# Mathematical Analysis

This textbook is dedicated to 'Goke' for providing the opportunity to formulate ideas on the teaching of mathematical analysis and to Rita for help and encouragement with the writing of the book.

# Mathematical Analysis

## R Maude
School of Mathematics, University of Leeds

# Edward Arnold

First published in Great Britain 1986 by
Edward Arnold (Publishers) Ltd, 41 Bedford Square, London WC1B 3DQ

Edward Arnold, 3 East Read Street, Baltimore, Maryland 21201, USA

Edward Arnold (Australia) Pty Ltd, 80 Waverley Road, Caulfield East, Victoria 3145, Australia

**British Library Cataloguing in Publication Data**
Maude, R.
  Mathematical analysis.
  1. Calculus
  I. Title
  515    QA303

  ISBN 0 7131 3529 8

Text set in 10/12pt Times Digiset
by J. W. Arrowsmith Ltd, Bristol

# Preface

On first meeting mathematical analysis it is almost always found difficult. There are basically two ways of trying to deal with this. Some books avoid as many as possible of the hard ideas, and hence also some of the fundamental concepts. They concentrate on giving the student a facility with techniques. The other way is to present the fundamental ideas fully, but to try to make the presentation easy. The central idea may be dealt with either as the concept of 'limit', or as that of 'continuity'. Whichever form is used, it has a quite daunting logical complexity; but it is classically expressed symbolically in two or three lines. In recent years many authors have tried to separate out the various parts of this complex of ideas. This book takes the process of separation a good deal further. It is a method which has been tried out on various groups of students. It has been thought useful to treat the case of complex numbers along with the real case, until differentiation is reached. There would be no problems in using the book for a course solely on real analysis. But there is some repetition in treating the two cases, and one can make use of this. The repetition in a first course in analysis should help to foster familiarity with the ideas. The alternative is to wait until the student has mastered elementary real analysis and then to go over essentially the same ground with complex numbers.

So after Chapter 0, which presents the requisite set theory, Chapter 1 describes both the real and the complex fields. Chapter 2 discusses sets of numbers, and in particular intervals. The basic ideas of analysis are to be discussed in terms of intervals. Thus the completeness axiom is given in a form which, in effect, says that real intervals can always be described by end points. Chapter 3 begins the discussion of continuity in a special situation where things are particularly simple. The continuity of monotone functions is described. The chapter then points the way to further development by defining open sets. In Chapter 4 the nature of an open set is considered. We meet the logically complex form 'for all ... there exists ...', in Theorem 4.1. This is one of the main logical difficulties in the concept of continuity. Here it occurs in the simple geometric form of a description of an open set. For many people this makes it much easier to picture. Every point of the open set is in some open interval contained in the open set.

Then in Chapter 5 the remaining difficulties about continuity are concerned with the domains of functions. Mistakes in connection with domains are a major source of error in applications to the calculus. Having separated out the various ingredients of the idea of continuity, they have to be reassembled. In Chapter 6 this is done in connection with the notion of the limit of a function. The limit as set up step by step in the book is shown to be equivalent to the limit given by the classic definition.

The idea of a sequence is introduced in Chapter 7, by showing how one arises in the solution of a particular kind of equation. This method is applied in the special case of a calculation of $\sqrt{2}$. The limit of a sequence is defined in a way which shows it to be a special case of the limit of a function, as met in Chapter 6. The previous theory can then be applied and simplified. It is pointed out that only some sets of real numbers can be formed into sequences, the countable sets.

Differentiation is discussed in Chapter 8, using the results on continuity. This is taken as far as the finite Taylor series. It is shown that in some cases the remainder tends to 0. This fact is used, in Chapter 8, to motivate the study of series, and in particular of power series. To avoid confusions which often arise, it is stressed throughout that a series consists of two sequences, a sequence of terms and a sequence of partial sums.

Integration, discussed in Chapter 10, is treated by using coverings of the ordinate set by rectangles. Restrictions are placed on the coverings when this is needed to make proof easy. Thus simple properties are proved in a way which includes the Lebesgue integral. But most of the more difficult results are given in a form which applies to the Riemann integral. However, to give an inkling as to why it is worth studying the Lebesgue integral, the limit theorem for an increasing sequence of functions is proved. But for the sake of any reader for whom this proof is too difficult, a Riemann integral limit theorem is also given.

Finally, in Chapter 11, the results of Chapters 1–10 are put to work to derive the properties of the exponential, logarithmic, sine, and cosine functions; starting from simple differential equations.

Except for this last chapter, each chapter ends with a set of exercises designed to promote familiarity with the ideas of the chapter.

# Contents

# Reading this book

It is quite usual to find the beginnings of rigorous mathematical analysis more difficult than other parts of mathematics at the same level. In part this is due to the fact that the key notion, whether considered as that of limit or of continuity, is really a tightly knit complex of ideas. The approach taken in this book is to try to separate out these ideas, so that the difficulties are spread out over the first few chapters. It is hoped that this will make for a better understanding of the ideas.

When reading the book, do not allow yourself to get stuck at some point which you feel you do not understand. The nature of mathematics is to make statements and prove that they are true. The most important such statements are made as theorems, and the proof then follows. So on reading the statement of a theorem one may well be left thinking 'Why on earth should that be so?'; but when a proof is presented first there may be the feeling 'What is this leading to?'. Even when both the theorem and its proof have been read, it may not be clear why that particular result was considered important. Often it will only become clearer when the uses to which the theorem are put have eventually been met. The same comment applies to definitions. The reason for the exact form of a definition may not be apparent until the definition is seen in action. Indeed, in the development of mathematics, very often several different forms of a definition are in use until it becomes clear that one form works best in practice. So to facilitate continual reference back to theorems and definitions, they are each given a number: first the number of the chapter in which they occur, and then after a decimal point the number of the theorem or definition in the chapter.

The reader should always have a pencil and paper handy, to fill in details of routine calculations which have been omitted, and to draw diagrams which are helpful to him but have not been included. The exercises at the end of the chapters are not intended as any form of test paper—they are part of the learning process. The reader should be prepared to refer to definitions, to make sure he knows what the questions mean, and to refer to theorems, to see which ones can be used in answering the questions.

To try to aid understanding, some comments, illustrations and repetitions have been inserted into formal proofs and statements. To make it clear that these are not part of the formal work, they have been enclosed in square brackets.

# Introduction

There are many problems in which we start with an infinite amount of information, but none the less arrive at a single answer. The object of mathematical analysis is to analyse ways in which this is possible. There are other closely related branches of mathematics concerned with the same kind of question. What distinguishes mathematical analysis from them is that the infinite amount of information is in the form of numbers and relations between numbers. Suppose that we have a number $a_n$ for every positive whole number $n$. Then we have an infinite amount of information. However, we may be able to find a sum, $a = a_1 + a_2 + a_3 + \cdots$, of all these numbers. The problem of how this can be done, and in what cases it can be done, is a question in mathematical analysis. In this book we are not greatly concerned with what numbers are, but we do have to say something about their nature if we are to answer such questions.

The material is presented in a logical order: first come the fundamental ideas and basic tools, and then only gradually are the results which have applications approached. Many readers will have seen some of the applications and may wonder why so much discussion of the theory is useful. One possible answer would be that intellectual curiosity ought to make us interested in knowing why the results we use are true. For those whose main interest is in the use of mathematics, a better answer may be that we are dealing with topics where people's intuition often leads them astray. This can cause calculations to give wrong answers, and so can be quite catastrophic.

When the numbers $a_n = 1/2^n$, there is a perfectly good result

$$1 = \tfrac{1}{2} + \tfrac{1}{4} + \cdots + 1/2^n + \cdots$$

However, until the subject had been properly investigated the 'sum'

$$1 - 1 + 1 - 1 + \cdots + (-1)^n \cdots$$

was a puzzle. Is the sum 0, as $(1-1) + (1-1) + \cdots + [(-1)^{2n} + (-1)^{2n+1}] + \cdots = 0 + 0 + 0 + \cdots + 0 + \cdots$ clearly has sum 0? Or is its sum 1, as $1 + (-1+1) + (-1+1) + \cdots + [(-1)^{2n+1} + (-1)^{2n+2}] + \cdots = 1 + 0 + 0 + \cdots + 0 + \cdots$ clearly has sum 1? Again, we can find a definite 'sum' for

$$1 - \tfrac{1}{2} + \tfrac{1}{3} - \tfrac{1}{4} + \cdots + \frac{(-1)^{n+1}}{n} + \cdots = s \quad \text{(say)}$$

Suppose that we multiply by $\frac{1}{2}$, and insert some zeros which cannot alter the value of the sum,

$$0+\tfrac{1}{2}+0-\tfrac{1}{4}+0+\tfrac{1}{6}+\cdots=\tfrac{1}{2}s$$

If we add these two equations we get

$$1+0+\tfrac{1}{3}-\tfrac{1}{2}+\tfrac{1}{5}+0+\tfrac{1}{7}-\tfrac{1}{4}+\cdots=\tfrac{3}{2}s$$

Omitting the zeros, this is

$$1+\tfrac{1}{3}-\tfrac{1}{2}+\tfrac{1}{5}+\tfrac{1}{7}-\tfrac{1}{4}+\cdots=\tfrac{3}{2}s$$

These are just the original numbers added up in a different order; however, their 'sum' is different. In fact it is only the even terms $(-1)^{2n+1}/(2n)=-1/(2n)$ which have been moved. To each such term has been added $\frac{1}{2}[(-1)^{n+1}/n]=(-1)^{n+1}/(2n)$. This gives 0 if $n$ is odd and $-2/(2n)=-1/n$ if $n=2m$, say, is even. Thus every term $-1/(2m)$ still occurs somewhere.

It is of considerable use, when dealing with such 'sums', to know just how bad this phenomenon can be. In particular, are there any cases where we do not have to be careful about keeping the numbers in their original order?

Many readers will already be familiar with the differential and integral calculus. This is the main source of problems where we get answers which are just a single number starting from an infinite amount of numerical information. Thus the definite integral $\int_a^b f(x)\,\mathrm{d}x$ is a number, but $f(x)$ is a different number for each of the infinite set of values of $x$ between $a$ and $b$. Readers who have not yet met the calculus should pass directly to the last paragraph of this chapter. Those familiar with the calculus may wonder why their knowledge needs to be supported by a knowledge of mathematical analysis. There follow three examples of difficulties in calculus which can be resolved by the use of mathematical analysis.

When a curve is given as the graph of a function, so that for points on the curve we have $y=f(x)$, the derivative $\mathrm{d}y/\mathrm{d}x$ is defined so as to give the slope of the tangent to the curve. Hence if the curve has a corner, and so no uniquely defined tangent at the point, there is no number $\mathrm{d}y/\mathrm{d}x$ at that point. Thus $y=|x|$ has the graph shown in Fig. 1, with a right-angle corner at 0, so that $(\mathrm{d}/\mathrm{d}x)|x|$ cannot have a meaning at $x=0$. Now there is a rule for differentiating products of functions:

$$\frac{\mathrm{d}}{\mathrm{d}x}f(x)\,g(x)=\left(\frac{\mathrm{d}}{\mathrm{d}x}f(x)\right)g(x)+f(x)\left(\frac{\mathrm{d}}{\mathrm{d}x}g(x)\right)$$

So some people, when asked to differentiate $|x|^{1+a}$ for $a>0$, write $|x|^{1+a}=|x|\cdot|x|^a$ and, applying the product rule, say that as $|x|$ has no derivative at 0 so also $|x|^{1+a}$ can have no derivative at 0. However, if $a=1$ then $|x|^{1+a}=|x|^2=x^2$, and $(\mathrm{d}/\mathrm{d}x)x^2=2x$ for all $x$ including $x=0$. So there are some problems which do have answers, but which we might think to be insoluble unless we know the exact requirements of the rule for differentiating products.

Amongst the further uses of derivatives there are differential equations. These are equations involving derivatives, and their solutions are functions

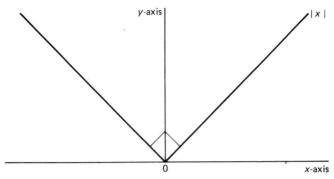

**Fig. 1**  Graph of |x|

usually containing arbitrary constants. One learns to expect that the number of constants will be the same as the highest order of derivative in the equation. Thus we might solve

$$\left(\frac{dy}{dx}\right)^2 = 4y$$

by

$$\frac{dy}{dx} = \pm 2\sqrt{y}$$

$$\frac{1}{2\sqrt{y}}\frac{dy}{dx} = \pm 1$$

$$\frac{dy}{2\sqrt{y}} = \pm \int dx$$

$$\sqrt{y} = \pm(x+c)$$

where $c$ is the arbitrary constant. Finally,

$$y = (x+c)^2$$

Since the derivative $dy/dx$ was only a first derivative, and there is one arbitrary constant $c$, it is tempting to think that this is the whole solution. Suppose, however, that $y = 0$ for all $x$. Then $dy/dx = 0$ for all $x$. This is a solution which cannot be obtained from $y = (x+c)^2$, whatever value of $c$ may be tried. Here again it would be useful to know conditions which will tell us when we have all the solutions to a differential equation. In particular, it would be nice to know when an equation with the highest order of derivative $n$ has no solutions other than those given by a form containing $n$ arbitrary constants.

The necessity to have precise conditions under which results hold can also be shown in connection with the integral. Suppose that we have a function defined by

$$f_n(x) = \begin{cases} nx & (0 \leqslant x \leqslant 1/n) \\ 2 - nx & (1/n < x < 2/n) \\ 0 & (2/n \leqslant x \leqslant 1) \end{cases}$$

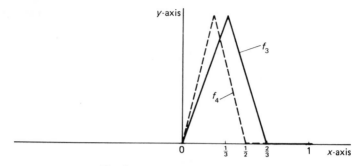

**Fig. 2**  The function $f_n$ for $n=3$ and 4

This function is shown in Fig. 2. Then $\int_0^1 f_n(x)\,dx = 1/n$ and so $\int_0^1 nf_n(x)\,dx = 1$, so that $\lim_{n\to\infty}\int_0^1 nf_n(x)\,dx = 1$. In addition, for any number $x$ between 0 and 1 and any sufficiently large $n$, we see that $f_n(x) = 0$, so that $\lim_{n\to\infty} nf_n(x) = 0$ and $\int_0^1 \lim_{n\to\infty} nf_n(x)\,dx = 0$. However, it is true and often used that

$$
\int_a^b \left( \lim_{N\to\infty} \sum_0^N a_n x^n \right) dx = \int_a^b \left( \sum_0^\infty a_n x^n \right) dx = \sum_0^\infty \left( \int_a^b a_n x^n \, dx \right)
$$

$$
= \lim_{N\to\infty} \sum_0^N \int_a^b a_n x^n \, dx
$$

In case the functions $f_n$ seem to be too artificial, readers who have sufficient technical skill with integrals and limits might like to show that

$$
\int_{-1}^1 \frac{(2m+1)!}{(m!)^2\,2^{2m+1}} (1-x^2)^m \, dx = 1
$$

but that

$$
\lim_{m\to\infty} \left\{ \frac{(2m+1)!}{(m!)^2\,2^{2m+1}} (1-x^2)^m \right\} = 0
$$

for all $x$ between $-1$ and 1, except for $x = 0$ where the numbers become arbitrarily large. So when can we change round the order of finding the limit and evaluating the integral?

As the techniques we use become more elaborate, it becomes less possible to rely upon our intuition. This is particularly true of calculations involving many variables and many limiting processes. Moreover, in order to have rigorously stated theorems, and logical arguments, to guide our intuition at the more advanced stages, we need to have followed the argument through the elementary stages. Indeed, because the early parts of mathematical analysis need to be properly understood, if the later stages are to be studied, these early parts are taken slowly in what follows. So it will be some time before the study begins to pay off with answers to the types of question suggested in this chapter.

# 0

# Sets, relations and functions

The information which we are given at the start of a problem, and from which we seek to find just one number, will usually be in the form of a set or several sets of numbers. A set is any collection, or grouping or lot, of different objects. The objects in our sets will be mathematical ones, numbers, points, lines, etc., but the notion of set is not confined to such things. The books on a bookshelf will form a set, or the cards one has been dealt in a hand of bridge. However, the members of a set do not have to be objects which are in some way similar. We could form a set from a shoe, a ship, a bit of sealing-wax, a cabbage, and a king. The essential point is that given any object it is either in the set or not in the set. If there is only a finite number of objects in a set then we can describe the set by giving a list of its members. We write {London, Lagos, Lima} for the set consisting of these three cities, or $\{1, 1^2, 2, 2^2\}$ for the set consisting of the three numbers 1, 2 and 4. Notice that we do not say that 1 is in the set twice; the number 1 is just in the set and we have said so twice. Sets will often be named by letters, and if we have a set represented by $A$ we write the fact that $a$ is in $A$ or is a member of $A$ by $a \in A$. The sign $\in$ is a modified form of the Greek letter $\varepsilon$. So we have that if $A = \{1, 1^2, 2, 2^2\}$ then $2 \in A$ but it is not true that $3 \in A$. This is written $3 \notin A$, and read '3 is not in $A$' or '3 does not belong to $A$'. The objects $a$ such that $a \in A$ are called the members of $A$ or the elements of $A$. It is clear that we should consider two sets $A$ and $B$ to be the same if whenever $a \in A$ then $a \in B$ and whenever $a \in B$ then $a \in A$, and in no other case, that is we write $A = B$ when $A$ and $B$ have the same members. Thus if Mr Smith has just three children, Peter, Ann and Rose, then the set of Mr Smith's children and the set {Peter, Ann, Rose} would be equal. Or again the set $\{1, 4, 9, 16\} = \{4^2, 3^2, 2^2, 1^2\}$. We can state this notion of equality of sets a little more concisely as

$A = B$ if and only if $a \in A$ implies and is implied by $a \in B$

Now the sets which are most important in this study have infinitely many members. So we cannot hope to be able to list them. In some cases we can indicate what the list must be. The set of all positive integers (an integer is a whole number) can be written $\{1, 2, 3, \ldots, n, \ldots\}$, or the set of positive integers which are squares as $S = \{1, 4, 9, \ldots, n^2, \ldots\}$. However, this is not always possible, and so we often define sets by the use of statements which are only true of members of the set. In the case of $S$ such a statement might be

$m = n^2$ and $n$ is a positive integer

Using this, $S$ is written as $\{m \,|\, m = n^2$ and $n$ is a positive integer$\}$, which is read as 'the set of all $m$ such that $m = n^2$ and $n$ is a positive integer'. More generally, if we write $p(a)$ to represent a statement about an object $a$ then $\{a \,|\, p(a)\}$ is the set of things $a$ for which $p(a)$ is true. Thus if $p(a)$ is the statement '$a$ is a child of Mr Smith', the set of Mr Smith's children is $\{a \,|\, p(a)\}$. Notice that sometimes $\{a: p(a)\}$ is used with exactly the same meaning. The set of even numbers could be written as $\{n \,|\, n = 2m$ and $m$ is an integer$\}$ or $\{n: n = 2m$ and $m$ is an integer$\}$.

A little care has to be used in deciding what statements $p$ can be used to define sets. The assumption that any statement will define a set can be shown to lead very quickly to a contradiction. The statement about sets, $A \notin A$, is a definite statement about the set $A$. For all the sets we have looked at it is true. Now, if possible, define a set $N$ by $N = \{A \,|\, A \notin A$ and $A$ is a set$\}$. Suppose it were true that $N \notin N$. Then putting $A = N$ in the definition, we see that this implies that $N \in N$. However, if $N \in N$ then by the definition $A \notin A$ has to be true for $A = N$, and so $N \notin N$. If $N$ were a set, both $N \in N$ and $N \notin N$ would be true. So $N$ is not a set. This is known, after Bertrand Russell, as the *Russell paradox*. Conditions under which $\{a \,|\, p(a)\}$ really does define a set have been extensively studied in mathematical logic. Fortunately, in the cases required in what follows, difficulties of this sort do not arise. Most of the sets which will be written in this form will be 'subsets' of some collection $S$ which is known or is assumed to be a properly defined set. A set $A$ is a subset of $S$ precisely if $a \in A$ implies that $a \in S$, so that $A$ is just part, or possibly all, of $S$. Thus the set of even numbers is a subset of the set of integers, or at any instant of time the set of all women mathematicians is a subset of the set of all mathematicians. Furthermore, the set of men first year students studying mathematics at a named university is a subset of the set of all first year students studying mathematics at that university. This still holds for those universities where there are no women first year students studying mathematics, so that the two sets are the same.

Our statement $p(a)$ will then usually be of the form $q(a)$ and $a \in S$. To see how this avoids the Russell paradox, consider

$N = \{A \,|\, A \notin A$ and $A \in S\}$

If $N \in N$ then both $N \notin N$ and $N \in S$. Since we cannot have both $N \in N$ and $N \notin N$, this is false. Thus $N \notin N$. Now this only implies that not both $N \notin N$ and $N \in S$. Since we are assuming $N \notin N$, it must be $N \in S$ which is false. So the conclusion is now only that $N \notin S$.

If $A$ is a subset of $B$ we write $A \subset B$ or $B \supset A$. Then $A \subset B$ is read '$A$ is contained in $B$' or '$A$ is a subset of $B$'. Also $B \supset A$ is read '$B$ contains $A$' or '$B$ has $A$ as a subset'. The following are immediate:

(i)   if $A \subset B$ and $B \subset A$ then $A = B$,
(ii)  if $A \subset B$ and $B \subset C$ then $A \subset C$.

Notice that the notation for subsets gives us a second way of writing $a \in A$. This is now the same as $\{a\} \subset A$.

Let us consider now the set of women first year students of mathematics. This can be defined in terms of two sets: the set $W$ of women students in the university in question, and the set $M$ of first year students studying mathematics. The set of women first year students of mathematics in the university is then $\{x \mid x \in W \text{ and } x \in M\}$. So in order to test whether a student is in the set, we first determine whether the student is male or female, and in the latter case find out if she is in her first year of studying mathematics. We make the following definition.

**Definition 0.1**  Given two sets $A$ and $B$, the set $A \cap B = \{x \mid x \in A \text{ and } x \in B\}$ and is called the *intersection* of the sets $A$ and $B$.

Then it follows that $A \cap B = B \cap A$ and that $A \cap B \subset A$ and $A \cap B \subset B$.

It has already been noted that there have been years and universities for which there were no women students in the first year studying mathematics. In such a case it would be found that whatever student $x$ was investigated the result was $x \notin M \cap W$. This sort of case is sufficiently interesting to give rise to another definition.

**Definition 0.2**  The set $\varnothing$ which is such that for all $x$, $x \notin \varnothing$ is called the *empty set*.

The symbol '$\varnothing$' is taken from the Scandinavian alphabets, but is sometimes referred to as a cancelled zero.

There is only one set $\varnothing$ because it is defined by stating what its members are—there are none!

Given any set $A$, it is always true that $\varnothing \subset A$, since if $x \in \varnothing$ then $x \in A$. This follows as there are no cases to be verified.

There is a second way of forming a new set from two given sets. Suppose that we have an equation

$$\sin ax \sin bx = 0$$

Then the set $X$ of solutions can be constructed from the fact that either $\sin ax = 0$ or $\sin bx = 0$. Hence

$$X = \{x \mid \sin ax = 0 \text{ or } \sin bx = 0\}$$

**Definition 0.3**  Given two sets $A$ and $B$, the set $A \cup B = \{x \mid x \in A \text{ or } x \in B\}$ and is called the *union* of the two sets $A$ and $B$.

Then it follows that $A \cup B = B \cup A$ and that $A \subset A \cup B$ and $B \subset A \cup B$.

**Theorem 0.1**  For any sets $A$, $B$ and $C$ we have:
(i)  $A \cap (B \cup C) = (A \cap B) \cup (A \cap C)$,
(ii)  $A \cup (B \cap C) = (A \cup B) \cap (A \cup C)$.

*Proof*

(i) $x \in A \cap (B \cup C)$ is the same as: $x \in A$ and $x \in B \cup C$. This, in turn, is the same as: $x \in A$, and $x \in B$ or $x \in C$. Now this is the same as $x \in A$ and $x \in B$, or $x \in A$ and $x \in C$, which holds precisely if $x \in A \cap B$ or $x \in A \cap C$, that is $x \in (A \cap B) \cup (A \cap C)$.

(ii) [We could prove this by the same sort of argument, but it is instructive to use a different approach.] For any $x$ and each of the sets $A$, $B$ and $C$ there are two possibilities: either $x$ belongs to the set or $x$ does not belong to the set. This gives $2^3 = 8$ cases in all. Now $x \in B \cap C$ only if $x \in B$ and $x \in C$, so that $x \in A \cup (B \cap C)$ if $x \in A$, whether $x \in B$ or $x \notin B$ and whether $x \in C$ or $x \notin C$, and also if $x \notin A$ but $x \in B$ and $x \in C$. This covers 5 cases.

In the remaining 3 cases $x \notin A \cup (B \cap C)$.

Now look at the 5 cases where $x \in A \cup (B \cap C)$. If $x \in A$ then $x \in A \cup B$ and $x \in A \cup C$, therefore $x \in (A \cup B) \cap (A \cup C)$, and if $x \notin A$ but $x \in B$ and $x \in C$ then again $x \in A \cup B$ and $x \in A \cup C$. So in each of these 5 cases $x \in (A \cup B) \cap (A \cup C)$.

However, if $x \notin A$ and $x \notin B$ then $x \notin A \cup B$, and so $x \notin (A \cup B) \cap (A \cup C)$; similarly, if $x \notin A$ and $x \notin C$ then $x \notin (A \cup B) \cap (A \cup C)$. Thus in the 3 remaining cases where $x \notin A$, that is when also $x \in B$ and $x \notin C$, when $x \notin B$ and $x \in C$, and when $x \notin B$ and $x \notin C$, we have $x \notin (A \cup B) \cap (A \cup C)$.

We can represent the second of these two arguments by a diagram, known as a *Venn diagram*. The members of the sets $A$, $B$ and $C$ are thought of as being represented by points in the plane, and $A$, $B$ and $C$ are drawn so as to allow for all the cases which can occur. In Fig. 0.1 the shaded area is $A \cup (B \cap C)$.

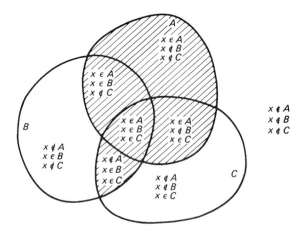

**Fig. 0.1** $A \cup (B \cap C) = (A \cup B) \cap (A \cup C)$

Given two sets $A$ and $B$, it is not possible to express all the different regions in the Venn diagram using only $\cap$ and $\cup$. It is necessary to be able to express the shaded area of Fig. 0.2.

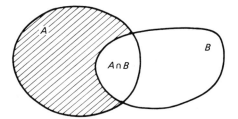

**Fig. 0.2**  $A\backslash B$

**Definition 0.4**  Given two sets $A$ and $B$, the set $A\backslash B = \{x\,|\,x \in A$ and $x \in B\}$ is called the *relative complement* of $B$ in $A$, and read '$A$ minus $B$'.

It is clear that the only case in which $A\backslash B = B\backslash A$ is the case $A = B$, since $A\backslash B$ and $B\backslash A$ must both be $\varnothing$, as $x \in A\backslash B$ implies that $x \notin B$ and so $x \notin B\backslash A$.

If $A \cap B = \varnothing$ we say that $A$ and $B$ are *disjoint*. For $A$ and $B$ disjoint sets we get both $A\backslash B = A$ and $B\backslash A = B$.

If $A$ is the set of all first year students studying mathematics and $B$ is the set of all men first year students studying mathematics then $A\backslash B$ is the set of all women first year students studying mathematics.

Now considering three sets $A$, $B$ and $C$, notice that $C\backslash(A \cup B) = (C\backslash A) \cap (C\backslash B)$, as $x \in C\backslash(A \cup B)$ is the same as $x \in C$ but $x \notin A$ and $x \notin B$, that is it is the same as $x \in C$ but $x \notin A$, and $x \in C$ but $x \notin B$, or $x \in (C\backslash A) \cap (C\backslash B)$. Similarly, $C\backslash(A \cap B) = (C\backslash A) \cup (C\backslash B)$.

If the intersection of $C$ with $(A \cap B)$ is formed then $(A \cap B) \cap C = A \cap (B \cap C)$, as $x \in (A \cap B) \cap C$ precisely if $x$ belongs to all of $A$, $B$ and $C$. So in forming the intersection of several sets, we do not need to insert brackets. In view of this, to fit the case when we have several sets $A_\alpha$ for different $\alpha$, the $\alpha$ forming a set $\Omega$, we make the following definition.

**Definition 0.5**  If $A_\alpha$ is a set for each $\alpha \in \Omega$ then the set

$$\bigcap_\Omega A_\alpha = \{x\,|\,x \in A_\alpha \text{ for all } \alpha \in \Omega\}$$

also written as $\bigcap \{A_\alpha \,|\, \alpha \in \Omega\}$ and sometimes as $\bigcap_{\alpha \in \Omega} A_\alpha$ or $\bigcap_\Omega A_\alpha$, is called the *intersection* of the $A_\alpha$. The notation $\bigcap_\alpha A_\alpha$ is also used.

When $\Omega$ is a set of consecutive integers $\{n_1, n_1 + 1, \ldots, m\}$, it is sometimes written $\bigcap_{n=n_1}^{m} A_n$. Thus $\bigcap_{n=1}^{3} A_n = A_1 \cap A_2 \cap A_3$.

Similar considerations apply to the union. $(A \cup B) \cup C = A \cup (B \cup C)$ is the set of all $x$ such that $x$ belongs to at least one of $A$, $B$ and $C$.

**Definition 0.6**  If $A_\alpha$ is a set for each $\alpha \in \Omega$ then the set

$$\bigcup_\Omega A_\alpha = \bigcup_{\alpha \in \Omega} A_\alpha = \bigcup \{A_\alpha \,|\, \alpha \in \Omega\} = \{x\,|\,x \in A_\alpha \text{ for some } \alpha \in \Omega\}$$

is called the *union* of the $A_\alpha$. This is also written as $\bigcup_\Omega A_\alpha$ and $\bigcup_\alpha A_\alpha$.

As for the case of the intersection, the notation $\bigcup_{n=n_1}^{m} A_n = A_{n_1} \cup A_{n_1+1} \cup \cdots \cup A_m$ is also used.

For the next result we need to use the fact that if it is not true that a whole collection of statements all hold, then at least one of them is false. We also need the fact that if it is not true that at least one of a collection of statements is true, then all the statements are false. Thus if I thought that John had the top marks in all his subjects, and I was told that this is false, it must be that in one (or possibly more) subject(s) he did not get the top mark. If, on the other hand, I thought that in some subject he achieved less than the top mark, and found this to be false, then he must have got the top mark in all his subjects.

**Theorem 0.2**   *The de Morgan laws*
(i)   $C \backslash (\bigcap_\Omega A_\alpha) = \bigcup_\Omega (C \backslash A_\alpha)$,
(ii)  $C \backslash (\bigcup_\Omega A_\alpha) = \bigcap_\Omega (C \backslash A_\alpha)$.

*Proof*
(i)   The statement

$$x \in C \backslash \left( \bigcap_\Omega A_\alpha \right)$$

is equivalent to

$x \in C$ but it is not true that $x \in A_\alpha$ all $\alpha$

which is the same as

$x \in C$ and for some $\alpha, x \notin A_\alpha$

that is

for some $\alpha, x \in C$ and $x \notin A_\alpha$

or     $x \in C \backslash A_\alpha$ some $\alpha$

which is

$$x \in \bigcup_\Omega (C \backslash A_\alpha)$$

(ii)  The statement

$$x \in C \backslash \left( \bigcup_\Omega A_\alpha \right)$$

is equivalent to

$x \in C$ but it is not true that $x \in A_\alpha$ some $\alpha$

which is the same as

$x \in C$ and for all $\alpha, x \notin A_\alpha$

that is

> for all $\alpha$, $x \in C$ and $x \notin A_\alpha$

which is

$$x \in \bigcap_\Omega (C \backslash A_\alpha)$$

Thus if all the sets under discussion are subsets of $C$ the operations of union and intersection are complementary.

We shall need the following result.

**Theorem 0.3**   If $A \cup B = C \cup D$ then $A \cup B = (A \cup C) \cup (B \cap D)$.

*Proof*   Since $A$, $B$, $C$ and $D$ are all subsets of $A \cup B$, we see that

$$(A \cup C) \cup (B \cap D) \subset A \cup B \tag{1}$$

Now if $x \in A \cup B$ but $x \notin A \cup C$ then $x \notin A$ and so $x \in B$. In addition, $x \notin C$, so that $x \in D$. Thus $x \in B \cap D$. So all $x \in A \cup B$ have $x \in (A \cup C)$ or $x \in B \cap D$, and

$$(A \cup C) \cup (B \cap D) \supset A \cup B \tag{2}$$

The equality now follows from eqs (1) and (2).

Now information is most often given not in the form of a simple set, but as a compound set called a relation. Thus the information might concern a set of passengers in a given aeroplane on a particular flight, and consist of the countries which each passenger had visited. For the set of positive integers it might consist of knowing the square roots of each integer. So the information comes in pairs consisting of one element $a$ of a set $A$ and a second element $b$ of a set $B$ giving information about $a$. The pair is written $\langle a, b \rangle$. In the first case it might be $\langle$Mr Smith, Japan$\rangle$; in the second case it might be $\langle 4, 2 \rangle$. Notice that the order of the members of the pair is important. In our second case $\langle 2, 4 \rangle$ would contain the statement that 4 was a square root of 2. So a pair is something more than a set with two elements: $\{a, b\} = \{b, a\}$ but $\langle a, b \rangle = \langle b, a \rangle$ only if $a = b$. It is possible to define pairs in terms of sets, but we will make use of the fact that the symbols $\langle a, b \rangle$ and $\langle b, a \rangle$ are different. Then the notion of *pair* is to be such that $\langle a, b \rangle = \langle c, d \rangle$ is the same as $a = c$ and $b = d$.

**Definition 0.7**   A *relation* $\rho$ between $A$ and $B$ is a set of pairs $\langle a, b \rangle$ with $a \in A$ and $b \in B$. $\langle a, b \rangle \in \rho$ is also written $a \rho b$.

In particular, between any two sets $A$ and $B$ there is the relation usually written $A \times B = \{\langle a, b \rangle \mid a \in A, b \in B\}$ and called the *Cartesian product* of $A$ and $B$. [This is a case where the set is not of the form $\{x \mid p(x) \text{ and } x \in S\}$; however, the '$a \in A$, $b \in B$' has the same effect as 'and $x \in S$' in preventing paradoxes.]

To see that the word 'product' is appropriate, notice that if $A$ has $n$ members and $B$ has $m$ members then $A \times B$ has $n \times m$ members. Table 0.1 illustrates this in the case where $n = 3$ and $m = 2$.

At the other extreme, there is always the empty relation $\emptyset$, so that $\langle a, b \rangle \notin \emptyset$ whatever $a \in A$ and $b \in B$. All relations between $A$ and $B$ are subsets of $A \times B$.

We shall need this fairly straightforward property of intersections of Cartesian products.

$$A = \{a, b, c\} \qquad B = \{d, e\}$$

| | | $A$ | |
|---|---|---|---|
| | $a$ | $b$ | $c$ |
| $d$ | $\langle a, d \rangle$ | $\langle b, d \rangle$ | $\langle c, d \rangle$ |
| $e$ | $\langle a, e \rangle$ | $\langle b, e \rangle$ | $\langle c, e \rangle$ |

$B\{$

**Table 0.1**   $A \times B$

**Theorem 0.4**   $\left( \bigcap_{\Omega} A_\alpha \right) \times \left( \bigcap_{\Omega} B_\alpha \right) = \bigcap_{\Omega} \left( A_\alpha \times B_\alpha \right)$.

*Proof*

$$z \in \bigcap_{\Omega} (A_\alpha \times B_\alpha)$$

if and only if

$$z \in (A_\alpha \times B_\alpha) \text{ all } \alpha \in \Omega$$

if and only if

$$z = \langle x, y \rangle \text{ where } x \in A_\alpha, \, y \in B_\alpha \text{ all } \alpha \in \Omega$$

if and only if

$$z = \langle x, y \rangle \text{ where } x \in \bigcap_{\Omega} A_\alpha, \, y \in \bigcap_{\Omega} B_\alpha$$

if and only if

$$z \in \left( \bigcap_{\Omega} A_\alpha \right) \times \left( \bigcap_{\Omega} B_\alpha \right)$$

Since it may be that only some of the members of the sets $A$ and $B$ occur in the pairs $\langle a, b \rangle$ of a relation $\rho$, we have the following definitions.

**Definition 0.8**   If $\rho$ is a relation between $A$ and $B$ then the *domain* of $\rho$ is the set

$$\mathcal{D}\rho = \{a \,|\, a \in A \text{ and } \langle a, b \rangle \in \rho\}$$

and the *range* of $\rho$ is the set

$$\mathscr{R}\rho = \{b \mid b \in B \text{ and } \langle a, b \rangle \in \rho\}$$

Thus if $\rho$ is the relation of marriage between sets $M$ and $W$ of all men and all women in the United Kingdom at 12 noon on 2 January 1985, the domain of $\rho$ is the set of all married men in the country at that time, and the range of $\rho$ is the set of all married women in the country at that time.

Notice that the only case in which either $\mathscr{D}\rho$ or $\mathscr{R}\rho$ is $\varnothing$ is when $\rho = \varnothing$, in which case both the domain and range of $\rho$ are empty.

Relations in general can be very difficult to handle. So all the ones we consider will have special properties which make them more tractable. Thus of particular interest are *equivalence* relations. These are relations $\rho$ between a set $A$ and itself, or on $A$, which are *reflexive*, that is $\langle a, a \rangle \in \rho$ or $a \rho a$ for all $a \in A$, *transitive*, that is if $a \rho b$ and $b \rho c$ then $a \rho c$, and *symmetric*, that is if $a \rho b$ then $b \rho a$. An equivalence relation $\rho$ on $A$ divides $A$ into equivalence classes $[a] = \{b \mid \langle a, b \rangle \in \rho\}$. For any two such classes $[a]$ and $[b]$, either $[a] = [b]$ or $[a] \cap [b] = \varnothing$. Since, if $c \in [a] \cap [b]$ and $x \in [b]$, we have $c \rho b$ as $c \in [b]$ and $b \rho x$ as $x \in [b]$, so $c \rho x$; also $c \in [a]$, so that $a \rho c$ and hence $a \rho x$ or $x \in [a]$. Thus if $A = \{a, b, c\}$ and $\rho = \{\langle a, a \rangle, \langle b, b \rangle, \langle c, c \rangle, \langle a, b \rangle, \langle b, a \rangle\}$ then $\rho$ is an equivalence relation and $[a] = [b] = \{a, b\}$ and $[c] = \{c\}$.

The relations we shall be most concerned with are functions. These are relations $\rho$ such that each $a \in \mathscr{D}\rho$ is related to just one $b \in \mathscr{R}\rho$. Thus the relation between the passengers in our aeroplane and the countries in which they were born would be a function.

**Definition 0.9** If the relation $\rho$ is such that whenever $\langle a, b \rangle \in \rho$ and $\langle a, c \rangle \in \rho$ then $b = c$, then $\rho$ is a *function* and we write $\rho(a)$ for the unique number $b$ such that $a \rho b$.

Thus any rule which applied to a member $a \in A$ produces a member $b \in B$ defines a function on $A$ to $B$. It is sometimes useful to think of a function as a 'black box' or computer which given any $a$ applies the rule, and produces the answer $b = \rho(a)$. Notice, however, that there may be many different rules which give the same function. Thus the two rules $b = 1 - a^2$ and $b = (1 + a)(1 - a)$, where $a$ is a number, give different computations, but always produce the same number $b$. Indeed many problems in mathematics are concerned with showing that two different rules do define the same function. Often we have a rule which is difficult to work, and wish to replace it by a rule which is easier to apply. A function can also be thought of as a table consisting of a list of members of $A$ and opposite each entry in the list the unique $b$ which corresponds to it. Such a list would be similar to a table of logarithms or a table of squares.

When the domain and range of a function $f$ both consist of numbers, a useful way of presenting such a listing is as the *graph* of the function. We set up rectangular coordinates, with the x-axis drawn across the paper and the y-axis at right angles to it. Then for each $x \in \mathcal{D}f$ we plot the point $\langle x, f(x) \rangle$, as shown in Fig. 0.3.

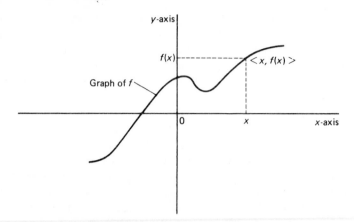

**Fig. 0.3**   The graph of a function f

Often functions can be defined by formulae. Thus $f(x) = 1 - x^2$ defines a function $f$ for all numbers $x$. But sometimes we may need to define a function by one formula for one part of its domain, and by a different formula for another part of its domain. We would write the definition of the function $g$, which has the value $-1$ for any negative $x$, but the value $+1$ for 0 and for all positive $x$, as

$$g(x) = \begin{cases} -1 & (x < 0) \\ 1 & (x \geqslant 0) \end{cases}$$

The graph of the function $g$ is shown in Fig. 0.4.

**Definition 0.10**   If $C$ is a set and $f$ is a function then $f(C)$ is the set

$$f(C) = \{b \mid \text{there is some } a \in C \text{ such that } f(a) = b\}$$

and is called the *image* of $C$ under $f$. If also $D$ is a set then the set $f^{-1}(D)$ is the set

$$f^{-1}(D) = \{a \mid f(a) = b \text{ for some } b \in D\}$$

and is called the *inverse image* of $D$ under $f$.

The notation $f^{-1}$ indicates that, whereas $f$ is thought of as starting from members of a set $A$ and producing members of a set $B$, in considering $f^{-1}(D)$ we are starting with members of $B$ and finding corresponding members of $A$. Suppose that $A = \{a, b, c\}$, $B = \{0, 1\}$ and $f = \{\langle a, 0 \rangle, \langle b, 1 \rangle, \langle c, 0 \rangle\}$. It is clear that $f$ is a function. Then $f(\{a, b\}) = B$, and $f^{-1}(\{0\}) = \{a, c\}$.

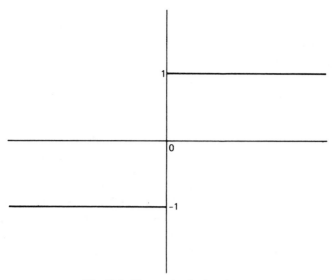

**Fig. 0.4** The graph of a function *g*

**Theorem 0.5**   If for each $\alpha \in \Omega$, $A_\alpha$ is a set and $f$ is a function then:
(i)   $f(\bigcup_\Omega A_\alpha) = \bigcup_\Omega f(A_\alpha)$,
(ii)   $f(\bigcap_\Omega A_\alpha) \subset \bigcap_\Omega f(A_\alpha)$.

*Note*   To see that equality need not hold in (ii), take $f$ to be the function just considered and let $A_1 = \{a, b\}$ and $A_2 = \{b, c\}$, so that $A_1 \cap A_2 = \{b\}$. Then $f(A_1 \cap A_2) = f(\{b\}) = \{1\}$, but $f(A_1) \cap f(A_2) = B \cap B = B = \{0, 1\}$.

*Proof*
(i)   $b \in f(\bigcup_\Omega A_\alpha)$ precisely if there is an $a \in \bigcup_\Omega A_\alpha$ for which $f(a) = b$; that is if there is an $A_\alpha$ and an $a \in A_\alpha$ such that $f(a) = b$; which is the same as there is an $A_\alpha$ such that $b \in f(A_\alpha)$; in other words $b \in \bigcup_\Omega f(A_\alpha)$.
(ii)   $b \in f(\bigcap_\Omega A_\alpha)$ precisely if there is an $a \in \bigcap_\Omega A_\alpha$ such that $f(a) = b$; that is if there is an $a \in A_\alpha$ for all $\alpha \in \Omega$ such that $f(a) = b$; which [is not the same as, but] does imply that $b \in f(A_\alpha)$ for all $\alpha \in \Omega$; in other words $b \in \bigcap_\Omega f(A_\alpha)$. [This could be true if $b = f(a_\alpha)$ with different $a_\alpha \in A_\alpha$ for different $\alpha$.]
   So we get $b \in f(\bigcap_\Omega A_\alpha)$ implies that $b \in \bigcap_\Omega f(A_\alpha)$, and hence $f(\bigcap_\Omega A_\alpha) \subset \bigcap_\Omega f(A_\alpha)$.

In this respect $f^{-1}$ behaves better than $f$.

**Theorem 0.6**   If for each $\alpha \in \Omega$, $B_\alpha$ is a set and $f$ is a function then:
(i)   $f^{-1}(\bigcup_\Omega B_\alpha) = \bigcup_\Omega f^{-1}(B_\alpha)$,
(ii)   $f^{-1}(\bigcap_\Omega B_\alpha) = \bigcap_\Omega f^{-1}(B_\alpha)$.

*Proof*

(i)   The proof of (i) follows exactly as in Theorem 0.5, as we did not have to use the fact that there is only one number $b = f(a)$ for a given $a$.

(ii)   $a \in f^{-1}(\bigcap_\Omega B_\alpha)$ precisely if there is a $b \in \bigcap_\Omega B_\alpha$ such that $f(a) = b$; that is if there is a $b \in B_\alpha$ for all $\alpha \in \Omega$ such that $f(a) = b$; which is the same as $a \in f^{-1}(B_\alpha)$ for all $\alpha \in \Omega$ [as there cannot be different numbers $b$, say $b_\alpha \in B_\alpha$, such that $f(a) = b_\alpha$]; in other words $a \in \bigcap_\alpha f^{-1}(B_\alpha)$.

The situation described by these theorems would arise if some university society decided to organize a jumble sale, and to collect just one item for the sale from each of its members. This would then generate a function $f$ such that if $a$ was a member of the society then $f(a)$ would be the item he or she gave for the sale. If the collection of items was organized by having a representative in each class, who collected the items from members of the society in his class, then there would be sets $A_\alpha$ of members of the society in class $\alpha$, and sets $f(A_\alpha)$ of items collected by the representatives. When the representatives bring all their items together for the sale we get $\bigcup_\Omega f(A_\alpha)$ as the set of items in the sale; while $\bigcup_\Omega A_\alpha$ would be the set of members of the society.

This particular function $f$ has a property which is often useful. Presumably no item was given by two different members of the society. We might have several members giving ball point pens, but then there would be several ball point pens. Thus for a given item $b$ there would be just one member of the society $a$ such that $b = f(a)$.

**Definition 0.11**   If the function $f$ is such that $f^{-1} = \{\langle b, a \rangle \mid \langle a, b \rangle \in f\}$ is also a function then $f$ is said to be *one–one* or (1–1) and $f^{-1}$ is the *inverse function* of $f$.

Notice that in such a case, as $f^{-1}$ is a function, $f^{-1}(C)$ is given a meaning by the first part of Definition 0.10 and also by the second part, but in both cases the set $f^{-1}(C)$ is the same.

**Theorem 0.7**   If $f$ is a one-one function and $a \in \mathcal{D}f$ and $b \in \mathcal{R}f$ then
(i)   $a = f^{-1}(f(a))$,
(ii)   $b = f(f^{-1}(b))$.

*Proof*
(i)   $\langle a, f(a) \rangle \in f$, therefore $\langle f(a), a \rangle \in f^{-1}$.
(ii)   $\langle f^{-1}(b), b \rangle \in f$.

The idea of combining two functions $f$ and $g$ in the form $f(g(a))$ leads to the final definition in this chapter.

**Definition 0.12**   If $f$ and $g$ are functions then $f \circ g$ is the function given by $f \circ g = \{\langle a, c \rangle \mid \langle g(a), c \rangle \in f\}$ and is called $f$ *composed with* $g$, or the *composition of $f$ and $g$*.

This is a function as if $c \neq c'$ the fact that $f$ is a function implies that at most one of $\langle g(a), c \rangle$ and $\langle g(a), c' \rangle$ can belong to $f$ [$g(a)$ is a definite number as $g$ is a function], so at most one of $\langle a, c \rangle$ and $\langle a, c' \rangle$ can belong to $f \circ g$. The function $f \circ g$ will always be defined, but will be the empty function $\varnothing$ unless $\mathscr{D}f \cap \mathscr{R}g \neq \varnothing$.

**Theorem 0.8**  If $f$, $g$ and $h$ are one-one functions such that $h = f \circ g$ then $h^{-1} = g^{-1} \circ f^{-1}$.

*Proof*  $\langle a, c \rangle \in h$ if and only if $\langle g(a), c \rangle \in f$. Thus $\langle c, a \rangle \in h^{-1}$ if and only if $\langle c, g(a) \rangle \in f^{-1}$. But $\langle g(a), a \rangle \in g^{-1}$, so that $\langle c, g(a) \rangle \in f^{-1}$ is the same thing as $\langle c, a \rangle \in g^{-1} \circ f^{-1}$.

This is really the same idea as that to reverse the process of coming into a room and sitting down, one first stands up and then leaves the room. One reverses sitting down first, then one reverses entering the room.

**Theorem 0.9**  If $f$, $g$ and $h$ are any three functions then $(f \circ g) \circ h = f \circ (g \circ h)$.

*Proof*  $\langle a, d \rangle \in (f \circ g) \circ h$ if and only if $\langle h(a), d \rangle \in f \circ g$, which holds precisely if $\langle g(h(a)), d \rangle = \langle g \circ h(a), d \rangle \in f$, but this is true if and only if $\langle a, d \rangle \in f \circ (g \circ h)$.

As a consequence of this, it is sufficient to write $f \circ g \circ h$ for either of the two expressions.

It is sometimes convenient to write a function almost as if it were identical with the set of its values, its range. Thus if the function $f$ is defined on a set $A$ we write $\{f(a)\}_{a \in A}$ and call it an *indexed set*. By this we mean that $f(a)$ and $f(b)$ are only to be identified and taken to be the same member of the set if $a = b$. If $f(a) = f(b)$ but $a \neq b$ then $f(a)$ and $f(b)$ are distinguished by their index, $a$ or $b$, as different members of the indexed set.

There are a few notations which are used to try to make some statements clearer, but which do not represent new ideas. Thus just as we sometimes write $A \supset B$ in place of $B \subset A$, we can write $A \ni a$ for $a \in A$, and read it as 'A has $a$ as a member'. When one statement implies another we sometimes write $p \Rightarrow q$ for $p$ implies $q$, or $q \Leftarrow p$ for $q$ is implied by $p$. When, as in the proof of Theorem 0.9, we are dealing with statements $p$ and $q$ such that $p$ if and only if $q$, we write $p \Leftrightarrow q$ or sometimes $p$ iff $q$, where iff is an abbreviation for 'if and only if'. Finally, $\exists$ is used for 'there is' or 'there exists', so that if $\exists a$ such that $a \in A$ then $A \neq \varnothing$. Sometimes the phrase 'we can find' is used for $\exists$. But if we see someone drop a needle into a haystack, we may be sure that there is a needle in the haystack without undertaking the proverbial problem of looking for it. In a similar way, $\exists a$ does not assert that there is any practical method of finding or determining $a$; it only says that there is such an object.

**Exercises**

**1**   Which of the following equations between sets are always true, and which are sometimes false? For the true equations an argument from a Venn diagram should suffice; for the other equations actual sets (for example, finite sets of numbers) for which the equations fail should be given.

(i)   $(A\backslash B)\cup C=(A\cup C)\backslash(B\cup C)$,

(ii)   $A\backslash B=A\backslash(A\cap B)$,

(iii)   $(A\cap B)\cup(B\cap C)\cup(C\cap A)=A\cup B\cup C$,

(iv)   $A\cup B=(A\cap B)\cup(A\backslash B)\cup(B\backslash A)$,

(v)   $A\cap B=(A\backslash B)\cap(B\backslash A)$.

**2**   Suppose that $S\subset X$ and $T\subset X$. Then some of the following statements express the same fact, and there are only 3 different situations described. Sort out the statements into three groups, each group containing statements expressing the same fact.

(i)   $S\subset T$,

(ii)   $S\cap T=\varnothing$,

(iii)   $S\cap T=S$,

(iv)   $S\cup T=X$,

(v)   $X\backslash T\subset S$,

(vi)   $S\subset X\backslash T$,

(vii)   $X\backslash S\subset T$,

(viii)   $(X\backslash T)\cap S=\varnothing$,

(ix)   $T\subset(X\backslash S)$.

**3**   The symmetric difference $A\,\Delta\,B$ is defined by $A\,\Delta\,B=(A\backslash B)\cup(B\backslash A)$. Show that $A\,\Delta\,B=(A\cup B)\backslash(B\cap A)$.

The unions and intersections in questions 4 and 5 are to be taken over all the positive integers.

**4**   Show that $\bigcap_{n=1}\{n,n+1,n+2,\ldots\}=\varnothing$, and that $\bigcap_{n=1}\{1,2,3,\ldots,n\}=\{1\}$.

**5**   Suppose that $A_1,A_2,\ldots,A_n,\ldots$ are sets. Show that for any other set $C$, $C\backslash\bigcup_{n=1}A_n=\bigcap_{n=1}(C\backslash A_n)$.

**6**   One of the relations $\rho_1,\rho_2,\rho_3$ and $\rho_4$ on the set $X=\{a,b,c,d,e\}$ is an equivalence relation; of the others, one is symmetric, one is reflexive and one is transitive. Determine which is which.

(i)   $\rho_1=\{\langle a,b\rangle,\langle b,c\rangle,\langle b,a\rangle,\langle d,e\rangle,\langle c,b\rangle,\langle e,d\rangle\}$,

(ii)   $\rho_2=\{\langle a,b\rangle,\langle b,c\rangle,\langle c,a\rangle,\langle a,a\rangle,\langle b,b\rangle,\langle c,c\rangle,\langle d,e\rangle,\langle e,d\rangle,\langle d,d\rangle,\langle e,e\rangle,$ $\langle b,a\rangle,\langle c,b\rangle,\langle a,c\rangle\}$,

(iii)   $\rho_3=\{\langle a,b\rangle,\langle b,c\rangle,\langle a,c\rangle,\langle c,d\rangle,\langle b,d\rangle,\langle a,d\rangle,\langle e,e\rangle\}$,

(iv)   $\rho_4=\{\langle a,e\rangle,\langle e,b\rangle,\langle a,a\rangle,\langle b,b\rangle,\langle d,e\rangle,\langle d,d\rangle,\langle e,e\rangle,\langle c,c\rangle\}$.

**7**   Find the domain and range of each of the relations $\rho_1,\rho_2,\rho_3$ and $\rho_4$ defined in question 6.

**8** Show that the relation $\subset$ defined on the collection $\{A \,|\, A \subset X\}$ of subsets of the fixed set $X$ is transitive and reflexive.

**9** Determine which of the following relations are functions:

(i) $r_1$ defined between $Y = \{a, b, c\}$ and $Z = \{g, h\}$ by $r_1 = \{\langle a, g \rangle, \langle b, h \rangle, \langle c, g \rangle\}$,

(ii) $r_2$ defined between the same $Y$ and $Z$ by $r_2 = \{\langle a, g \rangle, \langle b, h \rangle, \langle a, h \rangle, \langle c, g \rangle\}$,

(iii) $r_3$ defined on the set of positive, zero or negative integers by $r_3 = \{\langle n, m \rangle \,|\, n^2 + m^2 = 25\}$,

(iv) $r_4$ defined on the set of positive, zero or negative integers by $r_4 = \{\langle n, m \rangle \,|\, m = n^2\}$.

# 1

# The number systems and inequalities

Analysis is mainly concerned with problems where both the information to be obtained and the data are expressed in terms of numbers. It is not necessary to go deeply into the nature of numbers; it is sufficient to lay down a system of *axioms*, which state elementary facts about the numbers we use. If we base our arguments only on these axioms then our conclusions will be true of anything which satisfies the axioms, and so in particular they will be true of numbers whatever may be the real nature of numbers. We shall certainly want the following axioms to hold for our set $F$ of numbers.

**Field axioms**

(1)  (i)   For any two members $a$ and $b$ of $F$ there is a number written $a + b \in F$.
    (ii)  $a + b = b + a$ for all $a, b \in F$ (that is, all $a \in F$ and all $b \in F$).
    (iii) $(a + b) + c = a + (b + c)$ for all $a, b, c \in F$.
    (iv)  $\exists 0 \in F$ such that $a + 0 = a$ for all $a \in F$.
    (v)   To any $a \in F$ there corresponds an $x \in F$ such that $a + x = 0$; $x$ is written $-a$.

(2)  (i)   For any $a, b \in F$ there is a number written $a \cdot b \in F$.
    (ii)  $a \cdot b = b \cdot a$ for all $a, b \in F$.
    (iii) $(a \cdot b) \cdot c = a \cdot (b \cdot c)$ for all $a, b, c \in F$.
    (iv)  $\exists 1 \in F$ such that $1 \neq 0$ and $a \cdot 1 = a$ for all $a \in F$.
    (v)   To any $b \neq 0$, $b \in F$ there corresponds a $y \in F$ such that $b \cdot y = 1$; $y$ is written $b^{-1}$ or $1/b$.

(3)  $a \cdot (b + c) = a \cdot b + a \cdot c$ for all $a, b, c \in F$.

Any set $F$ satisfying these axioms is called a *field*. The axioms (ii) are *commutative laws*, axioms (iii) are *associative laws*, and axiom (3) is a *distributive law*. There can only be one 0 in $F$. To see this, suppose that $0' \in F$ also satisfied (1)(iv). Then $0 + 0' = 0$ but also $0 + 0' = 0' + 0$ by (1)(ii) and $0' + 0 = 0'$, so that $0 = 0'$. Again, to any $a \in F$ there corresponds only one $-a$. Thus if $a + x = 0$ and $a + y = y + a = 0$ then $y + (a + x) = y = (y + a) + x$ by (1)(iii) and $(y + a) + x = x$, so that $y = x$. In exactly the same way, there is only one $1 \in F$ and only one $b^{-1}$ corresponding to a given $b \in F$.

There are many fields. To see that our axioms do not yet ensure that the numbers are anything like what we expect numbers to be, consider a set $F$ with only two members. By (1)(iv) and (2)(iv) they will have to be called 0 and 1, although in this case they will not behave as 0 and 1 usually do. In any

field

$$0 \cdot a = a \cdot 0 = 0$$

as    $a \cdot (0+1) = a \cdot 1 = a = a \cdot 0 + a \cdot 1 = a \cdot 0 + a$

so    $a = a \cdot 0 + a$

and hence

$$0 = a + (-a) = (a \cdot 0 + a) + (-a) = a \cdot 0 + [a + (-a)] = a \cdot 0 + 0 = a \cdot 0$$

so    $0 = a \cdot 0$

Thus all the products between members of this particular $F$ are determined by the axioms. We have thus the multiplication table for $F$ shown as 'Table of $a \cdot b$', Table 1.1. Also for $+$ the axioms give $0+0=0$, $0+1=1+0=1$, and this leaves only $1+1$ to be defined. We define $1+1=0$, as to be in the field $F$ it must be either 0 or 1. Then we have an addition table shown as 'Table of $a+b$', Table 1.2. Since $a+b$ and $a \cdot b$ have thus been defined for all cases, (1)(i) and (2)(i) hold. By the symmetry of the tables, (1)(ii) and (2)(ii) hold. The tables have been constructed to ensure that (1)(iv) and (2)(iv) hold. To see that (1)(iii) holds, notice that both $(a+b)+c$ and $a+(b+c)$ are 1 if only one or all three of $a$, $b$ and $c$ are 1, and both are 0 if two or none of $a$, $b$ and $c$ are 1. Similarly, (2)(iii) can be seen to hold as $(a \cdot b) \cdot c$ and $a \cdot (b \cdot c)$ are both 0 unless $a = b = c = 1$ when they are both 1. The number $-a$ is 1 if $a = 1$ and 0 if $a = 0$, as $1+1=0$ and $0+0=0$, so that (1)(v) holds. $1/a$ is 1 if $a = 1$, so that (2)(v) holds. Finally, by taking the two cases $a = 0$ and $a = 1$, it is clear that (3) holds.

**Table 1.1**  Table of $a \cdot b$

**Table 1.2**  Table of $a + b$

Now the system of numbers which is our main concern is the system known as the *real numbers*. These numbers correspond to the geometric idea of a line, each number being a point on the line, and so the real numbers should have an order. Figure 1.1 shows a line with some of the numbers indicated. We should be able to say that one number $a$ is to the left of $b$, or that some third number $c$ is between $a$ and $b$. These two ways of describing the order can be expressed in terms of each other. So although the idea of a number $c$ being between two others, $a$ and $b$, is the one which is most basic to mathematical

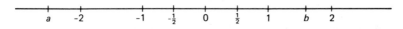

**Fig. 1.1**  The line of real numbers

analysis in the real field, the convenience of only having to use two numbers to say that $a$ is to the left of $b$ leads us to work in terms of this idea. We therefore lay down the following further axioms, to turn our field into an *ordered field*.

**Order axioms**
On the field $F$ there is a relation $<$ between members of $F$ such that:
 (i)   for all $a, b \in F$ one and only one of the following holds: $a = b$, $a < b$ or $b < a$,
 (ii)   if $a, b, c \in F$ and $a < b$ and $b < c$ then $a < c$,
 (iii)   if $a, b, c \in F$ and $a < b$ then $a + c < b + c$,
 (iv)   if $a, b, c \in F$ and $a < b$ and $0 < c$ then $a \cdot c < b \cdot c$.

Notice that $b > a$, read $b$ is greater than $a$, is often written for $a < b$, read $a$ is less than $b$. Also, in view of (i), if $b < a$ is false, which could be written $b \not< a$, then either $a = b$ or $a < b$, so this is written $a \leqslant b$ and read $a$ is less than or equal to $b$.

To see the effect of the order axioms on the possible nature of a field, some elementary facts about fields are needed. For the most part, standard algebraic results will be assumed, but here, since the results affect the outcome of the order axioms, the argument is given in detail.

$$-a = -1 \cdot a$$

as   $$a + (-1) \cdot a = 1 \cdot a + (-1) \cdot a = [1 + (-1)] \cdot a = 0 \cdot a = 0$$

$$(-1)^2 = 1$$

as   $$0 = (-1) \cdot (1 - 1) = (-1) \cdot 1 + (-1) \cdot (-1)$$

by (3). Adding 1 to both sides

$$1 = 1 + (-1) + (-1)^2 = (-1)^2$$

Hence if $x \in F$ then

$$x^2 = (-x)^2$$

as   $$(-x)^2 = [(-1) \cdot x]^2 = (-1) \cdot x \cdot (-1) \cdot x = (-1) \cdot (-1) \cdot x \cdot x = x \cdot x = x^2$$

**Theorem 1.1**   $0 < 1$.

*Proof*  By order axiom (i) just one of $0 = 1$, $1 < 0$ or $0 < 1$ holds, but field axiom (2)(iv) says that $0 \neq 1$, and so just one of $1 < 0$ or $0 < 1$ is true. [Consider whether $1 < 0$ is possible.] If $1 < 0$ then by order axiom (iii) with $c = -1$, $0 < -1$. Now by order axiom (iv) $0 \cdot 0 < (-1) \cdot (-1)$ or $0 < (-1)^2 = 1$. Thus in either of the two remaining cases $0 < 1$ [and so $1 < 0$ cannot hold].

From this theorem it follows that an ordered field must have infinitely many numbers. Order axiom (iii) gives that $0 < 1$ implies $1 < 1 + 1$. Hence by order axiom (ii) $0 < 1 + 1$. Thus $2 = 1 + 1$ is neither 0 nor 1, from order axiom (i). Similarly, if $3 = 2 + 1$ this is a further number of the field, as we get $1 < 2$ implies $2 < 2 + 1 = 3$ and so $3 > 2 > 1 > 0$. [We use this notation in view of order axiom (ii), since if $a > b$ and $b > c$ we have $a > c$, and $a > b > c$ expresses these three inequalities.] If it has been shown in this way that $n > n - 1 > \cdots > 1 > 0$ then it follows that $n + 1 > (n - 1) + 1 = n$ and so $n + 1$ is a number distinct from $0, 1, \ldots, n$. It is clear that as this process does not terminate, there are infinitely many numbers of the form $n$ in $F$. The set of all such numbers is the set of *natural numbers* and will be denoted by $\mathbb{N}$. The subset $\mathbb{N} \setminus \{0\}$ of $\mathbb{N}$ is the set of *positive integers* and will be written $\mathbb{N}'$. $F$ must also contain negative integers, as if $n \in \mathbb{N}'$ then $0 < n$, and adding $-n$ to both sides by order axiom (iii) gives $-n < 0$, so that $-n \notin \mathbb{N}$ by order axiom (i). Also $-n = (-1) \cdot n$, so that $-(-n) = (-1)^2 \cdot n = n$. Thus the function $f$ defined by $f(n) = -n$ is one-one, and there are infinitely many negative integers; there is just one corresponding to each positive integer. The set of *integers* $\{n \mid \text{either } n \in \mathbb{N} \text{ or } -n \in \mathbb{N}\}$ is denoted by $\mathbb{Z}$. However, $F$ must contain even more members because of field axiom (2)(v). If $n \in \mathbb{N}' = \mathbb{N} \setminus \{0\}$ there is a $1/n \in F$. Now since $n > 0$, if $1/n \le 0$ then $n \cdot 1/n = 1 \le 0$, so that $0 < 1/n$. Again if $1/n \ge 1$ then $n \cdot 1/n = 1 \ge n$, so that if $n \ne 1$ then $0 < 1/n < 1$ and $1/n \notin \mathbb{Z}$. In fact every number of the form $p/q$ where both $p, q \in \mathbb{Z}$ and $q \ne 0$ must be in $F$, but they are not all distinct and some belong to $\mathbb{Z}$. Thus $p/q = pr/(qr)$ with $r \in \mathbb{Z}$, and every $n \in \mathbb{Z}$ can be written $n/1$. The set of all numbers of the form $p/q$ is called the set of *rational numbers* (rational comes from ratio) and is denoted by $\mathbb{Q}$ (for quotient). We have $\mathbb{Z} \subset \mathbb{Q}$.

The question as to whether $\mathbb{Q} = F$ or whether there are *irrational numbers* in our ordered field is discussed later.

The order axioms can be expanded to give further rules for manipulating inequalities.

**Theorem 1.2**
($\alpha$)    If $a, b, c, d \in F$ and $a < b$ and $c < d$ then $a + c < b + d$.
($\beta$)    If $a, b, c \in F$ and $c < 0$ and $a < b$ then $b \cdot c < a \cdot c$.

*Proof*
($\alpha$)    By order axiom (iii) $a + c < b + c$ and $b + c < b + d$. So by order axiom (ii) $a + c < b + d$.
($\beta$)    By order axiom (iii) $c < 0 \Rightarrow c - c < -c$ [$c - c$ is, of course, $c + (-c)$], that is $0 < -c$. So by order axiom (iv) $(-c) \cdot a < (-c) \cdot b$, and using order axiom (iii) again to add $c \cdot a + c \cdot b$ to each side, $c \cdot a + c \cdot b - c \cdot a < c \cdot a + c \cdot b - c \cdot b$. Simplifying both sides by using the field axioms $c \cdot b < c \cdot a$.

**Theorem 1.3**    For all $a$ and $b$ in an ordered field:
($\alpha$)    if $a > 0$ and $b > 0$ then $a \cdot b > 0$,
($\beta$)    if $a > 0$ and $b < 0$ then $a \cdot b < 0$,
($\gamma$)    if $a < 0$ and $b < 0$ then $a \cdot b > 0$.

*Proof*

($\alpha$)   This is order axiom (iv) with $a$ put equal to 0 and $c$ put equal to $a$.

($\beta$)   This is Theorem 1.2($\beta$) with $c$ put equal to $b$ and $a$ put equal to 0 and $b$ put equal to $a$.

($\gamma$)   This is Theorem 1.2($\beta$) with $c$ put equal to $b$ and $b$ put equal to 0.

Notice that if $a > 0$ we say that $a$ is *positive* or $a$ is of *positive sign*, and if $a < 0$ we say that $a$ is *negative* or $a$ is of *negative sign*. So we can put ($\alpha$) and ($\gamma$) together and say that if $a$ and $b$ are of the same sign then $a \cdot b > 0$. A special, but important, case of this is as follows.

**Corollary to Theorem 1.3**   For all $x$ in an ordered field either $x^2 > 0$ or $x = 0$, that is $x^2 \geq 0$.

*Proof*   If $x > 0$ then $x^2 > 0$ by ($\alpha$) of the theorem; if $x < 0$ then $x^2 > 0$ by ($\gamma$) of the theorem; if $x = 0$ then $x^2 = 0$.

**Theorem 1.4**   For all $a$ and $b$ in an ordered field:

($\alpha$)   $a > 0 \Rightarrow 1/a > 0$,   $a < 0 \Rightarrow 1/a < 0$,

($\beta$)   $a > b > 0 \Rightarrow 1/a < 1/b$,   $b < a < 0 \Rightarrow 1/b > 1/a$.

*Proof*

($\alpha$)   If $a$ and $1/a$ are of opposite signs then $(-a)$ and $1/a$ are of the same sign, so that $(-a) \cdot 1/a > 0$, but $-a \cdot 1/a = (-1) \cdot 1 = -1 < 0$. Thus $a$ and $1/a$ are of the same sign.

($\beta$)   Since $a$ and $b$ are of the same sign in both cases, $a \cdot b > 0$. Hence from ($\alpha$) [which has just been proved] $1/(a \cdot b) > 0$. Then using order axiom (iv) gives that if $a > b$ then $a \cdot 1/(a \cdot b) > b \cdot 1/(a \cdot b)$. Thus we get $1/b > 1/a$.

We can now discuss how to express '$b$ is between $a$ and $c$' in terms of the relation '$<$'. The obvious way of doing this is to say that $b$ is between $a$ and $c$ iff one of $a < b < c$ or $c < b < a$. This is illustrated in Fig. 1.2. This can be reduced to a single inequality since $a < b < c$ is equivalent to $a - b < 0 < c - b$ and $c < b < a$ is equivalent to $c - b < 0 < a - b$. So $b$ is between $a$ and $c$ iff $a - b$ and $c - b$ are of opposite sign. Hence by Theorem 1.3 $(a - b) \cdot (c - b) < 0$, which is the same as $(a - b) \cdot (b - c) > 0$, gives a single inequality for '$b$ is between $a$ and $c$'. It is often useful to express the relation of $b$ to the pair $\langle a, c \rangle$ by using a parameter $t$ which obeys inequalities in a standard form.

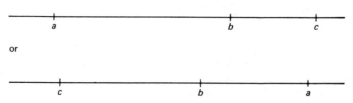

**Fig. 1.2**   $b$ between $a$ and $c$

Thus $b$ is between $a$ and $c$ iff $b = a + t \cdot (c - a)$ where $0 < t < 1$. Equivalently, $b$ is between $a$ and $c$ iff $b = c + \tau \cdot (a - c)$ where $0 < \tau < 1$. Since $a + t \cdot (c - a) = a \cdot (1 - t) + t \cdot c$ and $c + \tau \cdot (a - c) = c \cdot (1 - \tau) + \tau a$, these can also be written $b = a \cdot s + c \cdot t$ where $s + t = 1$ and $s, t > 0$.

To illustrate the rules for manipulating inequalities established in Theorems 1.2 to 1.4, we prove some inequalities. Two of these are of great importance later in mathematical analysis and are set out as theorems. For any two real numbers $a$ and $b$, by the Corollary to Theorem 1.3

$$(a - b)^2 \geq 0$$

Thus $\quad a^2 - 2 \cdot a \cdot b + b^2 \geq 0$

Hence

$$a^2 + b^2 \geq 2 \cdot a \cdot b$$

by order axiom (iii). Furthermore,

$$(a + b)^2 = a^2 + b^2 + 2 \cdot a \cdot b \geq 4 \cdot a \cdot b$$

from this we get that if $a, b > 0$

$$\frac{1}{a} + \frac{1}{b} = \frac{a + b}{a \cdot b} \geq \frac{4}{a + b}$$

using Theorem 1.4($\beta$). Hence

$$(a + b) \cdot \left( \frac{1}{a} + \frac{1}{b} \right) \geq 4$$

If $a, b < 0$ then by Theorem 1.4($\beta$)

$$\frac{a + b}{a \cdot b} \leq \frac{4(a + b)}{(a + b)^2} = \frac{4}{a + b}$$

Then a further use of Theorem 1.4($\beta$) gives

$$(a + b) \cdot \left( \frac{1}{a} + \frac{1}{b} \right) \geq 4$$

So this inequality holds provided only that both $a$ and $b$ are of the same sign. Similarly, for any four real numbers $a_1, a_2, b_1$ and $b_2$

$$(a_1 \cdot b_2 - a_2 \cdot b_1)^2 \geq 0$$

and so

$$a_1^2 \cdot b_2^2 + a_2^2 \cdot b_1^2 \geq 2 \cdot a_1 \cdot b_1 \cdot a_2 \cdot b_2$$

Adding $a_1^2 \cdot b_1^2 + a_2^2 \cdot b_2^2$ to both sides by order axiom (iii),

$$a_1^2 \cdot b_2^2 + a_2^2 \cdot b_1^2 + a_1^2 \cdot b_1^2 + a_2^2 \cdot b_2^2 = (a_1^2 + a_2^2) \cdot (b_1^2 + b_2^2)$$

$$\geq a_1^2 \cdot b_1^2 + 2 \cdot a_1^2 \cdot b_2^2 \cdot a_2^2 \cdot b_1^2 + a_2^2 \cdot b_2^2 = (a_1 \cdot b_1 + a_2 \cdot b_2)^2$$

So we have

$$(a_1^2 + a_2^2) \cdot (b_1^2 + b_2^2) \geq (a_1 \cdot b_1 + a_2 \cdot b_2)^2$$

In order to present some of these inequalities in a general form, we make use of the 'sigma' notation. If there are $n$ numbers $a_1, \ldots, a_n$ then their sum is denoted by $\sum_{i=1}^{n} a_i$. Thus $\sum_{i=1}^{3} a_i = a_1 + a_2 + a_3$. More generally, $\sum_{i=k}^{n} a_i = a_k + a_{k+1} + \cdots + a_n$, so that $\sum_{i=3}^{7} a_i = a_3 + a_4 + a_5 + a_6 + a_7$. Notice that $\sum_{i=1}^{n} a_i$ does not depend on a variable $i$. The $i$ in the formula helps to indicate how the sum is formed, but nowhere occurs in the sum $a_1 + a_2 + \cdots + a_n$ when it has been formed. The other point to be noticed is that the $\sum$ is one sided. It is only the expressions which come after the $\sum$ which are to be added. So, if we are to take a factor from one side of the $\sum$ to the other, we must be sure that it does not depend on $i$. Thus $2 \cdot \sum_{i=1}^{n} a_i = \sum_{i=1}^{n} 2 \cdot a_i$, but $a_i \cdot \sum_{i=1}^{n} a_i = a_i \cdot a_1 + a_i \cdot a_2 + \cdots + a_i \cdot a_n$, while $\sum_{i=1}^{n} a_i^2 = a_1^2 + a_2^2 + \cdots + a_n^2$. For this reason it is helpful to avoid using the letter, here $i$, denoting the variable of summation for any other purpose. There is a similar notation using a capital pi for products. Thus $a_1 \cdot a_2 \cdot a_3 \cdot a_4$ is written $\prod_{i=1}^{4} a_i$, or in general the product of $a_k, a_{k+1}, \ldots, a_n$ as $\prod_{i=k}^{n} a_i$. Both notations are similar to the $\bigcup$ and $\bigcap$ notations for sets. As for these notations, $\sum$ and $\prod$ are used when the set of numbers to be added or multiplied is not indexed by consecutive integers. Thus, for example,

$$\sum_{\substack{i=2, i \text{ even}}}^{10} a_i = a_2 + a_4 + a_6 + a_8 + a_{10}$$

and

$$\sum_{\substack{i,j=1, i>j}}^{n} a_i \cdot b_j = \sum_{j=1}^{n} \left( \sum_{i=j+1}^{n} a_i \cdot b_j \right)$$

$$= (a_2 \cdot b_1 + a_3 \cdot b_1 + \cdots + a_n \cdot b_1)$$
$$+ (a_3 \cdot b_2 + a_4 \cdot b_2 + \cdots + a_n \cdot b_2) + \cdots$$
$$+ (a_{n-1} \cdot b_{n-2} + a_n \cdot b_{n-2}) + a_n \cdot b_{n-1}$$

**Theorem 1.5   Cauchy's inequality**   For any real numbers $a_1, \ldots, a_N$ and $b_1, \ldots, b_N$

$$\left( \sum_{n=1}^{N} a_n \cdot b_n \right)^2 \leq \sum_{n=1}^{N} a_n^2 \cdot \sum_{n=1}^{N} b_n^2$$

*Note*   Once we have made sure that positive real numbers have square roots we will be able to write this in its customary form

$$\sum_{n=1}^{N} a_n \cdot b_n \leq \left( \sum_{n=1}^{N} a_n^2 \right)^{1/2} \cdot \left( \sum_{n=1}^{N} b_n^2 \right)^{1/2}$$

*Proof* By the Corollary to Theorem 1.3

$$(a_m \cdot b_n - a_n \cdot b_m)^2 \geqslant 0$$

for all $n$ and $m$. Hence

$$a_m^2 \cdot b_n^2 + a_n^2 \cdot b_m^2 \geqslant 2 \cdot a_n \cdot a_m \cdot b_n \cdot b_m = 2 \cdot a_n \cdot b_n \cdot a_m \cdot b_m$$

Keeping $m$ fixed we use Theorem 1.2($\alpha$) to add up the inequalities corresponding to $n = 1, \ldots, N$

$$a_m^2 \cdot \sum_{n=1}^{N} b_n^2 + b_m^2 \cdot \sum_{n=1}^{N} a_n^2 \geqslant 2 \cdot a_m \cdot b_m \cdot \sum_{n=1}^{N} a_n \cdot b_n$$

Now in the same way adding up these inequalities corresponding to $m = 1, \ldots, N$

$$\sum_{m=1}^{N} \left( a_m^2 \cdot \sum_{n=1}^{N} b_n^2 \right) + \sum_{m=1}^{N} \left( b_m^2 \cdot \sum_{n=1}^{N} a_n^2 \right)$$

$$= \sum_{n=1}^{N} b_n^2 \cdot \sum_{m=1}^{N} a_m^2 + \sum_{n=1}^{N} a_n^2 \cdot \sum_{m=1}^{N} b_m^2$$

$$\text{(as neither } \sum_{n=1}^{N} b_n^2 \text{ nor } \sum_{n=1}^{N} a_n^2 \text{ depends on } m)$$

$$= 2 \cdot \sum_{n=1}^{N} a_n^2 \cdot \sum_{n=1}^{N} b_n^2 \qquad \text{(as } \sum_{i=1}^{M} \alpha_i = \sum_{j=1}^{M} \alpha_j \text{ in all cases)}$$

$$\geqslant 2 \cdot \sum_{m=1}^{N} a_m \cdot b_m \cdot \sum_{n=1}^{N} a_n \cdot b_n$$

$$= 2 \cdot \left( \sum_{n=1}^{N} a_n \cdot b_n \right)^2$$

As $2 > 0$, on dividing both sides by 2 we get the result.

**Theorem 1.6  Minkowski's inequality**  If $a_1, \ldots, a_N$ and $b_1, \ldots, b_N$ are any real numbers then

$$\left( \sum_{n=1}^{N} (a_n + b_n)^2 \right)^{1/2} \leqslant \left( \sum_{n=1}^{N} a_n^2 \right)^{1/2} + \left( \sum_{n=1}^{N} b_n^2 \right)^{1/2}$$

provided that the square roots exist.

*Note* We shall arrive later at the situation where we know that as the sums are each positive or zero, they have a unique positive or zero square root.

*Proof* [First we notice that we can take positive square roots of an inequality.] If $0 < A < B$ then $A^2 < A \cdot B$ by Theorem 1.2($\beta$) and $A \cdot B < B^2$ by the same theorem. Hence $A^2 < A \cdot B < B^2$ and $A^2 < B^2$. Thus if $A^2 \leqslant B^2$ and $A$ and $B$ are both greater than or equal to 0 then it cannot be true that $A > B$ since this would give $A^2 > B^2$, so it must be that $A \leqslant B$.

From Theorem 1.5

$$2 \cdot \left( \sum_{n=1}^{N} a_n \cdot b_n \right) \leqslant 2 \cdot \left( \sum_{n=1}^{N} a_n^2 \right)^{1/2} \cdot \left( \sum_{n=1}^{N} b_n^2 \right)^{1/2}$$

Hence

$$\sum_{n=1}^{N} a_n^2 + 2 \cdot \left( \sum_{n=1}^{N} a_n \cdot b_n \right) + \sum_{n=1}^{N} b_n^2$$

$$\leqslant \sum_{n=1}^{N} a_n^2 + 2 \cdot \left( \sum_{n=1}^{N} a_n^2 \right)^{1/2} \cdot \left( \sum_{n=1}^{N} b_n^2 \right)^{1/2} + \sum_{n=1}^{N} b_n^2$$

$$= \left[ \left( \sum_{n=1}^{N} a_n^2 \right)^{1/2} + \left( \sum_{n=1}^{N} b_n^2 \right)^{1/2} \right]^2$$

so that

$$\sum_{n=1}^{N} (a_n + b_n)^2 \leqslant \left[ \left( \sum_{n=1}^{N} a_n^2 \right)^{1/2} + \left( \sum_{n=1}^{N} b_n^2 \right)^{1/2} \right]^2$$

On taking positive square roots the theorem results.

Minkowski's inequality is the triangle inequality for Euclidean space of $N$ dimensions. For convenience, Fig. 1.3 illustrates this in two dimensions, but let us consider it analytically in three dimensions. In three dimensions $\sqrt{(a_1^2 + a_2^2 + a_3^2)}$ is the distance from the point $(0, 0, 0)$ to the point $(a_1, a_2, a_3)$, $\sqrt{(b_1^2 + b_2^2 + b_3^2)}$ is the distance from the point $(a_1, a_2, a_3)$ to the point $(a_1 + b_1, a_2 + b_2, a_3 + b_3)$, and $\sqrt{[(a_1 + b_1)^2 + (a_2 + b_2)^2 + (a_3 + b_3)^2]}$ is the distance from the point $(0, 0, 0)$ to the point $(a_1 + b_1, a_2 + b_2, a_3 + b_3)$. In this case, therefore, the square roots are the lengths of the sides of the triangle with vertices $(0, 0, 0)$, $(a_1, a_2, a_3)$ and $(a_1 + b_1, a_2 + b_2, a_3 + b_3)$, and the inequality states that the length of the last of these three sides is less than or equal to the sum of the lengths of the other two sides.

In the case $N = 1$ Theorem 1.6 is

$$[(a + b)^2]^{1/2} \leqslant [a^2]^{1/2} + [b^2]^{1/2}$$

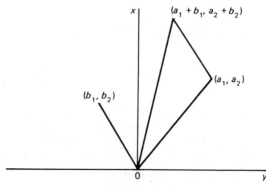

**Fig. 1.3**  $N = 2$

**Definition 1.1**  We write $|x| = \sqrt{(x^2)}$ and call $|x|$ the *modulus* of $x$ or mod $x$. An alternative name for $|x|$ is the *absolute value* of $x$.

Thus

$$|x| = \begin{cases} x & (x \geqslant 0) \\ -x & (x \leqslant 0) \end{cases}$$

The modulus of $x$ is the distance of the point $x$ on the real line from the point 0. The Minkowski or triangle inequality for $N = 1$ can now be written

$$|a + b| \leqslant |a| + |b|$$

Figure 1.4 shows the graph of the modulus as a function of $x$.

The following are immediate consequences of the definition:

(i)   $|x| \geqslant 0$ all real numbers $x$,
(ii)  $|a \cdot b| = |a| \cdot |b|$ all real numbers $a$ and $b$,
(iii) $|a| = |-a|$ all real numbers $a$, so that, for example, $|a - b| = |b - a|$,
(iv)  $|x| < a \Leftrightarrow -a < x < a$ all real numbers $x$ and $a$,
(v)   $|x| > a \Leftrightarrow x < -a$ or $x > a$ all real numbers $x$ and $a$.

Notice the difference between these last two: $|x| < a$ implies that both $-a < x$ and $a > x$ hold, but $|x| > a$ implies only that one of $x < -a$ or $x > a$, and it is clear that both cannot hold.

In this book two fields are in continual use. The first field is the *real field* denoted by $\mathbb{R}$, on which the discussion has so far centred. This is an ordered field, which is thought of as representing by its members the points on a line, and so its members will sometimes be called points and sometimes numbers. A further axiom is introduced later which determines whether there are numbers in $\mathbb{R}$ not in $\mathbb{Q}$, and if there are such numbers how they are related

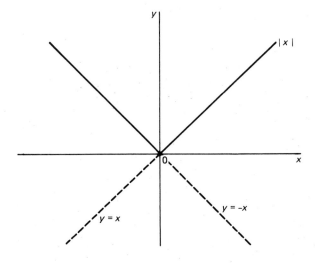

**Fig. 1.4**  The graph of mod $x$

to $\mathbb{Q}$. For the moment, however, we shall consider that we have finished the introduction of $\mathbb{R}$, and in particular the product $a \cdot b$ will now usually be written as $ab$, although the dot will occasionally be used for clarity in a complicated expression. The second field is the *complex field* $\mathbb{C}$. This is not an ordered field, but contains $\mathbb{R}$ and so can be described in terms of this ordered field. In many contexts it is necessary to have a solution to the equation $x^2 = -1$. This is not possible in $\mathbb{R}$, as in any ordered field $x^2 \geqslant 0$ and $-1 < 0$. $\mathbb{C}$ is the smallest field with $\mathbb{R} \subset \mathbb{C}$ such that there is an $i \in \mathbb{C}$ with $i^2 = -1$.

It has yet to be shown that there is such a field $\mathbb{C}$. Should one exist then for $\alpha$, $\beta$, $\gamma \in \mathbb{R}$ the equation $\alpha + i \cdot \beta = \gamma$ would imply $i = (\gamma - \alpha)/\beta$ or $\beta = 0$. However, $(\gamma - \alpha)/\beta$ is in the real field, so that $[(\gamma - \alpha)/\beta]^2 \geqslant 0$, and so is not $-1$. Thus $i \neq (\gamma - \alpha)\beta$ and so $\beta = 0$; that is if $\beta \neq 0$, $\alpha + i \cdot \beta \notin \mathbb{R}$. From this it follows that if $\alpha$, $\beta$, $\alpha_1$, $\beta_1 \in \mathbb{R}$, so that $\alpha + i \cdot \beta$, $\alpha_1 + i \cdot \beta_1 \in \mathbb{C}$, then $\alpha + i \cdot \beta = \alpha_1 + i \cdot \beta_1$ or $\alpha - \alpha_1 + i \cdot (\beta - \beta_1) = 0$ implies, since 0 is real, that $\beta = \beta_1$ and so $\alpha = \alpha_1$. Thus every pair $\langle \alpha, \beta \rangle$ of real numbers determines a complex number $\alpha + i \cdot \beta$, and no two pairs determine the same number. To see if there are any other numbers which must be in $\mathbb{C}$, we check those field axioms which assert the existence of members of the field. If $z_1 = x_1 + i \cdot y_1$ and $z_2 = x_2 + i \cdot y_2$ with $x_1$, $x_2$, $y_1$, $y_2 \in \mathbb{R}$ then (1)(i) says that $z_1 + z_2 \in \mathbb{C}$. Using (1)(ii) and (iii) and (3), $z_1 + z_2 = x_1 + x_2 + i \cdot (y_1 + y_2)$. Field axiom (1)(iv) says that $0 \in \mathbb{C}$, but $0 = 0 + i \cdot 0$, since we have seen that $a \cdot 0 = 0$ for all $a$ in the field. Next comes (1)(v), according to which there is a $-z_1 \in \mathbb{C}$ such that $z_1 + (-z_1) = 0$. Clearly, $-x_1 + i \cdot (-y_1) = -z_1$, as again using (1)(ii) and (iii), $x_1 + i \cdot y_1 + [-x_1 + i \cdot (-y_1)] = x_1 - x_1 + i \cdot (y_1 - y_1) = 0 + i \cdot 0 = 0$. By 2(i) we should have $z_1 \cdot z_2 \in \mathbb{C}$. Now

$$
\begin{aligned}
z_1 \cdot z_2 &= (x_1 + i \cdot y_1) \cdot (x_2 + i \cdot y_2) \\
&= x_1 \cdot (x_2 + i \cdot y_2) + i \cdot y_1 \cdot (x_2 + i \cdot y_2) \\
&= x_1 x_2 + i \cdot x_1 y_2 + i \cdot y_1 x_2 + i \cdot y_1 \cdot i \cdot y_2 \\
&= (x_1 x_2 - y_1 y_2) + i \cdot (x_1 y_2 + x_2 y_1)
\end{aligned}
$$

The axiom (2)(iv) makes $1 \in \mathbb{C}$ and $1 = 1 + i \cdot 0$. There remains axiom 2(v). This requires that if $z_1 \neq 0$ there should be a $z_1^{-1} \in \mathbb{C}$ such that $z_1 \cdot z_1^{-1} = 1$. It is easy to calculate that if $z_1^{-1}$ is of the form $u + i \cdot v$ with $u$, $v \in \mathbb{R}$ then $u = x_1/(x_1^2 + y_1^2)$ and $v = -y_1/(x_1^2 + y_1^2)$. Both $u$ and $v$ are defined provided that not both $x$ and $y$ are 0, that is if $z_1 \neq 0$. It is sufficient here to check that

$$
z_1 \cdot \left( \frac{x_1}{x_1^2 + y_1^2} + i \cdot \frac{-y_1}{x_1^2 + y_1^2} \right) = 1
$$

that is

$$
\begin{aligned}
(x_1 + i \cdot y_1) &\cdot \left( \frac{x_1}{x_1^2 + y_1^2} + i \cdot \frac{-y_1}{x_1^2 + y_1^2} \right) \\
&= \frac{x_1^2}{x_1^2 + y_1^2} + i \cdot \frac{-y_1 x_1}{x_1^2 + y_1^2} + i \cdot \frac{x_1 y_1}{x_1^2 + y_1^2} + i^2 \cdot \frac{-y_1^2}{x_1^2 + y_1^2} = \frac{x_1^2 + y_1^2}{x_1^2 + y_1^2} = 1
\end{aligned}
$$

Thus $\mathbb{C}$ (if it exists) consists of just one number $a + i \cdot b$ corresponding to each pair $\langle a, b \rangle$ of real numbers. This yields the following way of showing that $\mathbb{C}$ does exist. We define the field $\mathbb{C}$ to consist of all pairs $\langle a, b \rangle$ of real numbers, with operations $+$ and $\cdot$ defined by

$$\langle a, b \rangle + \langle c, d \rangle = \langle a + c, b + d \rangle$$

and $\quad \langle a, b \rangle \cdot \langle c, d \rangle = \langle ac - bd, ad + bc \rangle$

It is then sufficient to verify the axioms (1)(ii) and (iii), (2)(ii) and (iii), and (3). This can be done for each of these axioms by working out each side of the equation, and using the properties of the real numbers to show that the resulting pairs are the same. The calculations are quite straightforward but a little long, so the verification, together with an outline of the verification that the axioms for the real numbers can be satisfied, will be left until further techniques are available. It remains only to identify the numbers of the form $\langle x, 0 \rangle$ with the real numbers $x$, so as to have $\mathbb{R} \subset \mathbb{C}$.

Since $\mathbb{C}$ is $\mathbb{R} \times \mathbb{R}$ with suitably defined operations, $\mathbb{C}$ can be represented in the plane. A line, called the real axis, is drawn corresponding to all $z \in \mathbb{C}$ of the form $z = x + i \cdot 0$. A line at right angles, called the imaginary axis, is drawn through the point corresponding to 0. This represents the numbers of the form $0 + i \cdot y$. Then $x + i \cdot y$ is represented by the fourth corner of the rectangle whose other three corners are the points representing $0$, $x + i \cdot 0$ and $0 + i \cdot y$. The construction is shown in Fig. 1.5. If it is wished to make the distinction between $\mathbb{C}$ and the plane as representing $\mathbb{C}$ in this way then the representation is called the *Argand diagram*. But it is usually more convenient to identify $\mathbb{C}$ and the plane, and to speak of the complex plane, and of complex numbers as points in the plane. This is in line with our treatment of the real numbers and the real line. The use of the word 'imaginary' for numbers of the form $i \cdot y$ with

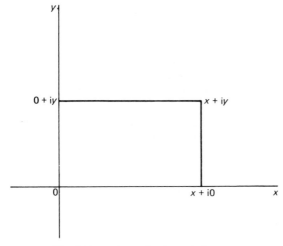

**Fig. 1.5** $x + iy$ on the Argand diagram

negative squares is a relic of the time when the existence of complex numbers was still being established, and people said 'let us imagine that such numbers exist, and see what happens'. What happened was that it was shown that the field $\mathbb{C}$ does indeed exist. For any complex number $z = x + i \cdot y$ with $x$ and $y$ real we write $\text{Re}(z) = x$ and call it the *real part* of $z$ and $\text{Im}(z) = y$ and call it the *imaginary part* of $z$. Notice that $\text{Im}(z)$ is the real number $y$, as this is usually the number that is needed. It might sound better for the imaginary part of $z$ to be an imaginary number, but we wish $\text{Re}(z)$ and $\text{Im}(z)$ to form the pair of real numbers which corresponds to $z$.

It is often convenient to think of a number $z$, not in terms of $\text{Re}(z)$ and $\text{Im}(z)$, but in terms of the line drawn from 0 to $z$. Thus we get a picture of $z_1 + z_2$ as being constructed by drawing a line parallel to and of the same length as that representing $z_2$, but starting from $z_1$ instead of from 0. The other end of this line is then $z_1 + z_2$. Equally we can draw a line parallel to and of the same length as that representing $z_1$, but starting from $z_2$. The picture of these constructions is given in Fig. 1.6.

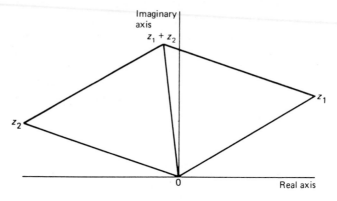

**Fig. 1.6**  The addition of complex numbers

The length of the line representing $z$, or the distance of the point $z$ from 0, is written $|z|$ and called the *modulus* of $z$. Thus $|z| = \sqrt{(x^2 + y^2)}$, where the positive square root is to be taken. The existence of the square root will be a problem which we will discuss in connection with the real numbers. Notice that if $z = x + i \cdot 0$ then $|z| = \sqrt{(x^2)}$, which agrees with the previous definition of the modulus of the real number $x$. If $z_1 = x_1 + i \cdot y_1$ and $z_2 = x_2 + i \cdot y_2$ with $x_1$, $x_2$, $y_1$ and $y_2$ real then

$$z_1 \cdot z_2 = x_1 x_2 - y_1 y_2 + i \cdot (x_1 y_2 + x_2 y_1)$$
$$|z_1 \cdot z_2| = \sqrt{[(x_1 x_2 - y_1 y_2)^2 + (x_1 y_2 + x_2 y_1)^2]}$$
$$= \sqrt{[x_1^2 x_2^2 + y_1^2 y_2^2 + x_1^2 y_2^2 + x_2^2 y_1^2]}$$
$$= \sqrt{[(x_1^2 + y_1^2)(x_2^2 + y_2^2)]} = |z_1| \cdot |z_2|$$

Thus $|z_1 \cdot z_2| = |z_1| \cdot |z_2|$, and also by repeated applications of this $|z^n| = |z|^n$. Furthermore, in view of Theorem 1.6, it follows from the definition of the modulus that

$$|z_1 + z_2| \leqslant |z_1| + |z_2|$$

If we look at this result in the Argand diagram we see why the name triangle inequality is appropriate. Figure 1.7 illustrates this.

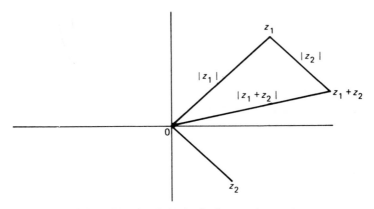

**Fig. 1.7** The triangle inequality for complex numbers

The notation of the modulus enables us to write a formula which describes a circle in the complex plane. The set $\{z \mid |z - a| = r\}$, where $a$ is a complex number and $r$ is a real positive number, is a circle centred at $a$ with radius $r$. This set can be thought of as a curve. The word *circle* will be used for such a set; the word 'circumference' is also used. Figure 1.8 illustrates this way of writing a circle. The sets $\{z \mid |z - a| < r\}$ and $\{z \mid |z - a| \leqslant r\}$ will be called *disks*, the former being called an *open disk* and the latter a *closed disk*. So a disk can be thought of as a set bounded by, or enclosed by, a circle.

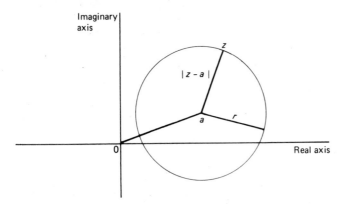

**Fig. 1.8** The circle $|z - a| = r$

One further notation used in connection with $\mathbb{C}$ is required at this stage. Geometrically it represents a reflection in the real axis. If $z = x + i \cdot y$ with $x$ and $y$ real then the *complex conjugate* of $z$ written $\bar{z}$ is $x - i \cdot y$. Usually there is no danger of confusion and it is just called the conjugate of $z$. The three numbers $\mathrm{Re}(z)$, $\mathrm{Im}(z)$ and $|z|$ can be expressed by using the conjugate, as follows.

$$\mathrm{Re}(z) = x = \tfrac{1}{2} \cdot [(x + i \cdot y) + (x - i \cdot y)] = \tfrac{1}{2}(z + \bar{z})$$

$$\mathrm{Im}(z) = y = \frac{1}{2i} \cdot [(x + i \cdot y) - (x - i \cdot y)] = \frac{1}{2i} \cdot (z - \bar{z})$$

$$|z| = \sqrt{(x^2 + y^2)} = \sqrt{[(x + i \cdot y) \cdot (x - i \cdot y)]} = \sqrt{(z \cdot \bar{z})} \quad \text{or} \quad \bar{z} \cdot z = |z|^2$$

It is often useful when calculating the conjugate of a number given by a complicated expression to notice the following:

(i)   $\overline{z_1 + z_2} = \overline{z_1} + \overline{z_2}$, as $x_1 + x_2 - i \cdot (y_1 + y_2) = x_1 - i \cdot y_1 + x_2 - i \cdot y_2$,
(ii)  $\overline{z_1 - z_2} = \overline{z_1} - \overline{z_2}$,
(iii) $\overline{z_1 \cdot z_2} = \overline{z_1} \cdot \overline{z_2}$, as $(x_1 x_2 - y_1 y_2) - i \cdot (x_1 y_2 + y_1 x_2) = (x_1 - i \cdot y_1) \cdot (x_2 - i \cdot y_2)$
since $(-i \cdot y_1) \cdot (-i \cdot y_2) = -y_1 y_2$,
(iv)  $\overline{z^{-1}} = (\bar{z})^{-1}$, as $\overline{z \cdot z^{-1}} = \bar{1} = 1$, and so by (iii) $\bar{z} \cdot \overline{(z^{-1})} = 1$ which is the equation that defines $(\bar{z})^{-1}$. Hence to find the conjugate of any expression which is built up from numbers $z_1, \ldots, z_n$ by additions, subtractions, multiplications and divisions, all we need to do is replace each of $z_1, \ldots, z_n$ by its conjugate. For instance, $\overline{[(z_1^2 + z_2)/(z_2^2 + z_3)]} = (\overline{z_1}^2 + \overline{z_2})/(\overline{z_2}^2 + \overline{z_3})$. Notice that $\bar{\bar{z}} = z$, and that $\bar{z} = z^{-1}$ iff $|z| = 1$, as $|z| = 1$ is equivalent to $z \cdot \bar{z} = 1$.

From now on the dot in the multiplication of complex numbers will regularly be omitted.

## Exercises

**1**   Suppose that $F$ is a field, and the set $P \subset F$ is such that:
(i)   for every $x \in F$ one and only one of $x \in P$, $-x \in P$ or $x = 0$ holds,
(ii)  if $x, y \in P$ then $x + y \in P$,
(iii) if $x, y \in P$ then $x \cdot y \in P$.
Also define the relation $<$ on $F$ by $x < y \Leftrightarrow y - x \in P$. Prove that with this relation of order, $F$ is an ordered field [that is, you are being asked to verify the order axioms].

**2**   Suppose now that $F$ is an ordered field and define the set $P$ by $P = \{x \mid x \in F \text{ and } x > 0\}$. Show that $P$ satisfies the conditions (i), (ii) and (iii) of question 1.

Notice that the results of questions 1 and 2 between them show that the set $P$ of positive numbers gives an alternative way of defining what is meant by an ordered field.

**3** Prove that for all $a \in \mathbb{R}$, $a^2/(a^4+1) \leqslant \frac{1}{2}$.

**4** Let $a, b > 0$ be real numbers and $n, m \in \mathbb{N}$. By considering $(a^n - b^n) \cdot (a^m - b^m)$, or otherwise, show that $a^{n+m} + b^{n+m} \geqslant \frac{1}{2}(a^n + b^n)(a^m + b^m)$. Deduce that $a^3 + b^3 \geqslant \frac{1}{4}(a+b)^3$.

**5** Prove that if $c_1, \ldots, c_N$ are any real numbers $(\sum_{n=1}^{N} c_n)^2 \leqslant N \sum_{n=1}^{N} c_n^2$.

**6** Find the real number $x$, either as a unique value or in terms of a set of inequalities for $x$, if
(i)   $|x-3| = 1$ but $(x-3) \neq 1$,
(ii)  $2 < |x+7| < 9$,
(iii) $2 < |x-7| < 9$,
(iv)  $|x^2 - 3x - 1| > 3$.

**7** Find the real and imaginary parts and the moduli of the following complex numbers: $1/(2+3i)$, $1/(2+i)$ and $(1+i)^2$ and, if $x$ and $y$ are real, of $(x-iy)/(x+iy)$ and $(x+iy)^3$.

**8** Prove that in the complex plane the set $\{z \mid |(z-1)/(z+1)| = 2\}$ is a circle.

**9** Prove that for $z_1, z_2 \in \mathbb{C}$, $||z_1| - |z_2|| \leqslant |z_1 - z_2|$.

# 2

# Sets of real numbers and sets of complex numbers

The situations to be analysed are ones in which there is an infinite amount of information. So it is necessary to consider not just real or complex numbers, but sets of real or complex numbers. The sets to be considered will usually have infinitely many members.

Sets of real numbers will be considered first. Such sets can be very complicated, as is shown by the following sets:

$$A = \{x \mid x = p/2^n \text{ with } p, n \in \mathbb{Z}\}$$

$$B = \{x \mid x \in \mathbb{R} \text{ and } |x - 1/n| < 1/n^4, n \text{ an integer and } |n| > 1\}$$

and $\quad C = \{x \mid x \in \mathbb{R} \text{ and } x^3 - x > 0\}$

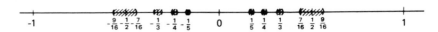

**Fig. 2.1**  The set $B$

The set $B$ is illustrated in Fig. 2.1 and the set $C$ is illustrated in Fig. 2.2. Investigating the set $C$ we have that if $x > 0$ then $x^3 - x > 0 \Leftrightarrow x^2 - 1 > 0 \Leftrightarrow x^2 > 1 \Leftrightarrow x > 1$, as $x > 0$. If $x = 0$ then $x^3 - x = 0$ and so $0 \notin C$. If $x < 0$ then

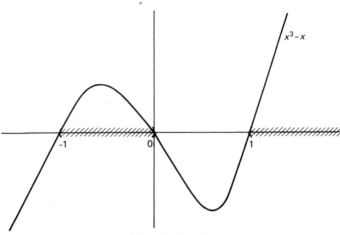

**Fig. 2.2**  The set $C$

$x^3 - x > 0 \Leftrightarrow x^2 - 1 < 0 \Leftrightarrow x^2 < 1 \Leftrightarrow x > -1$, as $x < 0$. Therefore $C = \{x \mid 1 < x\} \cup \{x \mid -1 < x < 0\}$. It is natural to treat $C$ by considering $\{x \mid 1 < x\}$ and $\{x \mid -1 < x < 0\}$ separately. The thing which makes such sets natural units for discussion is that they each consist of one piece, in the sense that given any two points of the set everything in between is also in the set. 'We can get from one point to another in the set without going outside.'

The set $B$ could also be treated by looking at the sets $\{x \mid |x - 1/n| < 1/n^4\}$ separately. They each have this same property. So we make a definition based on this property.

**Definition 2.1(i)** A set $I$ of real numbers is called an *interval* iff whenever both $\alpha, \beta \in I$ then also $x \in I$ for all $x$ such that $\alpha < x < \beta$.

Figure 2.3 illustrates this definition. If $\alpha_1, \beta_1 \in I$, but $\beta_1 < \alpha_1$, we can take $\alpha = \beta_1$ and $\beta = \alpha_1$. Thus it would give an equivalent definition to say that $x \in I$ for all $x$ between $\alpha$ and $\beta$.

**Fig. 2.3** Definition of interval

**Examples of intervals**

(i) $\{x \mid a \leq x \leq b\}$ is an interval since if $a \leq \alpha \leq b$ and $a \leq \beta \leq b$ then $\alpha < x < \beta$ implies that $a \leq \alpha < x < \beta \leq b$, so that $a \leq x \leq b$ and $x$ is in the set. This interval is called a bounded closed interval and is denoted by $[a, b]$.

(ii) $\{x \mid a < x < b\}$ is an interval by a precisely similar argument. It is denoted by $(a, b)$ and is called a bounded open interval.

(iii) $\{x \mid x < b\}$ is an interval. Again this can be proved by a similar argument. It is denoted by $(\cdot, b)$ and is called a semi-infinite open interval.

The notation for these intervals is systematic. The condition that $x$ be in such an interval is expressed by at most two inequalities on $x$. If there is an inequality saying that $x$ must be less than, or less than or equal to $b$, this is denoted by $b$ with a bracket to the right. The form of the bracket depends on whether the inequality is '$<$', in which case a round bracket ')' is used, or whether it is '$\leq$', in which case a square bracket ']' is used. Similarly, an inequality saying that $x$ is larger than $a$, or is at least as large as $a$, is denoted by $a$ with a bracket to the left. The same rule determines the form of the bracket. The absence of an inequality is denoted by a dot with a round bracket. We thus have the following 9 possible intervals of this form:

$$
\begin{array}{ccc}
(\cdot, \cdot) & (\cdot, b) & (\cdot, b] \\
(a, \cdot) & (a, b) & (a, b] \\
[a, \cdot) & [a, b) & [a, b]
\end{array}
\qquad (A)
$$

The intervals $(a, b]$ and $[a, b)$ are sometimes called half closed or half open. The intervals $(\cdot, b)$, $(\cdot, b]$, $(a, \cdot)$ and $[a, \cdot)$ are all called semi-infinite, and $(\cdot, \cdot)$ is the whole real field. These five types are all unbounded intervals, while the remaining four are all bounded intervals.

Various forms of this notation are in use. The principle difference is in what brackets are used to indicate the two types of inequality '<' and '≤'. Thus some books use a square bracket facing outwards to indicate '<', and would write $]a, b[$ where we have $(a, b)$. This has the advantage of suggesting that $a$ and $b$ are not in the interval, but can be confusing to read if two intervals of this kind are written near to each other. Also most books use the sign '∞' where '·' is used here. But there is no number ∞ in the real field, and it is best to avoid using this symbol in places which might suggest it was a number.

These nine types of interval $A$ are all expressed in terms of end points $a$ and $b$. So we have a formal definition of the term 'end point'.

**Definition 2.1(ii)**  If $I$ is an interval such that $c \in I$, and if there is a point $a$ such that $a$ is between $c$ and $x$ implies that $x \notin I$ while for all $x$ between $c$ and $a$, $x \in I$, then $a$ is an *end point* of $I$. If $a \leq c$ then $a$ is a *lower* or *left* end point of $I$ and if $c \leq a$ then $a$ is an *upper* or *right* end point of $I$.

If it is necessary to check whether or not a given set is an interval, it is sometimes easier to use the condition given by the following theorem than to use the definition directly.

**Theorem 2.1**  The non-empty set $I$ of real numbers is an interval if and only if there is some fixed point $c \in I$ such that for every $\gamma \in I$, $c < x < \gamma$ and $\gamma < x < c$ each imply $x \in I$ [that is if $x$ is between $c$ and any $\gamma \in I$ then $x \in I$].

*Proof*
(i)  It is first shown that $I$ is an interval only if the condition holds. If $I$ is an interval then for all $a$, $c \in I$, $b$ between $a$ and $c$ implies that $b \in I$. In particular, this is true for the fixed point $c$.
(ii)  In the second place it is shown that $I$ is an interval if the condition holds. Assume that there is a fixed point $c \in I$ such that if $\gamma \in I$ and $x$ is between $c$ and $\gamma$ then $x \in I$. [It must be shown that $I$ is an interval.] Take any $a, d \in I$ and $b$ between $a$ and $d$ so that $b = a + t(d - a)$ for some $t$ with $0 < t < 1$. Then $(b - a)(1 - t) = (d - b)t$. Hence, as $1 - t > 0$ and $t > 0$, $a - b$ and $d - b$ are of opposite signs and, unless $b = c$, one of them is of the same sign as $b - c$. Defining $\tau$ as follows:

$$\tau = \begin{cases} (b-c)/(a-c) & ((a-b)(b-c) > 0) \\ (b-c)/(d-c) & ((d-b)(b-c) > 0) \end{cases}$$

gives

$$(b-c)(1-\tau) = \begin{cases} (a-b)\tau & ((a-b)(b-c) > 0) \\ (d-b)\tau & ((d-b)(b-c) > 0) \end{cases}$$

Then $1 - \tau$ and $\tau$ are of the same sign. If they were both negative this would imply that $1 < 0$. So they are both positive and $0 < \tau < 1$. Therefore either $b = c + \tau(a - b)$ or $b = c + \tau(d - b)$ with $0 < \tau < 1$, that is either $b = c$ or $b$ is between $a$ and $c$ or $b$ is between $d$ and $c$. In all these cases the assumption implies that $b \in I$. Hence $I$ is an interval.

In view of this theorem, instead of having to test every pair of numbers $\langle \alpha, \beta \rangle$ with $\alpha, \beta \in I$ to see if $I$ is an interval, we can take only one fixed number $c$ and test the pairs $\langle c, \gamma \rangle$ for all $\gamma \in I$. We might, for instance, prove that $[0, 1)$ is an interval by taking $c = 0$. Then if $\gamma \in [0, 1)$, that is $0 \leqslant \gamma < 1$, we have for all $x$ between 0 and $\gamma$, $0 < x < \gamma < 1$. Hence $0 < x < 1$ and so $x \in [0, 1)$. Without the theorem we should have to put the argument in the following way: if $\alpha < \beta$ and $\alpha, \beta \in [0, 1)$ then $0 \leqslant \alpha < \beta < 1$, so that if $\alpha < x < \beta$ then $0 \leqslant \alpha < x < \beta < 1$ and hence $0 < x < 1$ and so $x \in [0, 1)$.

There are certain ways in which intervals can be combined to give further intervals.

**Theorem 2.2** If $\{I_\alpha\}$ is a non-empty set of intervals such that there is a fixed number $c$ with $c \in I_\alpha$ for all $I_\alpha$ of the set then $\bigcup_\alpha I_\alpha$ is an interval which contains $c$.

*Proof* For any number $\gamma \in \bigcup_\alpha I_\alpha$ it follows from the definition of the union that there is an index $a$ for which $\gamma \in I_a$. So by Theorem 2.1 [this is really the same thing as using the definition of an interval] if $x$ is between $c$ and $\gamma$ then $x \in I_a$ and hence $x \in \bigcup_\alpha I_\alpha$. Thus by Theorem 2.1 used in the opposite direction $\bigcup_\alpha I_\alpha$ is an interval. Finally, $c \in \bigcup_\alpha I_\alpha$, as $c \in I_\alpha$ for at least one $\alpha$ since $\{I_\alpha\}$ contains some interval as a member.

Thus
(i)   $(-1, 2] \cup [-3, 0] = [-3, 2]$ where $0 \in (-1, 2]$ and $0 \in [-3, 0]$,
(ii)  $\bigcup_{i \in \mathbb{N}'} [0, 1 - 1/i] = [0, 1)$ where $0 \in [0, 1 - 1/i]$ all positive integers $i$.

**Theorem 2.3** If $\{I_\alpha\}$ is any set of intervals then $\bigcap_\alpha I_\alpha$ is an interval.

*Proof* If both $\gamma, \delta \in \bigcap_\alpha I_\alpha$ then both $\gamma, \delta \in I_\alpha$ for each $I_\alpha$ of the set. Since each $I_\alpha$ is an interval it follows from the definition of an interval that if $\gamma < x < \delta$ then $x \in I_\alpha$ for each $I_\alpha$ of the set, that is $x \in \bigcap_\alpha I_\alpha$. So that if both $\gamma, \delta \in \bigcap_\alpha I_\alpha$ and $\gamma < x < \delta$ then $x \in \bigcap_\alpha I_\alpha$. This is precisely the statement that $\bigcap_\alpha I_\alpha$ is an interval.

Thus
(i)   $[-1, 0) \cap (0, 1] = \emptyset$, and the empty set is an interval,
(ii)  $\bigcap_\Omega [0, b)$ where $\Omega = \{b \mid b^2 > 2 \text{ and } b > 0\}$ is an interval.
Now for any two sets of real or complex numbers we can define two further ways of combination making use of the operations of addition and multiplication.

**Definition 2.2**    If $E$ and $F$ are any two sets either both of real or both of complex numbers then the sets $E + F$ and $E \cdot F$ are as follows:

$$E + F = \{z \mid z = x + y \text{ for some } x \in E \text{ and some } y \in F\}$$

and    $$E \cdot F = \{z \mid z = x \cdot y \text{ for some } x \in E \text{ and some } y \in F\}$$

It is then natural also to write $E - F$ for the set

$$E + (\{-1\} \cdot F) = E + \{z \mid -z \in F\} = \{z \mid z = x - y, x \in E, y \in F\}$$

Notice the danger of confusing $E \cdot F$ with the purely set theoretic Cartesian product $E \times F$ which is a set of pairs. There is also a danger of confusing $E - F$ with $E \backslash F$. Thus, for example, if $E = \{1, 2, 3, 4\}$ and $F = \{2, 4\}$ then $E - F = \{1-2, \ 2-2, \ 3-2, \ 4-2, \ 1-4, \ 2-4, \ 3-4, \ 4-4\} = \{-3, \ -2, \ -1, \ 0, \ 1, \ 2\}$, whereas $E \backslash F = \{1, 3\}$.

**Theorem 2.4**    If the set $I$ of real numbers is an interval and $a$ is any real number then both $\{a\} + I$ and $\{a\} \cdot I$ are intervals.

*Proof*    If $x, y \in \{a\} + I$ and also $x < y$ then $x - a, \ y - a \in I$ and $x - a < y - a$. Thus for any $z$ for which $x < z < y$ we have $x - a < z - a < y - a$ and, as $I$ is an interval, $z - a \in I$. By the definition of $\{a\} + I$ if $z - a \in I$ then $z \in \{a\} + I$. So $\{a\} + I$ is an interval.

Again if $a \neq 0$ and if $x, y \in \{a\} \cdot I$ and we assume that $x < y$ then $x/a, y/a \in I$. Also for any $z$ for which $x < z < y$ we have $z/a$ is between $x/a$ and $y/a$. [The direction of the inequalities will depend on the sign of $a$: if $a > 0$ then $x/a < z/a < y/a$ and if $a < 0$ then $y/a < z/a < x/a$.] So as $I$ is an interval, $z/a \in I$ and hence $z \in \{a\} \cdot I$. Thus $\{a\} \cdot I$ is an interval. In the remaining case when $a = 0$, $\{a\} \cdot I = \{z \mid 0y = z \text{ for some } y \in I\} = \{0\} = [0, 0]$, unless $I = \varnothing$ when $\{a\} \cdot I = \varnothing$ in any case.

**Theorem 2.5**    If both $I$ and $J$ are intervals then the set $I + J$ is also an interval.

*Proof*    [In view of Theorem 2.4, we can simplify the proof by taking intervals which both contain 0.] First, if either $I$ or $J$ is empty then $I + J$ is empty and $\varnothing$ is an interval. So we suppose that $I$ and $J$ both contain numbers. Let $x \in I$ and $y \in J$, and consider $I' = I - \{x\}$ and $J' = J - \{y\}$. Then $I' + J' = I + J - \{x + y\}$, as if $z \in I' + J'$ it can be written $z = x' + y'$ where $x' \in I'$ and $y' \in J'$ and so it can be written $z = x_1 - x + y_1 - y = x_1 + y_1 - (x + y)$ where $x_1 \in I$ and $y_1 \in J$. Now $I'$ and $J'$ are both intervals by Theorem 2.4 and if we can prove that $I' + J'$ is an interval then $I + J = I' + J' + \{x + y\}$ is an interval.

We have $0 \in I'$ and $0 \in J'$ and so $0 \in I' + J'$. If $w \in I' + J'$ then there must be $u \in I'$ and $v \in J'$ such that $w = u + v$. Suppose that $s$ is between 0 and $w$. [Then it must be shown that $s \in I' + J'$.] Now $s = su/w + sv/w$, but $0 < s/w < 1$, so that $(s/w)u$ is between 0 and $u$, and $(s/w)v$ is between 0 and $v$. Thus $su/w \in I'$ and $sv/w \in J'$, so that $s \in I' + J'$. Thus $I' + J'$ is an interval.

The question arises as to whether every interval is of one of the nine forms
A. It would clearly make intervals easier to visualize and some proofs simpler
if we could always express an interval in terms of end points. In the ordered
field $\mathbb{Q}$ of rational numbers there are intervals which cannot be defined by
end points. The set $\{x \mid 0 \leqslant x, x^2 \leqslant 2\}$ is such an interval. It is an interval because
it is the set

$$\bigcap_{\Omega} [0, b) \text{ where } \Omega = \{b \mid b > 0, b^2 > 2\}$$

which we have seen to be an interval. In fact if there is a number $\alpha$ such that
$\alpha^2 = 2$ then $\bigcap_{\Omega} [0, b) = [0, \alpha]$. Moreover it is not too difficult to see that there
cannot be any number $\beta$ with $\beta^2 \neq 2$ such that $\bigcap_{\Omega} [0, b)$ is either $[0, \beta]$ or
$[0, \beta)$. This will be proved in the course of establishing Theorem 2.8. But the
ordered field $\mathbb{Q}$ does not contain any number $\alpha$ with $\alpha^2 = 2$. The next theorem
proves this result.

**Theorem 2.6** There is no number $r \in \mathbb{Q}$ such that $r^2 = 2$.

*Proof* We first note that if $n \in \mathbb{N}$ and $n^2$ is even then $n$ is even. Suppose that
$n = 2m + 1 \in \mathbb{N}$ is odd. Then $n^2 = (2m + 1)^2 = 4m^2 + 4m + 1 = 2(2m^2 + 2m) + 1$,
where $2m^2 + 2m \in \mathbb{N}$. If $n$ is odd then its square is odd. [Actually it is obvious
also that if $n$ is even then $n^2$ is even, but what we use is the fact that if $n^2$ is
even then $n$ is even.]

Next, suppose that there were some $r \in \mathbb{Q}$ with $r^2 = 2$. As $r^2 = (-r)^2$, we may
assume that $r > 0$. Then we could write $r = p/q$ with $p, q \in \mathbb{N}$. We would have
$2 = p^2/q^2$ and hence $p^2 = 2q^2$. Thus $p^2$ would be even, and so $p$ would be even.
Write $p = 2p_1$. Then $(2p_1)^2 = 4p_1^2 = 2q^2$, so that $2p_1^2 = q^2$. Hence $q^2$ and so $q$
would be even. Write $q = 2q_1$. We get $r = p_1/q_1$.

Now $2^p > p$ and $2^q > q$, so that by dividing out by, at most the smaller of $p$
and $q$, factors of 2, we can get $r = p_2/q_2$ with at least one of $p_2$ and $q_2$ odd.
However, we saw that generally if $r = p/q$ then both of $p$ and $q$ are even. So
$p_2$ and $q_2$ have to be both even. We have a contradiction, and it must be that
there is no such $r$.

The full situation is as follows: if $r \in \mathbb{Q}$ then $r^2 \in \mathbb{N}$ iff $r \in \mathbb{N}$. This can be
proved by essentially the same argument as was used in Theorem 2.6. We need
to use the fact that if $p$ and $k$ are integers with no common factors and $p^2$ is
a multiple of $k$ then $p$ is a multiple of $k$. This is an easy consequence of the
fact that any integer can be resolved into a product of prime factors in an
essentially unique way. We shall find, however, in Theorem 2.10 that if we
once have a number $\sqrt{2}$ then there are plenty of other numbers which cannot
be in $\mathbb{Q}$. So we will not stop here to develop the moderate amount of algebra
needed to establish the unique factorization. There are many books on algebra
which do this, and a very full discussion of the factorization properties in $\mathbb{Z}$
is given in *Rings, Fields and Groups* by R B J T Allenby, published by Edward
Arnold. The result required is Theorem 1.5.1 in that book.

It is easy to see that there are many positive integers which are not squares of integers. We can just go through $\mathbb{N}'$ calculating the squares and noticing which positive integers get missed out. In particular, as $1^2 = 1$, $2^2 = 4$, $3^2 = 9$ and as for any $n \geq 4$ also $n^2 \geq 16 > 9$, the numbers 2, 3, 5, 6, 7 and 8 cannot be the squares of any integers. Thus 2, 3, 5, 6, 7 and 8 are not the squares of any rational numbers.

Hence if all intervals in $\mathbb{R}$ are to be of one of the forms $A$, there must be numbers $\alpha \in \mathbb{R}$ which are not rational. We will call such numbers *irrational* numbers. Since it is not immediately clear what other ways there may be of finding intervals which would need irrational end points, we ensure the existence of enough irrational numbers by adopting an axiom to the effect that all intervals have end points. However, it cannot say exactly this as $(\cdot, \cdot)$ has no end points.

**Axiom C (completeness)**    In the field $\mathbb{R}$ if $I$ is an interval and $c \in I$ but $d \notin I$ then there is an end point $a$ of $I$, and $a$ is between $c$ and $d$.

This axiom is equivalent to the assumption that every interval on the real line is of one of the nine forms $A$. For if there is no point $c \in I$ then $I = \varnothing$ and we can write $\varnothing$ in the form $(1, -1) = [1, -1]$, there being no numbers $x$ such that $1 \leq x \leq -1$. If there is a number $c \in I$ but no number $d \notin I$ then $I = (\cdot, \cdot)$. If there are both numbers $c \in I$ and $d \notin I$, and $c < d$, then by Axiom C, $I$ has an upper end point $b$. Also $b$ may or may not belong to $I$. If $c \in I$ and $d_1 \notin I$ but $d_1 < c$ then $I$ has a lower end point $a$ which may or may not belong to $I$. There are no other possibilities, and these cases give the forms $(\cdot, b)$, $(\cdot, b]$, $(a, \cdot)$ and $[a, \cdot)$ and, if both $d$ and $d_1$ exist, $(a, b)$, $[a, b)$, $(a, b]$ and $[a, b]$. It is straightforward to check that none of the forms $A$ is in conflict with the axiom.

Axiom C completes our description of the real numbers by axioms, and therefore also completes our description of the complex numbers.

It is often important to know whether or not the end points of an interval belong to the interval. So the following classification of intervals is made.

**Definition 2.3**
(i)    If $I$ is an interval and for all end points $a$ of $I$, $a \in I$ then $I$ is called a *closed* interval.
(ii)    If $I$ is an interval and for all end points $a$ of $I$, $a \notin I$ then $I$ is called an *open* interval.
(iii)    If $I$ is an interval which has one end point $a_1 \in I$ and a second end point $a_2 \notin I$ then $I$ is called a *half open* or a *half closed* interval.

So that intervals of the four forms $(\cdot, \cdot)$, $(\cdot, b]$, $[a, \cdot)$ and $[a, b]$ are closed; intervals of the four forms $(\cdot, \cdot)$, $(\cdot, b)$, $(a, \cdot)$ and $(a, b)$ are open; intervals of the two forms $(a, b]$ and $[a, b)$ are half open or half closed.

Notice that the intervals $(\cdot, \cdot)$ and $\varnothing = (1, -1) = [1, -1]$ are the only ones which are both open and closed.

Also if the intervals of Theorem 2.5 are both bounded and open and $I = (a, b)$ and $J = (c, d)$ then $I + J = (a + c, b + d)$ is bounded and open. If, however, they are both bounded and closed with $I = [a, b]$ and $J = [c, d]$ then $I + J = [a + c, b + d]$ is bounded and closed. There are similar results with unbounded intervals.

If $I$ is an open interval and $c \in I$ then there are numbers $a$ and $b$ such that $a < c < b$ and $a, b \in I$. To see this, we can write down the numbers $a$ and $b$ in terms of the end points of $I$. If $I$ has an upper end point $\beta$ we can take $b = \frac{1}{2}(c + \beta)$, and if $I$ has no upper end point we can take $b = c + 1$. Similarly, if $\alpha$ is a lower end point of $I$ then put $a = \frac{1}{2}(\alpha + c)$, and if there is no lower end point put $a = c - 1$.

**Theorem 2.7**    If $I$ is a non-empty open interval and $J$ and $K$ are also non-empty open intervals with $J \cap K = \varnothing$ then $I \neq J \cup K$.

*Proof*    If $I = J \cup K$ then, as there are numbers $a \in J$ and $b \in K$ such that $b \notin J$, there is an end point $c$ of $J$ between $a$ and $b$. As $J$ is open, $c \notin J$, but $c \in I$ and so we must have $c \in K$. However, as we have just noted, the fact that $K$ is open now implies that $\exists \alpha, \beta$ with $\alpha < c < \beta$ and $\alpha, \beta \in K$. This contradicts the fact that all points between $a$ and $c$ are in $J$.

To see that Axiom C does in fact ensure the existence of square roots, we prove the following theorem.

**Theorem 2.8**    In any ordered field if the interval $\{x \mid 0 \leqslant x, x^2 \leqslant \gamma\}$ has an upper end point $\alpha$ then $\alpha^2 = \gamma$ and $\{x \mid 0 \leqslant x, x^2 \leqslant \gamma\} = [0, \alpha]$. In particular, in the real field every positive number has a unique positive square root.

*Proof*    Suppose that $\beta \in (0, \alpha]$ and $\beta^2 < \gamma$. [We have to show that $\beta$ is not the end point $\alpha$. To do this, we choose a number $\beta + \varepsilon > \beta$ such that $\beta + \varepsilon \in [0, \alpha)$.] Let $\varepsilon$ be any number satisfying $\varepsilon > 0$, $\varepsilon < \beta$ and $\varepsilon < (\gamma - \beta^2)/(3\beta)$, this last number is positive as $\beta^2 < \gamma$. Such a number could be $\frac{1}{2} \min\{\beta, (\gamma - \beta^2)/(3\beta)\}$. [Here the notation $\min E$ is used to indicate the smallest number in the set $E$.] Then

$$(\beta + \varepsilon)^2 = \beta^2 + 2\beta\varepsilon + \varepsilon^2 < \beta^2 + 2\beta\varepsilon + \beta\varepsilon \qquad (\text{as } \beta > \varepsilon)$$

$$= \beta^2 + 3\beta\varepsilon$$

$$< \beta^2 + \gamma - \beta^2 \qquad (\text{as } 3\beta\varepsilon < \gamma - \beta^2)$$

$$= \gamma$$

Suppose that $\beta_1^2 > \gamma$. [We have to show again that $\beta_1$ is not an end point of $[0, \alpha]$ and so is not $\alpha$. As before, we do this by showing that there are numbers between $\beta_1$ and the end point $\alpha$.] Choose $\varepsilon_1 > 0$ such that $\varepsilon_1 < (\beta_1^2 - \gamma)/(2\beta_1)$, again $\varepsilon_1 = (\beta_1^2 - \gamma)/(4\beta_1)$ will do, and $\beta_1^2 - \gamma > 0$. Then

$$
\begin{aligned}
(\beta_1 - \varepsilon_1)^2 &= \beta_1^2 - 2\beta_1 \varepsilon_1 + \varepsilon_1^2 \\
&> \beta_1^2 - 2\beta_1 \varepsilon_1 \quad (\text{as } \varepsilon_1^2 > 0) \\
&> \beta_1^2 - (\beta_1^2 - \gamma) \quad (\text{as } \varepsilon_1 < (\beta_1^2 - \gamma)/(2\beta_1)) \\
&= \gamma
\end{aligned}
$$

Thus neither $\alpha^2 < \gamma$ nor $\alpha^2 > \gamma$, so that $\alpha^2 = \gamma$, and $\{x \mid 0 \leqslant x, x^2 \leqslant \gamma\} = [0, \alpha]$. [It might have been $[0, \alpha)$.]

In the field $\mathbb{R}$ we have that the interval $\{x \mid 0 \leqslant x, x^2 \leqslant \gamma\}$ must have an upper end point $\alpha$. This follows as for all $\gamma > 0$, 0 is in the interval, but since $(\gamma + 1)^2 = \gamma^2 + 2\gamma + 1 > \gamma^2$, the number $\gamma + 1$ is not in the interval. Also by the above $0 < \alpha$ and $\alpha^2 = \gamma$. To prove the uniqueness of the positive square root of $\gamma$, suppose also that $0 < \alpha_1$ and $\alpha_1^2 = \gamma$. Then $\alpha - \alpha_1 = (\alpha^2 - \alpha_1^2)/(\alpha + \alpha_1) = 0$, that is $\alpha = \alpha_1$.

Axiom C not only ensures the existence of a lot of numbers; it also implies some limit to what numbers there can be in $\mathbb{R}$. In particular, $(\cdot, \cdot)$ cannot have end points and so it cannot be possible to have real numbers larger than all the rational numbers.

**Theorem 2.9   Archimedean Property**   In the real field if $0 < a < b$ then there is an integer $n$ such that $na > b$.

*Proof*   Consider the set $A = \bigcup_{n \in \mathbb{N}} [0, na]$. $A$ is an interval by Theorem 2.2, as each interval $[0, na]$ contains 0. If there is no integer $n$ for which $na > b$ then $b \notin A$, as $b \notin [0, na]$ for all $n \in \mathbb{N}$. So by Axiom C the interval $A$ has an end point $c$ between 0 and $b$. Hence if $0 \leqslant x < c$ then $x \in A$ and so $x \in [0, ma]$ for some $m \in \mathbb{N}$. In particular, $c - a < c$ and, as $a \in [0, a] \subset A$, we have $c \geqslant a$, so that $c - a \geqslant 0$ and the inequalities imply that $c - a \in A$. Thus there is an $n_0 \in \mathbb{N}$ such that $c - a \in [0, n_0 a]$, and so $c - a \leqslant n_0 a$. As $a > 0$, this gives $c < (n_0 + 2)a$. [It could be that $(n_0 + 1)a = c$, which does not lead to the contradiction] but $[0, (n_0 + 2)a] \subset A$, so that $(n_0 + 2)a \in A$ and $c$ is not an end point of $A$. Thus we have a contradiction unless there is an $n$ for which $na > b$. [The proof is illustrated in Fig. 2.4.]

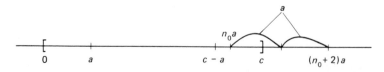

**Fig. 2.4**   Proof of the Archimedean Property

*Applications*

(i)   Given any real number $x$, there is an integer $n > x$. If $x \leq 1$ take $n = 2$, otherwise let $a = 1$ and $b = x$ in the above theorem.

(ii)   Given any real number $\delta > 0$, $\exists$ an integer $n > 0$ such that $1/n < \delta$. All we have to do is to find an $n \in \mathbb{N}$ such that $n > 1/\delta$.

**Theorem 2.10**   If $(a, b)$ is a non-empty interval of real numbers then there is a rational number $r$ and an irrational number $\alpha$ such that $r, \alpha \in (a, b)$.

*Proof*

(i)   We first consider the case in which both $a$ and $b$ are rational. Take $r = \frac{1}{2}(a+b)$ and $\alpha = a + \sqrt{2}[\frac{1}{2}(b-a)]$. Then $a < r < b$ and $0 < \sqrt{2} < 2$, so that $a < \alpha < a + (b-a) = b$. Now $r$ is clearly rational. Also $\alpha$ is irrational, since if $\alpha$ were rational then $\sqrt{2} = 2(\alpha - a)/(b - a)$ would be rational. For suppose that $\alpha = p/q$, $a = p_1/q_1$ and $b = p_2/q_2$ where all of $p$, $q$, $p_1$, $q_1$, $p_2$ and $q_2$ are integers, then

$$\sqrt{2} = 2(pq_1q_2 - p_1qq_1)/(p_2qq_1 - p_1qq_2)$$

(ii)   Now if $a$ and $b$ are any real numbers $0 \leq a < b$ we can find $a_1$ and $b_1$ both rational with $a_1 < b_1$ and $[a_1, b_1] \subset (a, b)$. By the second application of the Archimedean Property there is an integer $n$ such that $0 < 1/n < \frac{1}{2}(b-a)$. Then by the Archimedean Property itself there is an integer $m$ such that $m/n > a$. Therefore there is a finite set $S$ of integers $S = \{1, 2, \ldots, k\}$ such that $l/n \leq a$ if $l \in S$ [that is, there is an integer $k$ which is the largest integer such that $k/n \leq a$]. Let $a_1 = (k+1)/n$ and $b_1 = (k+2)/n$. They are clearly both rational and $a < a_1 < b_1$. Also as $a_1 - 1/n \leq a$, we get

$$a_1 + \frac{1}{n} = b_1 \leq a + \frac{2}{n} < a + (b - a) = b$$

by choice of $n$.

If $a < 0 < b$ then we can find $a_1$ and $b_1$ by using the interval $(0, b)$.

If $a < b \leq 0$ then we can find $-a_1$ and $-b_1$ by using the interval $(-b, -a)$.

(iii)   Given any interval $(a, b)$ with $a < b$ by (ii), there are rational numbers $a_1$ and $b_1$ with $a_1 < b_1$ such that $[a_1, b_1] \subset (a, b)$. Hence by (i) there is a rational number $r$ and an irrational number $\alpha$ both in $(a_1, b_1)$ and therefore both in $(a, b)$.

Now let us consider what kinds of sets of complex numbers should play the same role as intervals of real numbers. Since $\mathbb{C}$ is not an ordered field, we have to decide what should be meant by saying that $\zeta$ is between $w$ and $z$. Perhaps the most natural way of defining '$\zeta$ is between $w$ and $z$' is to say that this means $\zeta = w + t(w - z)$ for some real number $t$ with $0 < t < 1$; in other words, to make '$\zeta$ is between $w$ and $z$' mean that $\zeta$ is on the straight line segment between the points $w$ and $z$. Sets $C$ such that if $w, z \in C$ then also any point of the form $w + t(w - z) \in C$ where $0 < t < 1$ are called *convex sets*.

Convex sets can be used in the complex plane in much the same way as intervals on the real line are used. However, if $L_1$ is the line segment from $-1$ to i and $L_2$ is the line segment from i to 1, $L_1$ and $L_2$ are both convex but $L_1 \cup L_2$ contains $-1$ and 1 but not 0 and so is not convex.

The definition for '$\zeta$ is between $w$ and $z$' which leads to the sets in the complex plane which are known as intervals is as follows: $\zeta$ is between $w$ and $z$ iff Re($\zeta$) is between Re($w$) and Re($z$), and Im($\zeta$) is between Im($w$) and Im($z$). The intervals defined on this basis will be convex sets, but they are special convex sets which in some ways are technically easier to handle.

**Definition 2.4**   A set $I$ of complex numbers is called an *interval* iff whenever both $w, z \in I$ then also $\zeta \in I$ for all $\zeta$ such that both

$$[\text{Re}(z) - \text{Re}(\zeta)][\text{Re}(\zeta) - \text{Re}(w)] \geq 0$$

and   $[\text{Im}(z) - \text{Im}(\zeta)][\text{Im}(\zeta) - \text{Im}(w)] \geq 0$

that is the real part of $\zeta$ is between the real parts of $z$ and $w$ or equal to one of them, and the imaginary part of $\zeta$ is between the imaginary parts of $z$ and $w$ or equal to one of them.

The possibility of equality between Re($\zeta$) and Re($z$) or Re($w$), and the corresponding equality between the imaginary parts, is included so as to make the next theorem hold.

**Theorem 2.11**   The set $I$ of complex numbers is an interval iff there are sets $I_r$ and $I_i$ of real numbers which are intervals and such that $I = I_r \times I_i$.   .

*Proof*
(i)   If $I_r$ and $I_i$ are two real intervals and $I_r \times I_i$ is thought of as a subset of $\mathbb{C}$ then $z \in I_r \times I_i$ implies that Re($z$) $\in I_r$ and Im($z$) $\in I_i$, and $w \in I_r \times I_i$ implies that Re($w$) $\in I_r$ and Im($w$) $\in I_i$. Then if $\sigma$ is between Re($z$) and Re($w$), $\sigma \in I_r$, as $I_r$ is an interval. Also if $\sigma = $ Re($z$) or $\sigma = $ Re($w$) then $\sigma \in I_r$. Similarly, if $\tau$ is between Im($z$) and Im($w$) or is equal to one of them then $\tau \in I_i$. Hence $\zeta = \sigma + i\tau \in I_r \times I_i$ which is thus an interval of complex numbers.
(ii)   If $I$ is an interval of complex numbers let $I_r = \{x | \exists z \in I$ such that $x = \text{Re}(z)\}$ and $I_i = \{y | \exists z \in I$ such that $y = \text{Im}(z)\}$. Then for any two real numbers $x \in I_r$ and $y \in I_i$ there are complex numbers $z_1, z_2 \in I$ such that Re($z_1$) $= x$ and Im($z_2$) $= y$. So as $I$ is an interval, the number $z = x + iy$ with Re($z$) = Re($z_1$) and Im($z$) = Im($z_2$) is in $I$. Thus $I = I_r \times I_i$. Now for any real numbers $x, u \in I_r$ and $y, v \in I_i$ the complex numbers $z = x + iy$ and $w = u + iv$ belong to $I$. Thus for any $\sigma$ between $x$ and $u$ and any $\tau$ between $y$ and $v$, as $I$ is an interval, the complex number $\zeta = \sigma + i\tau$ belongs to $I$. Hence $\sigma \in I_r$ and $\tau \in I_i$, and so $I_r$ and $I_i$ are both intervals.

In view of this theorem, we have the following definitions.

**Definition 2.5**  If the complex interval $I = I_r \times I_i$ and:
  (i)   both $I_r$ and $I_i$ are bounded then $I$ is said to be *bounded,*
  (ii)  both $I_r$ and $I_i$ are open then $I$ is said to be *open,*
  (iii) both $I_r$ and $I_i$ are closed then $I$ is said to be *closed.*

There are no results for complex intervals analogous to Theorems 2.1 and 2.2. If we take $c = 0$ and the set $I = \{z = x + iy \mid 0 \leqslant x \leqslant 1, \ 0 \leqslant y \leqslant 1\} \cup \{z = x + iy \mid -1 \leqslant x \leqslant 0, \ -1 \leqslant y \leqslant 0\}$ it is clear (see Fig. 2.5) that the analogue to Theorem 2.1 would make $I$ an interval, and $I$ is not an interval. Also $I$ is the union of two intervals each containing the point 0, but $I$ is not an interval. So the analogue of Theorem 2.2 must be false. However, we do get the following results.

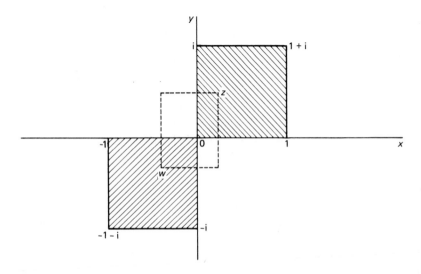

**Fig. 2.5**  The two shaded areas are intervals but their union is not an interval

**Theorem 2.12**  If $\{I_\alpha\}$ is any set of complex intervals then $\bigcap_\alpha I_\alpha$ is an interval.

*Proof*  Each $I_\alpha$ can be written as $I_\alpha = I_{r,\alpha} \times I_{i,\alpha}$ where $I_{r,\alpha}$ and $I_{i,\alpha}$ are real intervals. Then by Theorem 0.4

$$\bigcap_\alpha I_\alpha = \bigcap_\alpha (I_{r,\alpha} \times I_{i,\alpha}) = \left(\bigcap_\alpha I_{r,\alpha}\right) \times \left(\bigcap_\alpha I_{i,\alpha}\right)$$

By Theorem 2.3 each of the intersections is a real interval and so their Cartesian product is again a complex interval.

**Theorem 2.13**  If the set $I$ of complex numbers is an interval and $a$ is any complex number then $\{a\} \cdot I$ is an interval.

*Proof*  Let $a = \alpha + i\beta$. Then if $z = x + iy$ we get $az = \alpha x - \beta y + i(\alpha y + \beta x)$. Hence if $I = I_r \times I_i$ where $I_r$ and $I_i$ are real intervals then

$$\{a\} \cdot I = (\{\alpha\} \cdot I_r - \{\beta\} \cdot I_i) \times (\{\alpha\} \cdot I_i + \{\beta\} \cdot I_r)$$

Now each of $\{\alpha\} \cdot I_r$, $\{\alpha\} \cdot I_i$, $\{\beta\} \cdot I_r$ and $\{\beta\} \cdot I_i$ is a real interval by Theorem 2.4. Hence by Theorem 2.5 $(\{\alpha\} \cdot I_r - \{\beta\} \cdot I_i)$ and $(\{\alpha\} \cdot I_i + \{\beta\} \cdot I_r)$ are real intervals. Thus by Theorem 2.11 $I$ is a complex interval.

**Theorem 2.14**  If the sets $I$ and $J$ of complex numbers are intervals then the set $I + J$ is also an interval.

*Proof*  We may write $I = I_r \times I_i$ and $J = J_r \times J_i$ where $I_r$, $I_i$, $J_r$ and $J_i$ are real intervals. Then

$$I + J = \{z \mid z = w + \zeta \text{ for some } w \in I, \zeta \in J\}$$

$$= \{x + iy \mid x = u + \sigma, y = v + \tau, u \in I_r, \sigma \in J_r, v \in I_i, \tau \in J_i\}$$

$$= (I_r + I_i) \times (J_r + J_i)$$

but $I_r + I_i$ and $J_r + J_i$ are intervals by Theorem 2.5, so that $I + J$ is an interval.

Now let us look at how some properties of more general sets of real or complex numbers can be discussed in terms of intervals.

**Definition 2.6**
(i)   If $E$ is either a set of real numbers or a set of complex numbers and there is a bounded interval $I$ such that $E \subset I$, then $E$ is said to be *bounded*. Notice that we can always take the interval $I$ to be closed.
(ii)   If $E$ is a set of real numbers and there is a semi-infinite interval $[a, \cdot)$ such that $E \subset [a, \cdot)$, then $E$ is said to be *bounded below*, and $a$ is called a *lower bound for E*.
(iii)   If $E$ is a set of real numbers and there is a semi-infinite interval $(\cdot, b]$ such that $E \subset (\cdot, b]$, then $E$ is said to be *bounded above*, and $b$ is called an *upper bound for E*.

Since a bounded interval $[a, b] = (\cdot, b] \cap [a, \cdot)$, a set of real numbers $E$ is bounded precisely if it is both bounded above and bounded below. The situation for complex sets is a little more complicated, and so we will not go beyond Definition (i) for such sets.

The number $a$ is a lower bound for a set $E$ of real numbers iff for all $x \in E, a \leq x$; the number $b$ is an upper bound for $E$ iff for all $x \in E, x \leq b$. Now if $b$ is just an upper bound for $E$ there may be a smaller one, e.g. if $E = \{x \mid 0 < x^2 < 1\}$, 2 is an upper bound for $E$ but $1 < 2$ and 1 is an upper bound for $E$. It is clear that the smaller the upper bound, the more it tells us about the set.

**Definition 2.7** If $s$ is an upper bound for a set $E$ of real numbers and $s \leqslant b$ for all numbers $b$ which are upper bounds for $E$, then $s$ is said to be the *supremum* of $E$, written $s = \sup E$, or $s$ is said to be the *least upper bound* of $E$. Alternative notations are $s = \text{l.u.b. } E = \overline{\text{bound}} E = \overline{\text{bd}} E$, and $\sup_E f(x)$ for $\sup\{f(x) \mid x \in E\}$.

Similarly, if $l$ is a lower bound for a set $E$ of real numbers and $l \geqslant a$ for all numbers $a$ which are lower bounds for $E$, then $l$ is said to be the *infimum* of $E$, written $l = \inf E$, or $l$ is said to be the *greatest lower bound* of $E$. Alternative notations are $l = \text{g.l.b. } E = \underline{\text{bound}} E = \underline{\text{bd}} E$, and $\inf_E f(x)$ for $\inf\{f(x) \mid x \in E\}$.

If they exist for a particular set $E$, $\sup E$ and $\inf E$ are uniquely defined and so the definite article (the) in the definitions is justified. The uniqueness follows as if $s_1$ and $s_2$ were both suprema for $E$, since they would both then be upper bounds for $E$, $s_1 \leqslant s_2$ and $s_2 \leqslant s_1$, so that $s_1 = s_2$. Similarly, if $l_1$ and $l_2$ were both infima for $E$, since $l_2$ would be a lower bound for $E$, $l_2 \leqslant l_1$ and since $l_1$ would be a lower bound for $E$, $l_1 \leqslant l_2$, so that $l_1 = l_2$.

**Definition 2.8**
(i)   If $\inf E$ exists and $\inf E \in E$ we write $\min E$ for $\inf E$.
(ii)  If $\sup E$ exists and $\sup E \in E$ we write $\max E$ for $\sup E$.

The notations $\min E$ and $\max E$ are most frequently used for sets $E$ with a finite number of members. In this case $\min E$ and $\max E$ are always defined. This usage has been mentioned in passing in the proof of Theorem 2.8.

It is useful to have a condition for the existence of the supremum and infimum.

**Theorem 2.15**   If the set of real numbers $E \neq \varnothing$ and is bounded above then $\sup E$ exists.

If the set of real numbers $E \neq \varnothing$ and is bounded below then $\inf E$ exists.

*Proof* [As the proof is similar in the two cases, we consider only the first.] We know that for some $b$, $E \subset (\cdot, b]$. Consider the set $A = \bigcap_{E \subset (\cdot, b]} (\cdot, b]$, that is the intersection of all semi-infinite intervals of this sort which contain $E$. This is an interval by Theorem 2.3. Now if $b$ is such that $E \in (\cdot, b]$ then $b + 1 \notin A$ and, as $E$ is not empty, $\exists c \in E$ and hence $c \in A$. So by Axiom C there is a number $s$, $c \leqslant s \leqslant b + 1$, such that $s$ is the upper end point of $A$. Now $E \subset A$, as $(\cdot, b]$ is used in forming $A$ only if $E \subset (\cdot, b]$. So $s$ is an upper bound for $E$. If $b$ is any upper bound for $E$ then $(\cdot, s) \subset A \subset (\cdot, b]$, so that $s \leqslant b$, that is $s = \sup E$. [It follows that $s \in A$, as $s \in (\cdot, b]$ for all $(\cdot, b] \supset E$, so that $A = (\cdot, s]$.]

The second result follows by an exactly similar argument.

We had as one of the consequences of Axiom C the existence of square roots of positive real numbers. The corresponding result for complex numbers can be deduced directly from this, without having to discuss complex intervals.

**Theorem 2.16**   Every complex number other than 0 has exactly two square roots.

*Proof*   Let $w = u + iv$ and $z = x + iy$ with $u, v, x, y \in \mathbb{R}$. Then $z$ is a square root of $w$ iff

$$z^2 = x^2 - y^2 + 2ixy = w = u + iv$$

iff    $x^2 - y^2 = u$ and $xy = \tfrac{1}{2}v$

iff    $(v^2/4y^2) - y^2 = u$   or   $y = 0$ and $x^2 = u$

iff    $v^2/4 = y^4 + uy^2$   or   $v = y = 0$, $u \geqslant 0$ and $x = \pm\sqrt{u}$

iff    $(y^2)^2 + u(y^2) - (v^2/4) = 0$   or   $v = y = 0$, $u \geqslant 0$ and $x = \pm\sqrt{u}$

iff    $y^2 = \tfrac{1}{2}[-u \pm \sqrt{(u^2 + v^2)}]$   or   $v = y = 0$, $u \geqslant 0$ and $x = \pm\sqrt{u}$

iff    $y = \pm\dfrac{1}{\sqrt{2}}\sqrt{[-u + \sqrt{(u^2 + v^2)}]}$ as $y$ is real, and $x = v/2y$

        or   $v = 0$, $u \geqslant 0$, $y = \pm\dfrac{1}{\sqrt{2}}\sqrt{[-u + \sqrt{(u^2 + v^2)}]}$ and $x = \pm\sqrt{u}$

iff    $y = \pm\dfrac{1}{\sqrt{2}}\sqrt{[-u + \sqrt{(u^2 + v^2)}]}$ and $x = v/2y$

        or   $x = \pm\sqrt{u}$ if $v = 0$ and $u \geqslant 0$ so that $y = 0$

Hence, unless both $u$ and $v$ are 0, $w$ has two square roots.

**Exercises**

**1**   Prove that if $I$, $J$ and $K$ are real intervals and $I \cap J \neq \varnothing$ and $J \cap K \neq \varnothing$ then $I \cup J \cup K$ is an interval. Give an example of this situation where $I \cup K$ is not an interval.

**2**   If $I = (a, b]$ and $J = [c, d]$, find $I + J$.

**3**   For what values of $r$ is the set of real numbers $\{x \,||x - 1||x - 2| < r\}$ an interval? [*Hint*: for a formal proof you can consider the three cases $x \leqslant 1$, $1 \leqslant x \leqslant 2$ and $2 \leqslant x$, and use the result of question 1.]

**4**   Find the sets $[1, 2] \cdot (\tfrac{1}{2}, 3]$ and $[-3, 1] \cdot [-1, 2]$. Are they intervals?

**5**   Find the set $\{z \,|\, z = x + iy, \ x = 1, \ 0 \leqslant y \leqslant 1\} \cdot \{z \,|\, z = x + iy, \ y = 0, \ 0 \leqslant x \leqslant 1\}$. Is this set an interval?

**6**   If $I$ is an interval and $a \in I$ is not an end point of $I$, show that there are numbers $b, c \in I$ such that $b < a < c$.

**7**   Show that if the definition of open interval is used in the field $\mathbb{Q}$ then there are bounded open intervals $I$ and $J$ in $\mathbb{Q}$ such that $I \cap J = \varnothing$ but $I \cup J$ is an open interval.

**8**  If the sets $S$ and $T$ of real numbers are such that $S \neq \emptyset$ and $T$ is bounded, show that $S \subset T \Rightarrow \sup S \leqslant \sup T$ and $\inf S \geqslant \inf T$.

**9**  If $U$ and $V$ are bounded non-empty sets of real numbers, express $\sup(U \cup V)$ and $\inf(U \cup V)$ in terms of $\sup U$, $\sup V$, $\inf U$ and $\inf V$.

**10**  With $U$ and $V$ as in question 9, are there similar expressions for $\sup(U \cap V)$ and $\inf(U \cap V)$? Give examples to illustrate all the cases in your answer.

# 3

# Functions

It very often happens that the information which we have to analyse is not given just as a set of real or complex numbers, but occurs in the form of a function. Thus in describing the speed of a satellite in orbit around the earth, we might state only the set of its speeds taken over the whole orbit. This would be an interval of real numbers. A great deal more would be said about the way it describes the orbit by giving its speed at each instant of time, that is by giving a function $f$ whose value $f(t)$ is the speed of the satellite at time $t$. Both $t$ and $f(t)$ would be real numbers. To take a different sort of example, for a club or society with a fixed annual subscription there is a function which relates the subscription income to the number of members. This would be an integer valued function of integers. However, it is functions with domains and ranges in $\mathbb{R}$ or $\mathbb{C}$ which are of such widespread importance that they are among the main objects of study in mathematical analysis.

In this book we shall mainly discuss functions for which either both the domain and range are in $\mathbb{R}$ or both are in $\mathbb{C}$. A function $f$ such that $\mathscr{D}f, \mathscr{R}f \subset \mathbb{R}$ will be called a *real* function, and a function $g$ such that $\mathscr{D}g, \mathscr{R}g \subset \mathbb{C}$ will be called a *complex* function. If it is wished to state that $\mathscr{R}f \subset \mathbb{R}$ but that $\mathscr{D}f$ may be in either of the fields $\mathbb{R}$ or $\mathbb{C}$, it will be said that $f$ is *real valued*. Correspondingly $f$ is said to be *complex valued* if $\mathscr{R}f \subset \mathbb{C}$ and the domain is in either field. The word function will be taken to refer to a function which may be of any of these sorts.

Sometimes a function will be defined by a formula, e.g. $f$ defined by $f(x) = 1/x$. In such a case if $\mathscr{D}f$ is not specified it will be assumed to be the set of all the numbers in the field, for which the formula makes sense. In this case if $f$ is to be a real function then $\mathscr{D}f = \{x \mid x \in \mathbb{R}, x \neq 0\}$. However, we shall also require to discuss functions such as $g$ defined by $g(x) = 1/x$ for $x > 0$. Here by definition $\mathscr{D}g = \{x \mid x \in \mathbb{R}, x > 0\}$. After the notion of differentiation has been discussed it will be seen that $g$ arises as the derivative of the function log. In most of the particular cases we shall consider, the domains of functions will be intervals or unions of intervals. However, where appropriate, results will be established without restricting $\mathscr{D}f$.

For many purposes the notion of real function is too wide. Consider a function $f$ defined for all real numbers by

$$f(x) = \begin{cases} 1 & (x = p/q\,;\,p, q \in Z\,;\,p, q \text{ no common factors}\,;\,q \text{ odd}) \\ -1 & (x = p/q\,;\,p, q \in Z\,;\,p, q \text{ no common factors}\,;\,q \text{ even}) \\ 0 & (x \text{ irrational}) \end{cases}$$

This defines a function with domain the whole of $\mathbb{R}$. For each irrational $x \in \mathbb{R}$ the value of the function is 0. The rational numbers have been divided into two classes. It is easy to check that it is not possible to have four integers $p$, $q$, $p_1$ and $q_1$ with no integer which divides exactly into both $p$ and $q$, and also no integer which divides exactly into both $p_1$ and $q_1$, with $q$ even, $q_1$ odd, and $p/q = p_1/q_1$. So the value of $f$ at a rational number is either 1 or $-1$, according to which of the two forms is possible. However, the function $f$ is very difficult to do anything with. For example, we cannot really draw a graph for $f$—Fig. 3.1 illustrates this difficulty. Also suppose that we were trying to find such a function by taking physical measurements. The readings would be confusing. Two different values of $x$, such as $\frac{103}{1000}$ and $\frac{103}{1001}$, might seem the same to our measurements but give quite different values to the function, in this case $f(\frac{103}{1000}) = -1$ and $f(\frac{103}{1001}) = 1$. By Theorem 2.10 there is some point in between at which $f$ takes the value 0.

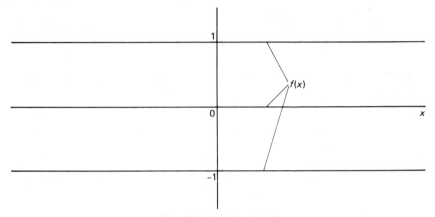

**Fig. 3.1** The function $f$

This function $f$ may seem a bit artificial, but there are in common use methods of constructing functions which can produce functions similar to $f$. So let us try to find conditions which are wide enough to include most of the functions we require to use, and at the same time narrow enough to give the functions special properties which will enable us to work with them. We would also expect the conditions to be such as to apply to most of the functions that the reader is likely to be familiar with.

One simple and useful restriction on a real function $f$ is to ask that it preserve inequalities; that is if $x$ and $z$ are any two real numbers with $x < z$ then we require that $f(x) < f(z)$. We can think of this as saying that as the point $x$ travels from left to right along the $x$-axis the value $f(x)$ travels upwards along the $y$-axis.

**Definition 3.1** If $f$ is a real valued function and the set $E \subset \mathcal{D}f$ then:
(i)   if $x < y \Rightarrow f(x) < f(y)$ for all $x, y \in E$, $f$ is said to be *strictly increasing* on $E$,

(ii)   if $x<y\Rightarrow f(x)\leqslant f(y)$ for all $x, y\in E$, $f$ is said to be *weakly increasing* on $E$,

(iii)   if $x<y\Rightarrow f(x)>f(y)$ for all $x, y\in E$, $f$ is said to be *strictly decreasing* on $E$,

(iv)   if $x<y\Rightarrow f(x)\geqslant f(y)$ for all $x, y\in E$, $f$ is said to be *weakly decreasing* on $E$.

A function which is either strictly increasing or strictly decreasing is called *strictly monotone*. A function which is either weakly increasing or weakly decreasing is called *weakly monotone*.

There are some minor variations in usage. Weakly increasing functions are sometimes called non-decreasing, and weakly decreasing functions are then called non-increasing. Also since it is clear that if a function is strictly monotone then it must be weakly monotone as well, if a function is said to be monotone this is taken to mean that it is weakly monotone.

An important feature of strictly monotone functions which can make them easy to use is the existence of an inverse.

**Theorem 3.1**    If the real function $f$ is strictly monotone on the set $E\subset \mathcal{D}f$ then there is an inverse function $f^{-1}$ with $\mathcal{D}f^{-1}=f(E)$ and $\mathcal{R}f^{-1}=E$.

*Proof*   Suppose that $x, y\in E$ and $f(x)=f(y)$. Then as either $x<y$ or $x>y$ would imply an inequality between $f(x)$ and $f(y)$, it must be that $x=y$. So to every $z\in f(E)$ there corresponds just one number $x\in E$ such that $f(x)=z$. The function $f^{-1}$ is the function which for each $z\in f(E)$ takes this corresponding $x$ as its value.

A second important feature of strictly monotone functions is that the relation of one number to two others of being between them is preserved under the mapping by such a function.

**Theorem 3.2**    If the real function $f$ is strictly monotone on a set $E$ and $a$, $b$, $c\in E$ then $b$ is between $a$ and $c$ iff $f(b)$ is between $f(a)$ and $f(c)$, that is

$$(a-b)(b-c)>0\Leftrightarrow[f(a)-f(b)][f(b)-f(c)]>0$$

*Proof*   Suppose that $f$ is increasing. Then for all $a, b\in E$, $a<b\Leftrightarrow f(a)<f(b)$. So $f(a)-f(b)$ is of the same sign as $a-b$. As $b, c\in E$ also $f(b)-f(c)$ is of the same sign as $b-c$. Thus $[f(a)-f(b)][f(b)-f(c)]$ is of the same sign as $(a-b)(b-c)$.

If, on the other hand, $f$ is decreasing then for all $a, b\in E$, $a<b\Leftrightarrow f(a)>f(b)$. So $(-1)[f(a)-f(b)]$ is of the same sign as $a-b$. Also $(-1)[f(b)-f(c)]$ is of the same sign as $b-c$. Thus $(-1)^2[f(a)-f(b)][f(b)-f(c)]$ is of the same sign as $(a-b)(b-c)$. Now $(-1)^2=1$.

So in either case

$$(a-b)(b-c)>0 \Leftrightarrow [f(a)-f(b)][f(b)-f(c)]>0$$

## Examples of monotone functions

(i) $f$ given by $f(x)=x^2$ for $x \geqslant 0$. The graph for $f$ is shown in Fig. 3.2. $f$ is strictly increasing, as if $0 \leqslant x < y$ then $0 \leqslant x^2 < y^2$, and so there is a function $f^{-1}$ usually written $\sqrt{}$, so that $f^{-1}(x)=\sqrt{x}$, defined on $\mathcal{R}f = \{x \mid x \geqslant 0\}$ such that $\sqrt{x^2}=x$ if $x \geqslant 0$.

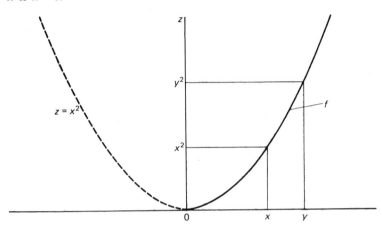

**Fig. 3.2**  *f*, the monotone function (i)

(ii) $g$ given by

$$g(x) = \begin{cases} 2x & (x>1) \\ x & (x \leqslant 1) \end{cases}$$

Figure 3.3 shows the graph of $g$. The function $g$ is strictly increasing, as if $x < y$ and if $y \leqslant 1$ then $g(x) = x < y = g(y)$, but if $x > 1$ then $g(x) = 2x < 2y = g(y)$, and finally if $x \leqslant 1$ with $y > 1$ then $g(x) = x \leqslant 1 < 2 \leqslant 2y = g(y)$.

(iii) $h$ given by

$$h(x) = \begin{cases} -(1+x) & (x \leqslant -1) \\ 0 & (-1 < x < 1) \\ 1-x & (1 \leqslant x) \end{cases}$$

Figure 3.4 shows the graph of $h$. The function $h$ is weakly decreasing, but is not strictly decreasing. It is easy to check that $h$ is weakly decreasing. Also we have that if both $x, y \in [-1, 1]$ then $h(x) = h(y) = 0$.

The function $g$ shows that even the class of strictly monotone functions can give rise to problems. Suppose that we are trying to determine $g$ from physical measurements. Two different points $x_1$ and $x_2$ could be so close to each other that they could not be distinguished by the measurements, although at the

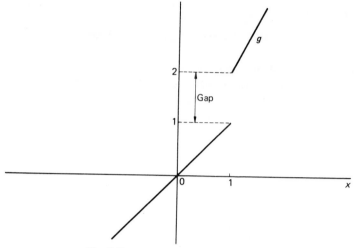

**Fig. 3.3**   *g*, the monotone function (ii)

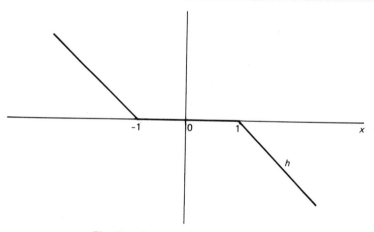

**Fig. 3.4**   *h*, the monotone function (iii)

same time it could be that $x_1 < 1 < x_2$. Then $g(x_1) < 1$ and $g(x_2) > 2$, so that at what appeared to be the same point the readings for the value of $g$ would be quite different. However, in the context of strictly monotone functions it is fairly easy to see what further restriction is needed. Functions whose ranges have gaps, like the gap in the range of $g$ between 1 and 2, must be excluded.

**Definition 3.2**   A function $f$ which is strictly monotone on an open interval $J$ is said to be *continuous on J* iff the image under $f$ of any open interval $I \subset J$ is an open interval $f(I)$.

The function $f^{-1}$ exists by Theorem 3.1, and so we get the following alternative description of continuity for strictly monotone functions.

**Theorem 3.3**  The strictly monotone function $f$ defined on an open interval $J$ is continuous on $J$ iff every open interval $I$ has as its image under $f^{-1}$ an open interval $f^{-1}(I)$.

*Notes*
(i)  We do not have to assume that $I \subset \mathscr{R}f$ because we have defined $f^{-1}(I) = \{x \,|\, f(x) = y \text{ for some } y \in I\}$.
(ii)  Because of the way in which $f^{-1}(I)$ is defined, the existence of this set does not need the existence of the function $f^{-1}$.

*Proof*
(i)  If $f$ is continuous then $f(J)$ is an open interval. Now $f^{-1}(I) = f^{-1}(I \cap f(J))$, so we can assume that $I \subset \mathscr{R}f = f(J)$.
  Consider any two numbers $\alpha, \beta \in f^{-1}(I)$ so that $f(\alpha), f(\beta) \in I$. Then for any $x$ between $\alpha$ and $\beta$, that is $(\alpha - x)(x - \beta) > 0$, we get by Theorem 3.2 $[f(\alpha) - f(x)][f(x) - f(\beta)] > 0$. So that since $I$ is an interval, $f(x) \in I$ and hence $x \in f^{-1}(I)$. Thus $f^{-1}(I)$ is an interval. Now if $\alpha$ is any number $a \in f^{-1}(I)$, as $f(\alpha) \in I$ and $I$ is an open interval, we have seen that $\exists a, b \in I$ such that $a < f(\alpha) < b$. But then $f^{-1}(a)$, $f^{-1}(b) \in f^{-1}(I)$ and $\alpha$ is between $f^{-1}(a)$ and $f^{-1}(b)$. Thus $\alpha$ is not an end point of $f^{-1}(I)$ and so $f^{-1}(I)$ is an open interval.
(ii)  Now assume that for every open interval $I$, $f^{-1}(I)$ is an open interval. [We have to show that $f$ is continuous.] Let $U \subset J$ be an open interval, and suppose that $\alpha, \beta \in U$. Then for $y_0$ any number between $f(\alpha)$ and $f(\beta)$, if $y_0$ did not belong to $f(U)$, it would follow that

$$f(U) = f(U) \cap [(\cdot, y_0) \cup (y_0, \cdot)]$$

as $(\cdot, y_0) \cup (y_0, \cdot) = \mathbb{R} \backslash \{y_0\}$. It would follow that

$$U = U \cap [f^{-1}((\cdot, y_0)) \cup f^{-1}((y_0, \cdot))] \tag{1}$$

However, $U \cap f^{-1}((\cdot, y_0))$ and $U \cap f^{-1}((y_0, \cdot))$ are open intervals by our assumption about $f$ and, as $f^{-1}(y_0)$ is between $\alpha$ and $\beta$, $\alpha$ is in one of these intervals and $\beta$ is in the other. Also $(\cdot, y_0) \cap (y_0, \cdot) = \varnothing$ gives $[U \cap f^{-1}((\cdot, y_0))] \cap [U \cap f^{-1}((y_0, \cdot))] = \varnothing$. Thus by Theorem 2.7 eq. (1) is not possible. So it must be that $y_0 \in f(U)$, that is $f(U)$ is an interval.
  Now for all $y_1 \in f(U)$, as $f^{-1}(y_1) \in U$ which is an open interval, there are numbers $a, b \in U$ with $a < f^{-1}(y_1) < b$. This means that $y_1$ is between $f(a)$ and $f(b)$, which are both in $f(U)$. So $y_1$ is not an end point of $f(U)$, which is thus an open interval.

Figures 3.5 and 3.6 illustrate the ways in which the condition of continuity, and the equivalent condition given by Theorem 3.3, prevent the occurrence of gaps in the range of $f$.

**Corollary to Theorem 3.3**  If the real function $f$ is strictly monotone and continuous on an open interval $I$ then $f^{-1}$ is strictly monotone and continuous on $f(I)$.

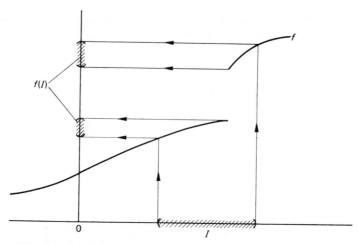

**Fig. 3.5**   $f(I)$ is not an interval; $f$ is not continuous by Definition 3.2

*Proof*   The fact that $f^{-1}$ is strictly monotone follows from writing the condition that $f$ be strictly monotone on $I$ as

$(a-b)[f(a)-f(b)]$ is of the same sign for all $a, b \in I$

For this can be written as

$[f^{-1}(\alpha)-f^{-1}(\beta)](\alpha-\beta)$ is of the same sign for all $\alpha, \beta \in f(I)$

So we can apply the definition of continuity to $f^{-1}$ and get that $f^{-1}$ is continuous on $f(J)$ iff for every open interval $I \subset f(J)$, $f^{-1}(I)$ is an open interval. Thus if $f$ is continuous on $J$ the theorem gives us that $f^{-1}$ is continuous on $f(J)$.

We see that for a strictly monotone continuous function, if we wish to measure $f(x_1)$, we can decide that there is an open interval $I \ni f(x_1)$ which will give us the accuracy we want (or perhaps only the accuracy that we can

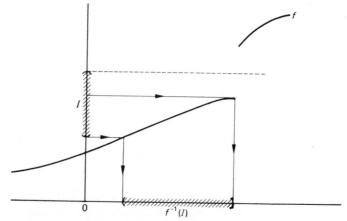

**Fig. 3.6**   $f^{-1}(I)$ is not open; $f$ is not continuous by Theorem 3.3

hope for). That is if we find a number in $I$ this will be taken as being near enough to $f(x_1)$. Thus the shorter the interval $I$, the greater is the accuracy. Then provided that we can measure $x$ so as to be sure that it is in $f^{-1}(I)$, we can measure $f(x_1)$ to the specified degree of accuracy.

If then continuity is such a desirable property for strictly monotone functions, we should seek to extend the definition of continuity to functions which need not be strictly monotone, or indeed monotone at all. Let us look at functions which are strictly increasing on one interval $I_1$ and strictly decreasing on a second interval $I_2$ with $I_1 \cap I_2 = \{c\}$ where $c$ is an end point of both $I_1$ and $I_2$. The functions $f$ and $g$ defined by

$$f(x) = \begin{cases} x & (x \leq 1) \\ 2 - x & (x > 1) \end{cases}$$

and $g(x) = x^2$ $(x \in \mathbb{R})$

are such functions. For $f$, $I_1 = (\cdot, 1]$ and $I_2 = [1, \cdot)$, and $c = 1$. For $g$, $I_1 = [0, \cdot)$ and $I_2 = (\cdot, 0]$, while $c = 0$. The graphs of $f$ and $g$ are indicated in Figs 3.7 and 3.8. If we take any open interval $I \ni c$ we find that $f(I)$ or $g(I)$ is not an open interval. Thus if $I = (a, b)$ and $a < 1 < b$ then $f(I) = (\min\{a, 2 - b\}, 1]$. If $I = (\alpha, \beta)$ and $\alpha < 0 < \beta$ then $g(I) = [0, \max\{\alpha^2, \beta^2\})$. Suppose that we take the view suggested by Theorem 3.3, and look at intervals $I$ to which the value of the function at $c$ belongs, we find that $f^{-1}(I)$ or $g^{-1}(I)$ is an open interval. Thus if $a < 1 = f(1) < b$ then $f^{-1}((a, b)) = (a, 2 - a)$. If $\alpha < 0 = g(0) < \beta$ then $g^{-1}((\alpha, \beta)) = (-\sqrt{\beta}, \sqrt{\beta})$. So in extending the definition of continuity we will consider the inverse images of open intervals.

However, if $I \subset \mathcal{R}f$ but $f(c) \notin I$ then $f^{-1}(I)$ consists of two disjoint open intervals. Thus if $a < b < 1$ then $f^{-1}((a, b)) = (a, b) \cup (2 - b, 2 - a)$, and if

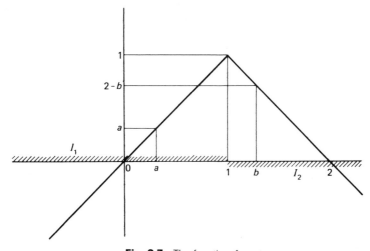

**Fig. 3.7** The function $f$

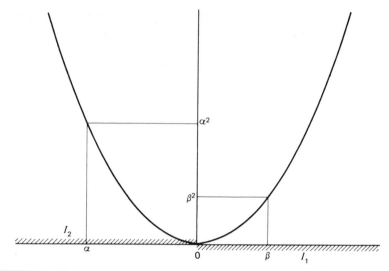

**Fig. 3.8** The function $g$

$0 < \alpha < \beta$ then $g^{-1}((\alpha, \beta)) = (-\sqrt{\beta}, -\sqrt{\alpha}) \cup (\sqrt{\alpha}, \sqrt{\beta})$. The solution to this problem is just to accept the situation. Open sets will be defined to be unions of open intervals, and then we will have that $f^{-1}(I)$ and $g^{-1}(I)$ are open sets. It is this that will be used to define continuity.

The function sin has yet to be defined or considered, but any reader who has a knowledge of it will see that the function $h$ defined by

$$h(x) = \begin{cases} -2n + x & (2n \leqslant x < 2n+1) \\ 2n + 2 - x & (2n+1 \leqslant x < 2n+2) \end{cases} \Bigg\} n \in \mathbb{Z}$$

imitates the behaviour of sin, at least when we are considering inverse images of open intervals. The function $h$ is illustrated in Fig. 3.9. Consider $I = (\frac{1}{3}, \frac{2}{3})$. Then $h^{-1}(I) = (\bigcup_{n \in \mathbb{Z}} (\frac{1}{3} + 2n, \frac{2}{3} + 2n)) \cup (\bigcup_{n \in \mathbb{Z}} (\frac{4}{3} + 2n, \frac{5}{3} + 2n))$. Thus $h^{-1}(I)$ is the union of an infinite collection of open intervals. It is in this way that $h$ imitates sin, as $\sin^{-1}(I)$ is the union of a similar collection of open intervals. As a result of these considerations, we make the following definitions.

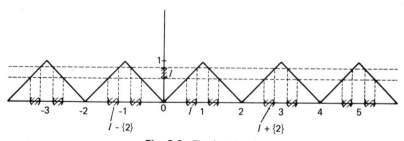

**Fig. 3.9** The function $h$

**Definition 3.3**  A set $E \subset \mathbb{R}$ is called an *open set* iff there is a collection of open intervals $\{I_\alpha\}$ such that $E = \bigcup_\alpha I_\alpha$.

A real function $f$ defined on an open interval $J$ is said to be *continuous* iff for every open interval $I$, $f^{-1}(I)$ is an open set.

Notice that these definitions are compatible with the previous ones. Thus an open interval is an open set as, for instance, $(a, b)$ is the union of the intervals of the collection $\{(a, b)\}$. Hence a strictly monotone function which was continuous by the previous definition is also continuous by the new definition. Also it follows from Theorem 3.2 that if $f$ is strictly monotone and $I$ is an interval then $f^{-1}(I)$ is an interval. Thus there are no strictly monotone functions which were not continuous by the previous definition, but are continuous by the new definition.

**Examples of open sets**
(i)  The empty set $\varnothing$ is an open set, either because $\varnothing = (a, a)$ or because $\varnothing$ is the union of an empty collection of open intervals.
(ii)  The open interval $(\cdot, \cdot)$ is an open set.
(iii)  For $a > 0$, $\{x \mid a < |x - b| < c\} = (b - c, b - a) \cup (b + a, b + c)$ is an open set.
(iv)  Define

$$I = \left( \frac{1}{3}, \frac{2}{3} \right) \quad I_{n_1} = \left( \frac{n_1}{3} + \frac{1}{9}, \frac{n_1}{3} + \frac{2}{9} \right) \quad I_{n_1,n_2} = \left( \frac{n_1}{3} + \frac{n_2}{9} + \frac{1}{27}, \frac{n_1}{3} + \frac{n_2}{9} + \frac{2}{27} \right)$$

and generally

$$I_{n_1,\ldots,n_k} = \left( \sum_{i=1}^{k} \frac{n_i}{3^i} + \frac{1}{3^{k+1}}, \sum_{i=1}^{k} \frac{n_i}{3^i} + \frac{2}{3^{k+1}} \right)$$

where $n_1, \ldots, n_k = 0, 1, 2$. Define also a set of the indices

$$\Omega = \{(n_1, n_2, \ldots, n_k) \mid k \in \mathbb{N}' \text{ and } n_1, \ldots, n_k = 0, 2\}$$

Then

$$U = \bigcup_\Omega I_{n_1,\ldots,n_k} \cup I$$

where $I$ corresponds to the case $k = 0$, is an open set.

The set $U$ is illustrated in Fig. 3.10. We see that it is not necessary to let $n_i$ take the value 1 as $I_{n_1,\ldots,n_k,1} \subset I_{n_1,\ldots,n_k}$. Also $I$ is the open middle third of the interval $[0, 1]$. $I_0$ and $I_2$ are the open middle thirds of the two closed intervals forming $[0, 1] \backslash I$. Then $I_{0,0}$, $I_{0,2}$, $I_{2,0}$ and $I_{2,2}$ are the middle thirds of the four closed intervals forming the set $[0, 1] \backslash (I \cup I_0 \cup I_2)$. So if the set $U$ is thought

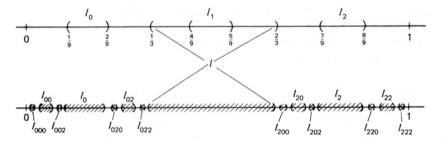

**Fig. 3.10**  The set $U$

of as being formed by successive steps of putting in the points of the intervals corresponding to the next integer $k$, then at each step the points $[0, 1]$ which are not yet in $U$ form a number of closed intervals, and the step consists of putting into $U$ their open middle thirds.

It should be noted that for none of the open intervals forming $U$ is there a next, or nearest, open interval of $U$. If we think of $U$ as being constructed by successive steps as described, then we see that at each step an open interval is put into $U$ between any two open intervals already in $U$. Alternatively, if $I_{n_1,\dots,n_l}$ and $I_{n'_1,\dots,n'_m}$ are any two of the intervals forming $U$, take $k > l, m$. It is then easy to choose $n''_1, \dots, n''_k$ in such a way that $I_{n''_1,\dots,n''_k}$ is between the other two open intervals. The set $C = [0, 1] \backslash U$ is known as the *Cantor ternary set* or *Cantor's discontinuum*. $C$ contains many points which are not end points of intervals forming $U$. The points 0 and 1 are such points. However, any $c \in C$ has the following in common with an end point: $c \notin U$ but for every $(a, b) \ni c$ we get $(a, b) \cap U \neq \emptyset$. All we need to do to find points of $U$ in $(a, b)$ is to take $k$ so large that $1/3^k < b - c$. Then there will be an interval with an index of the form $n_1, \dots, n_k$ which meets $(c, b)$, and even one with an index of the form $n_1, \dots, n_{k+1}$ which is contained in $(c, b)$. If $c = 1$ a similar argument applies using the interval $(a, c)$.

(v)  The sets $B$ and $C$ defined at the start of Chapter 2 (p. 32) are open sets as

$$B = \bigcup_{n \in \mathbb{Z} \backslash \{0\}} \left( \frac{1}{n} - \frac{1}{n^4}, \frac{1}{n} + \frac{1}{n^4} \right)$$

and    $C = (-1, 0) \cup (1, \cdot)$

$B$ is illustrated in Fig. 2.1 and $C$ in Fig. 2.2.

Whilst the set $U$ shows that in some respects open sets can be quite complicated, we shall see that the properties of open sets which we need to use are much simpler. The set $U$ can be regarded as a warning that it is not safe to assume that all properties of open sets are simple, and to proceed only on the basis of results which have been proved.

**Examples of continuous and discontinuous functions**

(i)   The function $h$ defined above, and illustrated in Fig. 3.9, is continuous. To see this, consider $I$ an open interval $I \subset [0, 1]$. Then

$$h^{-1}(I) = (I + \{2n\}) \cup (\{2n\} - I)$$

If now $I$ contains 1 but not 0 then that part of $h^{-1}(I)$ which is in $[0, 2)$ is

$$A = (I \cap (0, 1]) \cup ((\{2\} - I) \cap [1, 2))$$

and   $h^{-1}(I) = A + \{2n \mid n \in \mathbb{Z}\}$

Since $1 \in I$, $A$ is an open interval as illustrated in Fig. 3.11. If $I$ contains 0 but not 1 then the set

$$B = ((\{0\} - I) \cap (-1, 0]) \cup (I \cap [0, 1))$$

is an open interval and

$$h^{-1}(I) = B + \{2n \mid n \in \mathbb{Z}\}$$

The formation of the set $B$ is illustrated in Fig. 3.12. Finally, if $I \supset [0, 1]$ then $h^{-1}(I) = (\cdot, \cdot)$.

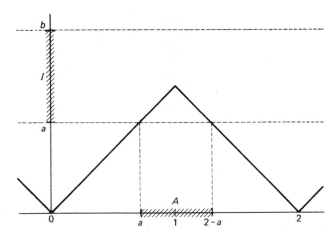

**Fig. 3.11**   The set $A$

(ii)   The function $k$ defined by

$$k(x) = \begin{cases} h(1/x) & (x \neq 0) \\ 0 & (x = 0) \end{cases}$$

is not continuous if its domain is taken to be any interval containing 0. Figure 3.13 shows the portion of the graph of $k$ near $x = 0$.

To see that $k$ is not continuous, we take $I = (-1, 1)$. Then all the values of $k$, except those taken at points of the form $1/(2n + 1)$, $n \in \mathbb{Z}$, are in $I$. Therefore $k^{-1}(I) = (\cdot, \cdot) \setminus \{1/(2n + 1) \mid n \in \mathbb{Z}\}$. Now if $k^{-1}(I)$ were open we would have

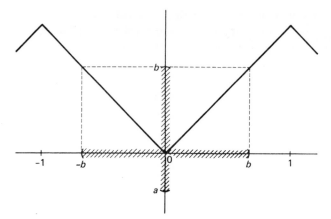

**Fig. 3.12**  The set $B$

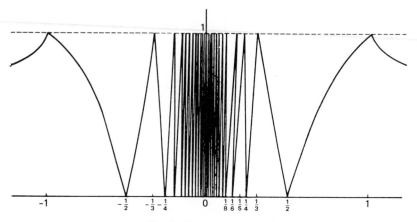

**Fig. 3.13**  The function $k$

$k^{-1}(I) = \bigcup_\alpha I_\alpha$ where each $I_\alpha$ was an open interval. So as $0 \in k^{-1}(I)$, there would be some open interval $(a, b) = I_\beta \subset k^{-1}(I)$ with $0 \in I_\beta$. However, there is an $n \in \mathbb{N}'$ such that $n > \frac{1}{2}[(1/b) - 1]$, so that $0 < 1/(2n+1) < b$. As $1/(2n+1)$ is not in $k^{-1}(I)$, there is no such interval $(a, b)$ and $k^{-1}(I)$ is not an open set.

Notice that we cannot get a continuous function from $k$ by changing the definition of $k(0)$. If $k(0)$ is taken to be any number in $[0, 1]$ then there is an interval $I \ni k(0)$ but not containing one of 0 or 1. Then $k^{-1}(I)$ contains 0 but not $1/(2n+1)$ for all $n \in \mathbb{Z}$, or if it is $0 \notin I$ then $1/(2n) \notin k^{-1}(I)$ for $n \in \mathbb{Z}\backslash\{0\}$. Suppose that we try taking $k(0)$ to be a number outside $[0, 1]$. Then there is an interval $I$ such that $k(0) \in I$ and $I \cap [0, 1] = \varnothing$. If $k(0) > 1$ take $I = (1, \cdot)$. If, on the other hand, $k(0) < 0$ take $I = (\cdot, 0)$. Then $k^{-1}(I) = \{0\}$ which is not an open set.

(iii)  The function $l$ defined by $l(x) = |x| \, k(x)$ is continuous. It is illustrated in Fig. 3.14. We see that the graph of $l$ consists of a number of straight line

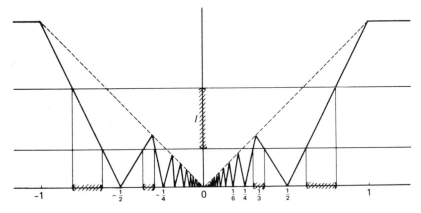

**Fig. 3.14** The function *l*

segments, and that if $a > 0$ there is only a finite number of these segments containing points for which $l(x) > a$. So if $I$ is an open interval not containing 0 then $l^{-1}(I)$ can be shown to be the union of a finite number of open intervals. This must be taken to include the possibility that $l^{-1}(I) = \varnothing$. If, on the other hand, $I$ is an open interval with $0 \in I$ then $I$ is either one of $(a, b)$ or $(\cdot, b)$ with $a < 0 < b$, in which case $(-b, b) \subset l^{-1}(I)$; or else $[0, \cdot) \subset I$, so that $l^{-1}(I) = (\cdot, \cdot)$. This follows as $|l(x)| \leqslant |x|$ all real $x$. Again it is easy to see that outside $(-b, b)$ the graph of $l$ consists of a finite number of straight line segments. Hence $l^{-1}(I)$ is the union of $(-b, b)$ and a finite number of other open intervals. Thus $l^{-1}(I)$ is open.

(iv)   The function $m$ illustrated in Fig. 3.15 and defined by

$$m(x) = \begin{cases} x & (x \neq 1) \\ 0 & (x = 1) \end{cases}$$

is not continuous if its domain is defined to be any open interval containing 1, as any open interval $I$ with $0 \in I$ and $1 \notin I$ has $m^{-1}(I) = I \cup \{1\}$. It is clear that as $1 \notin I$ this is not an open set.

The reader who is familiar with the properties of the function sin will see that as the function $h$ in certain ways imitates the behaviour of sin, so $k$ and $l$ imitate the behaviour of $\sin(1/x)$ and $x \sin(1/x)$. Thus the considerations by which it was shown that $k$ is not continuous can be used to show that if $k_1(x) = \sin(1/x)$ for $x \neq 0$ and $k_1(0) = a$ then, whatever number $a$ may be, $k_1$ is not continuous. Further considerations, similar to those used to show that $l$ is continuous, can be used to show that if $l_1(x) = x \sin(1/x)$ and $l_1(0) = 0$ then $l_1$ is continuous.

Now consider how to define open sets in the complex field $\mathbb{C}$. Since open intervals in $\mathbb{C}$ have been defined, there is no difficulty in seeing what the definition must be.

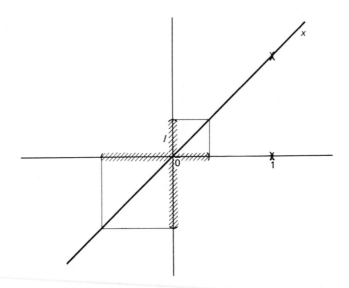

**Fig. 3.15**  The function $m$

**Definition 3.4**  A set $E \subset \mathbb{C}$ is called an *open set* (or a set *open in* $\mathbb{C}$) iff there is a collection of open intervals in $\mathbb{C}$, $\{I_\alpha\}$, such that $E = \bigcup_\alpha I_\alpha$.

As in the case of open sets in $\mathbb{R}$, it is clear that all open intervals are open sets. In particular, $\varnothing$ and $\mathbb{C}$ are open sets. But it may not be quite so obvious that any open disk, that is any set of the form $D = \{z \mid |z - a| < r\}$, is an open set. The number $d_z = r - |z - a|$ is the distance of the point $z \in D$ from the circle $\{z \mid |z - a| = r\}$. So the interval

$$I_z = \left\{ w \mid \text{Re}(z) - \frac{d_z}{\sqrt{2}} < \text{Re}(w) < \text{Re}(z) + \frac{d_z}{\sqrt{2}}, \ \text{Im}(z) - \frac{d_z}{\sqrt{2}} < \text{Im}(w) < \text{Im}(z) + \frac{d_z}{\sqrt{2}} \right\}$$

satisfies $z \in I_z \subset D$. This is shown in Fig. 3.16. From this it follows that $D \subset \bigcup_{z \in D} I_z \subset D$, so that $D = \bigcup_{z \in D} I_z$, and $D$ is an open set.

**Theorem 3.4**  A set $E \subset \mathbb{C}$ is an open set iff there is a collection $\{D_\alpha\}$ of open disks such that $E = \bigcup_\alpha D_\alpha$.

*Proof*
(i)  Suppose that $E = \bigcup_\alpha D_\alpha$ where the $D_\alpha$ are open disks. Then each $D_\alpha$ can be written as $D_\alpha = \bigcup_{z \in D} I_{\alpha,z}$ where $I_{\alpha,z}$ is an open interval for each $z \in D_\alpha$. Then writing $\Omega = \{\langle \alpha, z \rangle \mid z \in D\}$ we get $E = \bigcup_\Omega I_{\alpha,z}$. So $E$ is an open set.
(ii)  Suppose now that $E$ is an open set so that there is a collection $\{I_\alpha\}$ of open intervals with $E = \bigcup_\alpha I_\alpha$. Write $I_\alpha = (a_\alpha, b_\alpha) \times (c_\alpha, d_\alpha)$. Then if $z \in I_\alpha$ both

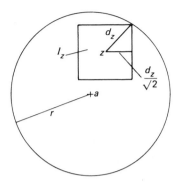

**Fig. 3.16** $z \in I_z \subset D$

$\text{Re}(z) \in (a_\alpha, b_\alpha)$ and $\text{Im}(z) \in (c_\alpha, d_\alpha)$. Let $r_{\alpha,z}$ be the smallest of the four numbers $\text{Re}(z) - a_\alpha$, $b_\alpha - \text{Re}(z)$, $\text{Im}(z) - c_\alpha$ and $d_\alpha - \text{Im}(z)$. Then the open disk $D_{\alpha,z} = \{w \mid |w - z| < r_{\alpha,z}\} \subset I_\alpha$. Thus, as $z \in D_{\alpha,z}$, we get $I_\alpha \subset \bigcup_{z \in I_\alpha} D_{\alpha,z} \subset I_\alpha$. Hence $I_\alpha = \bigcup_{z \in I_\alpha} D_{\alpha,z}$. So writing $\Omega' = \{\langle \alpha, z \rangle \mid z \in I_\alpha\}$ we get $E = \bigcup_{\Omega'} D_{\alpha,z}$.

We now extend the definition of continuity to functions whose domains or ranges may be in $\mathbb{C}$.

**Definition 3.5** If $f$ is a function with $\mathscr{D}f$ contained in either $\mathbb{R}$ or $\mathbb{C}$ and $\mathscr{R}f$ contained in either $\mathbb{R}$ or $\mathbb{C}$, and if $\mathscr{D}f$ is an open interval, then $f$ is said to be *continuous* iff for every open interval $I, f^{-1}(I)$ is an open set. The terms interval and open set have to be interpreted in the appropriate field.

Notice that the case of $\mathscr{D}f$ and $\mathscr{R}f \subset \mathbb{R}$ has not been excluded as in this case the definition is exactly the same as the one we already had.

Even if we are considering a function between $\mathbb{R}$ and $\mathbb{C}$ its graph requires to be drawn in three-dimensional space, one dimension of $\mathbb{R}$ and two dimensions of $\mathbb{C}$. The graph of a complex function would need four dimensions. So no examples of continuous or discontinuous functions for these cases are given here.

There are, of course, many other conditions similar to continuity which are studied or play some role in mathematical analysis, but continuity is by far the most important condition of its kind. The full reasons for its importance can only be seen by finding the results which follow from the definition, and noting the applications that these results have. So it is hoped that any reader who remains unconvinced of the significance of continuity will become convinced in the course of the next few chapters.

### Exercises

**1** If the function $f$ is strictly monotone, show that $f^{-1}$ is increasing if $f$ is increasing, and is decreasing if $f$ is decreasing.

**2**　For a real function $f$ defined and strictly monotone on the open interval $J$, show that if $f$ is continuous then for every $a \in J$, $\sup\{f(x)\,|\,f(a)>f(x)\}=\inf\{f(x)\,|\,f(x)>f(a)\}=f(a)$.

**3**　Show that if a real strictly monotone function $f$ defined on an open interval $J$ is not continuous then there is some open interval $I$ such that $I \neq \varnothing$, $f(J) \cap I = \varnothing$ and there are $a, b \in J$ such that for every $y \in I$, $f(a)<y<f(b)$.

**4**　Use the result of question 3 to show that if $f$ and $g$ are both real functions defined on an open interval $J$, both are strictly increasing, and $f$ is not continuous then the function $h$ given by $h(x)=f(x)+g(x)$ for $x \in J$ is not continuous.

**5**　Show that if $(a, b)$ is any open interval and $[\alpha_k, \beta_k]$ for $k=1,\ldots,n$ are closed intervals in $\mathbb{R}$ then $(a, b)\backslash\bigcup_{k=1}^{n}[\alpha_k, \beta_k]$ is an open set.

**6**　If $f$ is the real function defined by $f(x)=|x|$, prove that $f$ is continuous on $(\cdot,\cdot)$.

**7**　Prove that the annulus $\{z\,|\,a<|z|<b,\, z \in \mathbb{C}\}$ is an open set.

**8**　Use the result from question 7 to show that the real valued function $g$ with domain $\mathbb{C}$ defined by $g(z)=|z|$ is continuous on $\mathbb{C}$.

**9**　Prove that if $f$ is a real function $\mathscr{D}f=(-2,2)$ and $f(1/2n)=1$, $f(1/(2n+1))=-1$ for $n \in \mathbb{N}'$ then $f$ is not continuous.

**10**　Prove that if $g$ is a complex function $\mathscr{D}g=\{z\,||z|<2,\, z \in \mathbb{C}\}$ and $g(1/2n)=1$, $g(1/(2n+1))=-1$ for $n \in \mathbb{N}'$ then $g$ is not continuous.

**11**　Suppose that $w$ is a complex function continuous on $\mathbb{C}$ and that $w(z)=u(z)+iv(z)$ with $u(z)$, $v(z) \in \mathbb{R}$. Prove that $u$ and $v$ are real valued functions continuous on $\mathbb{C}$.

# 4

# Open sets

In this chapter the basic properties of general open sets are considered. This is in preparation for the study of continuous functions in Chapter 5. The first result is an alternative characterization of open sets, which is very often more convenient to use than the definition. This will be seen in the subsequent theorems when it is used in the proofs.

**Theorem 4.1**  A set $U$ of real or complex numbers is an open set if and only if every $x \in U$ belongs to some open interval $I_x$ such that $x \in I_x \subset U$.

*Proof*
(i)  The condition is true of any open set $U$, as $U$ is the union of a collection of open intervals $\{I_\alpha\}$. So $x \in U \Leftrightarrow x \in I_\alpha$ for some $\alpha$, but also $I_\alpha \subset U$.
(ii)  If the condition holds then for every $x \in U$ there is some open interval $I_x$ with $x \in I_x \subset U$, so that $U \subset \bigcup_{x \in U} I_x \subset U$ and hence $U = \bigcup_{x \in U} I_x$. Thus $U$ is open.

Part (i) of the proof shows that if $U$ is to·be open then it is necessary for the condition to hold. This part of the result can be used to show that certain sets are not open sets. Thus any closed or half closed interval $K$ on the real line with an end point $b \in K$ is not open. This follows as every open interval $I \ni b$ must contain points not in $K$. For example, $(\cdot, b]$ is not open as there is no open interval $(\alpha, \beta)$ with $\alpha < b < \beta$ and $(\alpha, \beta) \subset (\cdot, b]$ which would imply $\beta \leq b$. More generally, if $I$ is an open interval and $a$ is any number $a \notin I$ then $I \cup \{a\}$ is not open. This follows as if $a > x$ all $x \in I$ then any open interval to which $a$ belongs contains a point $b$ with $b > a$ and so $b \notin I \cup \{a\}$, and if $a < x$ all $x \in I$ then any open interval to which $a$ belongs contains a point $b$ with $b < a$ and so again $b \notin I \cup \{a\}$.
This result is precisely the one which is needed in Chapter 3 to see that the function $m$ is not continuous (p. 61). The argument which leads to this part of Theorem 4.1 had been considered informally in discussing the function $k$.

**Theorem 4.2**  The union of any collection of open sets either all in $\mathbb{R}$ or all in $\mathbb{C}$ is an open set.

*Proof*  Let the collection of open sets be $\{U_\alpha \mid \alpha \in \Omega\}$. Then by the definition of union, for every $x \in \bigcup_\Omega U_\alpha \; \exists$ an $a \in \Omega$ such that $x \in U_a$. Hence by

Theorem 4.1 there is an open interval $I$ with $x \in I \subset U_a$. As this implies $x \in I \subset \bigcup_\Omega U_\alpha$, Theorem 4.1 shows that $\bigcup_\Omega U_\alpha$ is an open set.

The use of Theorem 4.1 in this proof can be avoided by using instead an argument similar to that used in the proof of Theorem 3.4. The important feature of Theorem 4.2 is that there is no restriction of any sort on the number of open sets in the collection $\{U_\alpha | \alpha \in \Omega\}$. When we look at intersections of open sets the situation is different.

**Theorem 4.3** The intersection of any finite collection of open sets all in $\mathbb{R}$ is an open set.

*Proof* Let the open sets be $n$ in number and be $U_1, \ldots, U_n$. Write $U = U_1 \cap U_2 \cap \cdots \cap U_n$. Then $x \in U \Rightarrow x \in U_1$ and $x \in U_2$ and $\cdots$ and $x \in U_n$. So by Theorem 4.1 there exist open intervals $I_1, \ldots, I_n$ such that

$$x \in I_i \subset U_i$$

for $i = 1, \ldots, n$. Now by Theorem 2.3 the set $I = I_1 \cap I_2 \cap \cdots \cap I_n$ is an interval. Clearly then also $x \in I$. [To see that $I$ is open] as $\varnothing$ is open, suppose that $I \neq \varnothing$. Let $c \in I$ and $b_1, \ldots, b_k$ be the upper end points of those $I_i$ which have upper end points [so that $k \leqslant n$ and $b_k$ is not necessarily an end point of $I_k$]. Then for $x > c$, $x \in I$ iff $x < b_i$ for $i = 1, \ldots, k$. Writing $b = \min_i b_i$, this is the same as $x < b$, since if $x < b_i$ and $b_i < b_j$ then $x < b_j$. Hence either there is no $b_i$ and all $x > c$ have $x \in I$, or $I$ has an upper end point $b$ and $b \notin I$. Similarly, if $I$ has a lower end point $a$ then $a$ is the largest of the lower end points of the $I_i$ and $a \notin I$. [Thus, for instance, if $I_1 = (1, \cdot)$, $I_2 = (2, \cdot)$, $I_3 = (\cdot, 3)$ and $I_4 = (1, 4)$ then $b_1 = 3$, $b_2 = 4$, and $I_1 \cap I_2 = (2, \cdot)$ has no upper end point, while $I_1 \cap I_2 \cap I_3 \cap I_4 = (2, 3)$ where 3 is $\min\{b_1, b_2\}$.] So $I$ is an open interval. Thus if $x \in U$ there is an open interval $I$ with $x \in I \subset U$. So by Theorem 4.1 $U$ is open.

The restriction of finiteness on the collection $\{U_i | i = 1, \ldots, n\}$ is essential. If, for example, we take $U_i = (-1/i, 1/i)$ then

$$\bigcap_{i=1}^{n} (-1/i, 1/i) = (-1/n, 1/n)$$

is open, but

$$\bigcap_{i \in \mathbb{N}'} (-1/i, 1/i) = \{0\}$$

(as for every $x \neq 0$ there is some $i \in \mathbb{N}'$ with $1/i < |x|$) which is not open.

However, the restriction to the field $\mathbb{R}$ is only a matter of convenience. Theorem 4.3 can be used to prove the corresponding result in $\mathbb{C}$.

**Theorem 4.4** The intersection of any finite collection of open sets all in $\mathbb{C}$ is an open set.

*Proof* Let the open sets be $n$ in number and be $U_1, \ldots, U_n$. Write $U = \bigcap_{i=1}^{n} U_n$. Then $z \in U \Rightarrow z \in U_1$ and $z \in U_2$ and $\ldots$ and $z \in U_n$. So by Theorem 4.1 $\exists$ open intervals $I_1, \ldots, I_n$ such that

$$z \in I_j \subset U_j$$

for $j = 1, \ldots, n$. If we write $I = \bigcap_{j=1}^{n} I_j$, and also $I_j = I_r^{(j)} \times I_i^{(j)}$ where $I_r^{(j)}$ and $I_i^{(j)}$ are real intervals for $j = 1, \ldots, n$, then

$$I = \left( \bigcap_{j=1}^{n} I_r^{(j)} \right) \times \left( \bigcap_{j=1}^{n} I_i^{(j)} \right)$$

by Theorem 0.4. Now by Theorem 2.3 both of $\bigcap_{j=1}^{n} I_r^{(j)}$ and $\bigcap_{j=1}^{n} I_i^{(j)}$ are intervals, and by Theorem 4.3 they are open sets. Hence these two intersections are open intervals, and so $I$ is an open interval. Since also $z \in I$, we have that for all $z \in U$ there is an open interval $I$ such that $z \in I \subset U$. So by Theorem 4.1 $U$ is an open set.

The next theorem shows that in $\mathbb{R}$ the intervals which form an open set can be chosen in a particularly useful way. There is no corresponding result in terms of intervals in the field $\mathbb{C}$.

**Theorem 4.5** A set $U$ of real numbers is open if and only if there is a collection of open intervals $\{I_\alpha\}$ such that $I_\alpha \cap I_\beta = \emptyset$ if $\alpha \neq \beta$ and $U = \bigcup_\alpha I_\alpha$. The intervals $I_\alpha$ are called components of $U$.

*Proof* The sufficiency of the condition to show that $U$ is open is immediate from the definition of an open set. [The fact that the collection $\{I_\alpha\}$ of open intervals satisfies an extra condition not required in that definition has no effect on this conclusion.]

To prove the necessity of the condition [that is to show that if $U$ is open then there is necessarily such a collection $\{I_\alpha\}$ of open intervals], consider a point $x \in U$. By Theorem 4.1 there is some open interval $I \subset U$ such that $x \in I$. Let $A_x$ be the union of all such intervals. Then by Theorem 2.2 $A_x$ is an interval with $x \in A_x$. In addition, by Theorem 4.2 $A_x$ is an open set. So $A_x$ is an open interval. Now if for two such sets $A_x$ and $A_y$ we have $A_x \cap A_y \neq \emptyset$ then consider $z \in A_x \cap A_y$. As $A_x$ and $A_y$ are both open intervals and $z \in A_x$ and $z \in A_y$, we get $A_x \subset A_z$ and $A_y \subset A_z$. Since $A_z$ is an open interval with $x, y \in A_z$, it follows that $A_z \subset A_x$ and $A_z \subset A_y$. Hence $A_x = A_y = A_z$. The set of intervals $\{A_x\}$ can be written with just one name for each interval as $\{I_\alpha\}$. [Then $A_x = I_\alpha$ for all $x \in I_\alpha$.] The set $\{I_\alpha\}$ is now clearly the required collection of intervals.

**Theorem 4.6** If $I_\alpha$ is a component of the open set $U$ in $\mathbb{R}$ and $a$ is an end point of $I_\alpha$ then $a \notin U$.

*Proof* Since $I_\alpha$ is an open interval, $a \notin I_\alpha$. Suppose that there is a component $I_\beta$ of $U$ such that $a \in I_\beta$. Then as $\{a\} \cup I_\alpha$ is an interval [e.g. if $I = (a, b)$ then

$\{a\} \cup I_\alpha = [a, b)]$, $(\{a\} \cup I_\alpha) \cup I_\beta$ is an interval. However, $a \in I_\beta$, so that $(\{a\} \cup I_\alpha) \cup I_\beta = I_\alpha \cup I_\beta$ and so is open. Since $I_\alpha$ is a component of $U$, any open interval $J$ with $I_\alpha \subset J \subset U$ is $I_\alpha$. Hence $a \in I_\alpha \cup I_\beta = I_\alpha$, and this is false. Thus there is no component $I_\beta$ with $a \in I_\beta$ and so $a \notin U$.

Thus we might define an open set $U$ by $U = \bigcup_{n=1}^{5} I_n$ where $I_n = (n, n+5/(2n))$ for $n = 1, \ldots, 5$. Then the collection of intervals given by Theorem 4.5, the components of $U$, would be $\{(1, 3\frac{5}{6}), (4, 4\frac{5}{8}), (5, 5\frac{1}{2})\}$, since $(1, 3\frac{1}{2}) \cup (2, 3\frac{1}{4}) \cup (3, 3\frac{5}{6}) = (1, 3\frac{5}{6})$. The two collections of intervals are illustrated in Fig. 4.1.

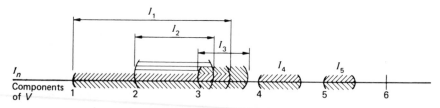

**Fig. 4.1**  The components of $U = \bigcup_{n=1}^{5} I_n$

Or again if we want an open set $U$ such that the set $G = \{x \mid x = \frac{1}{2}(1/n + 1/(n+1)), n \in \mathbb{N}'\}$ has $G \subset U$, we might construct $U$ from the intervals $I_* = (-\frac{1}{3}, \frac{1}{3})$, $I_n = (1/(n+1), 1/n)$ for $n \in \mathbb{N}'$. Then $G \subset I_* \cup \bigcup_{n \in \mathbb{N}'} I_n = U$ because $\frac{1}{2}(1/n + 1/(n+1)) \in I_n$ for $n \in \mathbb{N}'$, but also $U = I_* \cup I_1 \cup I_2$ and $I_*$, $I_1$ and $I_2$ are components of $U$. This holds as for $n > 3$ we have $I_n \subset I_*$.

Consider now a slightly different situation. Let $I_0 = (-\frac{3}{10}, \frac{3}{10})$ and write $V = \bigcup_{n \in \mathbb{N}} I_n$. As before, $G \subset V$. This time, however, let us select only enough of the intervals $I_n$ to preserve this inclusion, that is we put $V' = I_0 \cup I_1 \cup I_2$, so that as $I_n \subset I_0$ for $n > 3$, and also $\frac{1}{2}(\frac{1}{4} + \frac{1}{3}) = \frac{7}{24} \in I_0$, then $G \subset V'$. The sets $V$ and $V'$ are illustrated in Fig. 4.2. Essentially here we have the sets $I_n$ and wish to use them to form a set $V'$ such that $G \subset V'$. But it is also to our advantage if we can use only a finite number of the sets $I_n$. This situation will arise in Chapter 5. There, instead of $G$, we will have a closed interval $[a, b]$, and the sets from which we need to form an open set $V' \supset G$ will be open sets $U_\alpha$. The $U_\alpha$ arise as the inverse images of open intervals under a continuous function $f$. We require that $V'$ should be the union of such open sets, and it turns out to be important that we need only a finite number of the sets $U_\alpha$ to form such a set $V'$. The following theorem establishes the fact that a finite number does suffice.

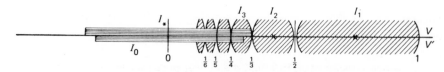

**Fig. 4.2**  The sets $V$ and $V'$

**Theorem 4.7  Heine–Borel theorem**  If $\{U_\alpha\}$ is a collection of open sets in $\mathbb{R}$ and $[a, b] \subset \bigcup_\alpha U_\alpha$, with $a \leqslant b$, then there is a finite collection of some of the $U_\alpha$ such that if this finite collection is written as $\{U_{\alpha_1}, U_{\alpha_2}, \ldots, U_{\alpha_n}\}$ then

$$[a, b] \subset \bigcup_{r=1}^{n} U_{\alpha_r}$$

*Proof*  Consider $I_x = [a, x]$ for $a \leqslant x \leqslant b$ and $x$ also such that there is some finite collection $\{U_{\beta_1}, \ldots, U_{\beta_k}\}$ of the $U_\alpha$ for which

$$I_x \subset \bigcup_{l=1}^{k} U_{\beta_l}$$

Let $A$ be the set of $x \in [a, b]$ for which there is such a finite collection. [Then what we must show is that $I_b$ is such an interval and $b \in A$.] Form $I = \bigcup_{x \in A} I_x$. Then, as $a \in I_x$ for all $x \in A$, we get by Theorem 2.2 that $I$ is an interval and $a \in I$, unless $I = \varnothing$. Clearly, as $I_x \subset [a, b]$ for all $x \in A$, then also $I \subset [a, b]$. Now $I$ is not empty as there is some $U_{\alpha'}$ such that $a \in U_{\alpha'}$, and hence $I_a = \{a\}$ is one of our intervals, that is $a \in A$. These two facts show that $I$ has an upper end point $c$.

Now $c \in [a, b] \subset \bigcup_\alpha U_\alpha$, and so there is a set $U_\gamma$ such that $c \in U_\gamma$. As the set $U_\gamma$ is open, there is a number $y_0$ with $a \leqslant y_0 \leqslant c$ and $y_0 \in U_\gamma \cap I$. Thus $y_0 \in I$ and there is an $x_0 \in A$ such that $y_0 \in [a, x_0]$. Hence there is a finite collection $\{U_{\beta_1}, \ldots, U_{\beta_k}\}$ of the $U_\alpha$ such that $[a, x_0] \subset \bigcup_{l=1}^{k} U_{\beta_l}$. Forming the union also with $U_\gamma$ we get

$$[a, c] \subset U_\gamma \cup \bigcup_{l=1}^{k} U_{\beta_l}$$

So that $I$ is contained in the union of a finite number of the sets $U_\alpha$ and we get:
(i)   $I = [a, c]$ because $c \in A$ and so $[a, c] \subset I$ [so that $I$ is the largest of the intervals $I_x$ for $x \in A$],
(ii)  also $c = b$ because if $c < b$ then $U_\gamma$ contains a point $x_1$ with $c < x_1 < b$, and so

$$[a, x_1] \subset U_\gamma \cup \bigcup_{l=1}^{k} U_{\beta_l}$$

giving $c < x_1 \in I$, contrary to the definition of $c$. [The situation just considered is illustrated in Fig. 4.3.]

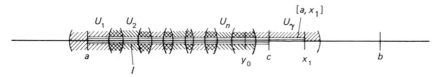

**Fig. 4.3**  Proof of Theorem 4.7

Hence

$$[a, b] \subset U_\gamma \cup \bigcup_{l=1}^{k} U_{\beta_l}$$

The required collection of $U_\alpha$ is now $\{U_{\beta_1}, \ldots, U_{\beta_k}, U_\gamma\}$.

The ideas associated with this theorem, of having sets $E$ and $U$ with $E \subset U$ but being more concerned with sets $U_\alpha$ such that $\bigcup_\alpha U_\alpha = U$, and of the possibility of forming $U' \supset E$ from only some of the sets $U_\alpha$, give rise to the following definitions.

**Definition 4.1**
(i)   If $E$ is any set and $\{U_\alpha\}$ is any collection of open sets, either all in $\mathbb{R}$ or all in $\mathbb{C}$, with $E \subset \bigcup_\alpha U_\alpha$ then $\{U_\alpha\}$ is called an *open covering of $E$* or a *covering of E by open sets*. Very often, when no other sort of covering is being considered, $\{U_\alpha\}$ is referred to simply as a covering of $E$.
(ii)   If $\{U_\alpha\}$ is an open covering of a set $E$, and $\{V_\beta\} \subset \{U_\alpha\}$ [so that each $V_\beta$ is a $U_\alpha$], and also $E \subset \bigcup_\beta V_\beta$ then $\{V_\beta\}$ is called a *subcovering of $E$*.
(iii)   If $E$ is a set in either $\mathbb{R}$ or $\mathbb{C}$ and is such that for every open covering $\{U_\alpha\}$ of $E$ there is an open subcovering $\{U_{\alpha_1}, \ldots, U_{\alpha_n}\}$ of $E$, consisting of a finite number of the sets of $\{U_\alpha\}$, then $E$ is said to be *compact*.

The use of the word 'covering' in Definition 4.1(i) has very direct connections with the non-mathematical use of the word. We might, for instance, cover the floor of a room with newspapers when we were about to redecorate the room. Then the set $E$ of points of the floor could be thought of as a set in $\mathbb{C}$, and $U_\alpha$ could be written for the set of points of the floor covered by a particular sheet of newspaper. To be sure that the floor was covered so that no paint could fall on it, we would need to have $E \subset \bigcup_\alpha U_\alpha$, that is $\{U_\alpha\}$ is a covering of $E$. If the sheets of newspaper were thrown down at random until the covering was achieved it is probable that several sheets of newspaper could then be picked up to be used elsewhere and still leave a covering of the floor. This would then give us a subcovering.

The important point to notice about Definition 4.1(iii) is that the finite subcovering must exist for *every* open covering of $E$. Thus although we saw that for the set $G = \{x \mid x = \frac{1}{2}(1/n + 1/(n+1)), n \in \mathbb{N}\}$ the collection of intervals $\{I_n \mid n \in \mathbb{N}\}$ formed an open covering, and $\{I_0, I_1, I_2\}$ formed a finite subcovering, $G$ is not compact. To prove that $G$ is not compact, we need only to notice that $\{I_n \mid n \in \mathbb{N}'\}$ is a covering of $G$ ($I_0$ is not needed) and that $\frac{1}{2}(1/n + 1/(n+1)) \notin I_m$ if $m \neq n$ and $m \neq 0$. Thus, except for $\{I_n \mid n \in \mathbb{N}'\}$ itself, there are no subcoverings of $G$. To omit even one interval $I_k$ would give a collection of $I_n$ which failed to cover $\frac{1}{2}(1/k + 1/(k+1))$.

In terms of the above definitions we can state the Heine–Borel theorem in the form: every closed bounded interval is compact. In this form the theorem is also true of sets in $\mathbb{C}$.

**Theorem 4.8   Heine–Borel theorem in $\mathbb{C}$**   If $I$ is a closed bounded interval in $\mathbb{C}$ then $I$ is compact.

*Proof*   Write $I = I_r \times I_i$ where $I_r$ and $I_i$ are real closed bounded intervals, and let $\{J^{(\alpha)}\}$ be any covering of $I$ by open intervals $J^{(\alpha)} = J_r^{(\alpha)} \times J_i^{(\alpha)}$, so that $J_r^{(\alpha)}$ and $J_i^{(\alpha)}$ are real open intervals. [Now for each $y \in I_i$ we can apply Theorem 4.7 to get a finite covering of $I_r \times \{y\}$ by a finite number of the $J^{(\alpha)}$.] The set $J^{(\alpha)}(y) = \{x \mid x + iy \in J^{(\alpha)}, x \in \mathbb{R}\}$ is $J_r^{(\alpha)}$ if $y \in J_i^{(\alpha)}$ and $\varnothing$ if $y \notin J_i^{(\alpha)}$. So $J^{(\alpha)}(y)$ is an open interval for all $y \in \mathbb{R}$, and $\{J^{(\alpha)}(y)\}$ is an open covering of $I_r$. Theorem 4.7 gives a finite subcovering $\{J^{(\alpha_1)}(y), \ldots, J^{(\alpha_n)}(y)\}$, where $\alpha_1, \ldots, \alpha_n$ and $n$ depend on $y$. The collection of corresponding intervals in $\mathbb{C}$, $\{J^{(\alpha_1)}, \ldots, J^{(\alpha_n)}\}$, is then a finite covering of $I_r \times \{y\}$. [To make the dependence on $y$ clear] call this collection of open intervals in $\mathbb{C}$, $\{K(1, y), \ldots, K(n(y), y)\}$. [But this collection covers not only $I_r \times \{y\}$, but also a whole interval containing this set.] Thus let $W(y) = \bigcap_{l=1}^{n(y)} J_i^{(\alpha_l)}$. Then $y \in W(y)$ and $I_r \times W(y) \subset \bigcup_{l=1}^{n(y)} K(l, y)$. [The relation between $W(y)$ and the sets $K(l, y)$ is shown in Fig. 4.4.] Also $W(y)$ is an open set by Theorem 4.3 [indeed it is an open interval]. Hence $\{W(y) \mid y \in I_i\}$ is an open covering of $I_i$ and so by Theorem 4.7 there is a finite subcovering $\{W(y_1), \ldots, W(y_k)\}$. Then

$$I = I_r \times I_i \subset I_r \times \bigcup_{m=1}^{k} W(y_m) = \bigcup_{m=1}^{k} I_r \times W(y_m)$$

$$\subset \bigcup_{m=1}^{k} \left( \bigcup_{l=1}^{n(y_m)} K(l, y_m) \right)$$

However, $\sum_{m=1}^{k} n(y_m)$ is finite, so that $\{K(l, y_m) \mid 1 \le m \le k, 1 \le l \le n(y_m)\}$ is a finite subcovering of $I$ by intervals of $J^{(\alpha)}$. The theorem holds for coverings of $I$ by open intervals.

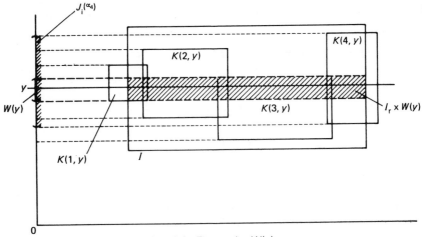

**Fig. 4.4**   The set $I_r \times W(y)$

Now if $\{U_\alpha\}$ is any open covering of $I$, write $U_\alpha = \bigcup_\beta I_\alpha^{(\beta)}$ where $I_\alpha^{(\beta)}$ are open intervals. Then $\{I_\alpha^{(\beta)}\}$ is a covering of $I$ by open intervals, as $I \subset \bigcup_{\alpha,\beta} I_\alpha^{(\beta)}$. Hence by what has already been proved there is a finite subcovering of $I$, $\{I_{\alpha_1}^{(\beta_1)}, \ldots, I_{\alpha_n}^{(\beta_n)}\}$. Now $I_{\alpha_l}^{(\beta_l)} \subset U_{\alpha_l}$. Hence

$$I \subset \bigcup_{l=1}^{n} I_{\alpha_l}^{(\beta_l)} \subset \bigcup_{l=1}^{n} U_{\alpha_l}$$

Thus $\{U_{\alpha_l}\}$ is the required finite subcovering.

In both $\mathbb{R}$ and $\mathbb{C}$ it is possible to describe exactly what sets are compact. The reader may find this result described as the Heine–Borel theorem. Since the main part of the work is already contained in the proofs of Theorems 4.7 and 4.8, this is not a very serious discrepancy in terminology.

In the proofs of several of the theorems in this chapter we have found corresponding to a point $x$ an open set $U_x$ such that $x \in U_x$. The sets $A_x$ in the proof of Theorem 4.5 are a case in point. This state of affairs occurs frequently enough to give rise to the following definitions.

## Definition 4.2

(i)   If the open set $U$ in $\mathbb{R}$ or $\mathbb{C}$ is such that $x \in U$ then $U$ is said to be a *neighbourhood* of $x$. In writing, neighbourhood is frequently abbreviated to nhd.

(ii)   If the open interval $I$ in $\mathbb{R}$ or $\mathbb{C}$ is such that $x \in I$ then $I$ is said to be an *interval neighbourhood* of $x$, and will be written $I(x)$.

(iii)   The set $\{w \mid |x - w| < \varepsilon, \; w \in F\}$ for $F = \mathbb{R}$ or $F = \mathbb{C}$ is called the *spherical* or *ball neighbourhood* of $x$ of radius $\varepsilon$, provided that $\varepsilon > 0$; it will be written $\mathcal{N}(x, \varepsilon)$.

In the case of $\mathbb{R}$ the neighbourhood $\mathcal{N}(x, \varepsilon) = (x - \varepsilon, x + \varepsilon)$ and so is also an interval neighbourhood. In the case of $\mathbb{C}$ this neighbourhood is a disk of radius $\varepsilon$ centred at $x$, but it is sometimes necessary to consider the interval neighbourhood

$$\{w \mid \operatorname{Re}(w) \in (\operatorname{Re}(x) - \varepsilon, \; \operatorname{Re}(x) + \varepsilon), \; \operatorname{Im}(w) \in (\operatorname{Im}(x) - \varepsilon, \; \operatorname{Im}(x) + \varepsilon)\}$$

The names spherical and ball neighbourhood arise from considering $\mathbb{R}$ and $\mathbb{C}$ as special cases of $n$-dimensional space, and $\mathcal{N}(x, \varepsilon)$ as a case of a set which in three dimensions is a solid sphere or ball.

## Exercises

**1**   Show that if $U \subset \mathbb{R}$ is any open set and $[\alpha_k, \beta_k]$ is a closed interval in $\mathbb{R}$ for $k = 1, \ldots, n$ then $U \backslash \bigcup_{k=1}^{n} [\alpha_k, \beta_k]$ is an open set.

**2**   Let $C$ be the Cantor ternary set and $U$ be the open set given under number (iv) of the Examples of open sets in Chapter 3 (p. 57). Show that if $\alpha \in C$ then the interval $I = (\alpha - \frac{1}{3}k, \; \alpha + \frac{1}{3}k)$ contains points of $U$ which belong to an interval (a component) $J$ of $U$ of length at least $\frac{1}{3}k$. Deduce that $C$ is not an open set. [Note that $J$ is not supposed to be contained in $I$.]

**3** For the real interval $[a, b]$, find an infinite collection of open intervals $\{(a_n, b_n)\}$ such that $\bigcap_{n \in \mathbb{N}} (a_n, b_n) = [a, b]$.

**4** For the real interval $[a, b)$, find an infinite collection of open intervals $\{(\alpha_n, \beta_n)\}$ such that $\bigcap_{n \in \mathbb{N}} (\alpha_n, \beta_n) = [a, b)$.

**5** If $z_0 = x_0 + iy_0$ with $x_0, y_0 \in \mathbb{R}$, let $T(z_0, s) = \{z \mid |x - \mathrm{Re}(z_0)| + |y - \mathrm{Im}(z_0)| < s$ and $z = x + iy, x, y \in \mathbb{R}\}$. Prove that $T(z_0, s)$ is an open set.

**6** Let $U$ and $V$ be open sets in $\mathbb{R}$ with $\{I_\alpha\}$ the collection of component intervals of $U$ and $\{J_\beta\}$ the collection of components of $V$, and $U \cap V = \emptyset$. Prove that $\{I_\alpha\} \cup \{J_\beta\}$ is the set of components of $U \cup V$.

**7** Prove that any set $\{a_1, \ldots, a_n\}$, consisting of a finite number of points $a_l \in \mathbb{R}$ or of points $a_l \in \mathbb{C}$, is compact.

**8** Prove that the set $F = \{0\} \cup \{x \mid x = 1/n, n \in \mathbb{N}'\}$ is compact.

**9** Show that $[1, 2) \subset \bigcup_{n \in \mathbb{N}'} (1 - 1/n, 2 - 1/n)$, but that there is no finite sub-covering of $[1, 2)$ by intervals of $\{(1 - 1/n, 2 - 1/n) \mid n \in \mathbb{N}'\}$.

**10** Show that $(0, 1) \subset \bigcup_{x \in (0,1)} (x - \delta, x + \delta)$ where $\delta$ is a fixed positive number. Find a finite subcovering of $(0, 1)$ by intervals of $\{(x - \delta, x + \delta) \mid x \in (0, 1)\}$.

**11** By considering $\{(1/n, 1 - 1/n) \mid n \in \mathbb{N}'\}$, or otherwise, show that $(0, 1)$ is not compact.

# 5

# Continuous functions

The properties of open sets established in Chapter 4 can now be used to yield basic results about continuous functions. In the process it will also emerge how the condition that $\mathcal{D}f$ should be an open interval can be dropped. The first theorem gives conditions under which continuity ensures the existence of solutions to certain equations.

**Theorem 5.1  The intermediate value theorem**  If $f$ is a real function continuous on the open interval $J$ and $[a, b] \subset J$ then for every number $\mu$ between $f(a)$ and $f(b)$ there is at least one number $\xi \in (a, b)$ such that $f(\xi) = \mu$.

*Proof*  The set $(\cdot, \cdot) \setminus \{\mu\}$ is the union of two disjoint semi-infinite open intervals $(\cdot, \mu)$ and $(\mu, \cdot)$. Since $\mu$ is between $f(a)$ and $f(b)$, it must be that $f(a)$ is in one of these intervals and $f(b)$ is in the other. Let $I_a$ be the interval of the pair for which $f(a) \in I_a$, and $I_b$ be the other interval so that $f(b) \in I_b$. As $f$ is continuous, $f^{-1}(I_a)$ is an open set containing $a$, and $f^{-1}(I_b)$ is an open set containing $b$. Also $(\cdot, \mu) \cap (\mu, \cdot) = \varnothing$ implies that $f^{-1}(I_a) \cap f^{-1}(I_b) = \varnothing$. Now by Theorem 4.5 there is a component $J$ of $f^{-1}(I_a)$ with $a \in J$. Also as $b \notin f^{-1}(I_a)$, so that $b \notin J$, there is an end point $\xi$ of $J$ with $a < \xi < b$. By Theorem 4.6 $\xi \notin f^{-1}(I_a)$. Also $\xi \notin f^{-1}(I_b)$ because if $J_0$ is a component of $f^{-1}(I_b)$ then $J \cap J_0 = \varnothing$, so that $J$ and $J_0$ are the components of the set $J \cup J_0$ and Theorem 4.6 implies that $\xi \notin J_0$. Hence $\xi \notin f^{-1}(I_a) \cup f^{-1}(I_b)$ which is the same as $f(\xi) \notin I_a \cup I_b$. So $f(\xi)$ must be the only number not in $I_a \cup I_b$, that is $f(\xi) = \mu$.

There may, of course, be any number of solutions to the equation $f(x) = \mu$ in $[a, b]$, but the proof proceeds by finding just one. If $f$ were monotone then there could be only the one solution. The proof has been arranged so that $\xi$ is the smallest of the numbers in $[a, b]$ for which $f(\xi) = \mu$.

**Corollary to Theorem 5.1**  If $f$ is a real function continuous on the open interval $J$ and $I$ is any interval $I \subset J$ then $f(I)$ is an interval.

*Proof*  If $\alpha, \beta \in f(I)$, let $a$ and $b$ be such that $f(a) = \alpha$ and $f(b) = \beta$. Then for any number $\mu$ between $\alpha$ and $\beta$ the theorem tells us that there is a $\xi$ between $a$ and $b$ such that $f(\xi) = \mu$, that is $\mu \in f(I)$ so that $f(I)$ is an interval.

It is natural to ask whether, perhaps, the image of a bounded interval is a bounded interval, and whether there are results about closed intervals having images which are closed intervals, or about open intervals having images which are open intervals. We have seen already in the discussion in Chapter 3 about the definition of continuity (p. 55) that a continuous function may map an open interval onto an interval which is either half closed or perhaps closed. By using the Heine–Borel theorem, Theorem 4.7, results about the image of a closed interval by a continuous function can readily be established.

**Theorem 5.2**   If the real function $f$ is continuous on the open interval $J$ then for any closed interval $[a, b] \subset J$ the set $f([a, b])$ is bounded.

*Proof*   Since $\bigcup_{n \in \mathbb{N}'} (-n, n) = (\cdot, \cdot)$, the collection of intervals $\{(-n, n) \mid n \in \mathbb{N}'\}$ is an open covering of every set of real numbers, and so in particular is an open covering of $f([a, b])$ [whether or not this is bounded]. Hence the collection $\{f^{-1}((-n, n)) \mid n \in \mathbb{N}'\}$ of sets $f^{-1}((-n, n))$, which are open in view of the continuity of $f$, is an open covering of $[a, b]$. So the Heine–Borel theorem, Theorem 4.7, implies that there is a finite subcovering $\{f^{-1}((-n_1, n_1)), \ldots, f^{-1}((-n_k, n_k))\}$. Since $[a, b] \subset \bigcup_{l=1}^{k} f^{-1}((-n_l, n_l))$, we have $f([a, b]) \subset \bigcup_{l=1}^{k} (-n_l, n_l) = (-N, N)$ where $N = \max\{n_l \mid l = 1, \ldots, k\}$. The interval $(-N, N)$ is a bounded interval so that $f([a, b])$ is bounded. [This proof is illustrated in Fig. 5.1.]

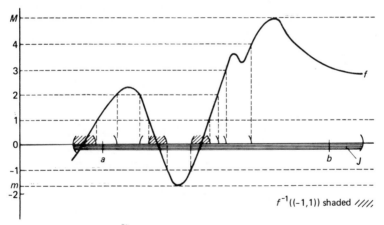

**Fig. 5.1**   Proof of Theorem 5.2

Now that $f([a, b])$ is known to be bounded, a covering can be chosen to fit it more exactly, and give a proof that $f([a, b])$ is closed.

**Theorem 5.3**   If the real function $f$ is continuous on the open interval $J$ then for any closed interval $[a, b] \subset J$ the set $f([a, b])$ is a closed interval.

*Proof*   We know that $f([a, b])$ is a bounded interval. Suppose that its end points are $m$ and $M$ [that is it is one of the four intervals $(m, M)$, $[m, M)$,

$(m, M]$ and $[m, M]]$. Then if $M \notin f([a, b])$ the collection of intervals $\{(m - x, M - x) | x > 0\}$ would be an open covering of $f([a, b])$. As $f$ is continuous, this would make $\{f^{-1}((m - x, M - x)) | x > 0\}$ an open covering of $[a, b]$. So the Heine–Borel theorem, Theorem 4.7, would imply that there was a finite subcovering of $[a, b]$; denote this by $\{f^{-1}((m - x_1, M - x_1)),$ $f^{-1}((m - x_2, M - x_2)), \ldots, f^{-1}((m - x_k, M - x_k))\}$. Hence the set $f([a, b]) \subset \bigcup_{l=1}^{k} (m - x_l, M - x_l)$, and the upper end point of $f([a, b])$ would be less than or equal to the largest of the numbers $M - x_1, \ldots, M - x_k$. Since $x_1, \ldots, x_k > 0$, this is $< M$, and so $M$ could not be the upper end point of $f([a, b])$. Thus it must be that $M \in f([a, b])$.

In a precisely similar way, it follows that if $m \notin f([a, b])$ then $\{(m + y, M + y) | y > 0\}$ would be an open covering of $f([a, b])$, and there would be a finite set of positive numbers $\{y_1, \ldots, y_{k'}\}$ such that $f([a, b]) \subset \bigcup_{l=1}^{k'} (m + y_l, M + y_l)$, and so $m$ could not be the lower end point of $f([a, b])$. Thus $m, M \in f([a, b])$ and $f([a, b]) = [m, M]$.

The number $M = \sup\{f(x) | x \in [a, b]\}$ and also $m = \inf\{f(x) | x \in [a, b]\}$. So the result of this theorem is often expressed by saying that $\exists \xi, \eta \in [a, b]$ such that $f(\xi) = \inf\{f(x) | x \in [a, b]\}$ and $f(\eta) = \sup\{f(x) | x \in [a, b]\}$. Thus $f(\xi)$ and $f(\eta)$ are, in fact, a minimum and a maximum of the values of $f$. The theorem is also sometimes stated in the form that a continuous function attains its bounds (least upper bound, and greatest lower bound) on a closed interval.

If we look at the conclusions of these three theorems, Theorems 5.1, 5.2 and 5.3, it can be seen that they are statements about those values of $f(x)$ for which $x \in [a, b]$. So that if a function $\varphi$ is defined only on $[a, b]$ by $\varphi(x) = f(x)$ if and only if $x \in [a, b]$, then the conclusions of the theorems would apply to $\varphi$. Thus, for example, $\varphi([a, b])$ is a closed interval. In order to describe the conditions on $\varphi$, it would be necessary to say that $\varphi$ could be extended to a function $f$ defined on an open interval $J \supset [a, b]$ and equal to $\varphi$ on $[a, b]$ such that $f$ was continuous. In other words, it must at least be true of $\varphi$ that the inverse image under $\varphi$ of any open interval $I$, $\varphi^{-1}(I)$, is the intersection of an open set with $[a, b]$. So the following definition is made.

**Definition 5.1** A function $f$ with $\mathscr{D}f$ a set $E$ is said to be *continuous on E* iff for every open interval $I$, $f^{-1}(I)$ is the intersection of some open set with $E$.

With this definition the following version of Theorem 5.3 can be stated.

**Theorem 5.4** If $f$ is a real function with domain the closed interval $[a, b]$, and $f$ is continuous on $[a, b]$, then $f([a, b])$ is a closed bounded interval.

*Proof* The proof follows exactly the proofs of Theorems 5.1, 5.2 and 5.3, except that when we consider inverse images of the open intervals $I_a$, $I_b$, $(-n, n)$, $(m - x, M - x)$ and $(m + y, M + y)$ we need an additional step to pass to open sets. If the open interval is $I$ then we have $f^{-1}(I) = [a, b] \cap U$ where

$U$ is an open set. Now it is the sets $U$ to which $\xi$ can be shown not to belong, or which provide the open covering of $[a, b]$ to which the Heine–Borel theorem can be applied. [The reader should check the remainder of each of the proofs to assure himself that they still hold.]

The definition of continuity can be extended even further. If $f$ is defined on a set $F$, but not necessarily continuous on $F$, it may be that there is a subset $E \subset F$ such that it is reasonable to say that $f$ is continuous on $E$. Following the previous definition we ask that for every open interval $I$ there should be an open set $U$ such that $f^{-1}(I) \cap E = U \cap E$. If this were all we asked then we would get, for instance, that $g$ defined by

$$g(x) = \begin{cases} -1 & (x < 0) \\ 1 & (x \geq 0) \end{cases}$$

was continuous on $(\cdot, 0)$ and on $[0, \cdot)$, but clearly not continuous on $(\cdot, \cdot) = (\cdot, 0) \cup [0, \cdot)$. So our definition will also require that $f^{-1}(I) \cap F \supset U \cap F$. With this condition the function $g$ can be seen not to be continuous on $[0, \cdot)$, see Fig. 5.2. Using this condition, we shall be able to derive a theorem to the effect that if $f$ is continuous on $E_1$ and continuous on $E_2$ then $f$ is continuous on $E_1 \cup E_2$.

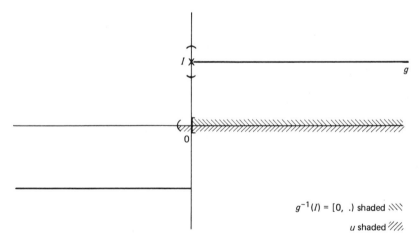

$g^{-1}(I) = [0, .)$ shaded

$u$ shaded

**Fig. 5.2** The function $g$

**Definition 5.2** A function $f$ defined on a set $F$ is said to be *continuous on the set $E \subset F$* iff for every open interval $I$, $f^{-1}(I) = (U \cap F) \cup G$ where $U$ is an open set and $G \cap E = \varnothing$. In view of the fact that $G \subset F$, this can also be written as $f^{-1}(I) = (U \cup G) \cap F$.

The set $G$ in no way adds to the continuity of $f$; it is merely that we can allow the existence of $G$ without affecting the continuity of $f$ on the set $E$.

The definition is consistent with our previous definition, as if $E = F$ then we have $G = \emptyset$, and what remains is that $f^{-1}(I)$ is the intersection of the open set $U$ with $E$. Also if $f$ is defined on $F$ and continuous on $E \subset F$ then the function $\varphi$ defined only on $F_1$, but with $\varphi(x) = f(x)$ for $x \in F_1$, is continuous on $E$, where $F_1$ is any set with $E \subset F_1 \subset F$. This holds in particular in the case $F_1 = E$.

**Examples to illustrate Definition 5.2**   It will be established later in this chapter that the function given by $x^2$ for all $x$ is continuous on every real interval.

Let $g$ be the function defined by $g(x) = x^2$ for $0 < x \leqslant 1$ and $h$ be the function defined by $h(x) = -x^2$ for $1 < x < 2$. Then $g$ is continuous on $(0, 1]$ and $h$ is continuous on $(1, 2)$. But if $f$ is defined by

$$f(x) = \begin{cases} g(x) & (x \in (0, 1]) \\ h(x) & (x \in (1, 2)) \end{cases}$$

with $f = (0, 2)$ then $f$ is not continuous on $(0, 2)$, although $f$ is continuous on $(0, 1) \cup (1, 2)$. Consider, for instance, $f^{-1}((0, 3)) = (0, 1]$ which is not the intersection of an open set with $(0, 2)$, but is the intersection of an open set with $(0, 1]$. The situation is illustrated in Fig. 5.3.

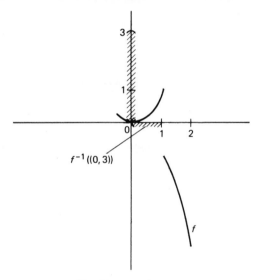

**Fig. 5.3**   The function $f$

**Theorem 5.5**   If the function $f$ is continuous on a set $F \subset \mathscr{D}f$ and $E \subset F$ then $f$ is continuous on $E$, and the function $\varphi$ which has $\varphi(x) = f(x)$ for $x \in E$ and is not defined for other $x$ is continuous on $E$.

*Proof*   For any open interval $I$ we have by the continuity of $f$ on the set $F$ that $f^{-1}(I) = (U \cap \mathscr{D}f) \cup G$ where $U$ is open and $G \cap F = \emptyset$. As $E \subset F$, this gives $G \cap E = \emptyset$. Thus $f$ is continuous on $E$.

Now $\varphi^{-1}(I) = f^{-1}(I) \cap E = (U \cap \mathscr{D}f \cap E) \cup (G \cap E) = U \cap E$, as $E \subset \mathscr{D}f$ and $G \cap E = \varnothing$. Since $U$ is an open set, this shows that $\varphi$ is continuous on $E = \mathscr{D}\varphi$.

**Theorem 5.6** If the function $f$ is defined on $F$ and is continuous on each of the sets $E_1, E_2 \subset F$ then $f$ is continuous on $E_1 \cup E_2$.

*Proof* For any open interval $I$ the continuity of $f$ on $E_1$ gives $f^{-1}(I) = (U_1 \cup G_1) \cap F$ where $U_1$ is open and $G_1 \cap E_1 = \varnothing$, and the continuity of $f$ on $E_2$ gives $f^{-1}(I) = (U_2 \cup G_2) \cap F$ where $U_2$ is open and $G_2 \cap E_2 = \varnothing$. Hence by Theorem 0.3

$$f^{-1}(I) = (U_1 \cup U_2) \cup (G_1 \cap G_2) \cap F$$

where $U_1 \cup U_2$ is open and $(G_1 \cap G_2) \cap (E_1 \cup E_2) = \varnothing$. Thus $f$ is continuous on $E_1 \cup E_2$.

Notice that both Theorem 5.5 and Theorem 5.6 are valid for functions $f$ with $\mathscr{D}f$ in $\mathbb{R}$ or in $\mathbb{C}$ and $\mathscr{R}f$ in $\mathbb{R}$ or in $\mathbb{C}$. The status of Theorems 5.1 to 5.4 for functions with one or both of their range and domain complex is discussed at the end of the chapter.

We now turn to some theorems which will allow us to construct a large collection of continuous functions from a few functions known to be continuous. The theorems also establish the continuity of a few particular functions, so that such a collection of continuous functions can be constructed. It will be seen that in the proofs of these theorems the set $G$ of Definition 5.2 plays no role. In fact the arguments can be simplified so as to be virtually the same as if our functions were defined and continuous on an open interval $J$ and the definition in use was Definition 3.5. To do this, we make the following definitions.

**Definition 5.3**
(i)  If the set $U \subset F$ is either in $\mathbb{R}$ or in $\mathbb{C}$ and is such that there is an open set $\hat{U}$ for which $U = \hat{U} \cap F$, then $U$ is said to be *open relative to $F$* or *open as a subset of $F$*.
(ii)  If the set $E \subset F$ is either in $\mathbb{R}$ or in $\mathbb{C}$ and the set $U \subset F$ can be written in the form $U = V \cup G$ where $V$ is open relative to $F$ and $G \cap E = \varnothing$, then $U$ is said to be *open in $E$ relative to $F$*.

The most important basic properties of open sets which carry over to relatively open sets are collected together in the next theorem.

**Theorem 5.7**
(a)  If $F$ is any subset of either $\mathbb{R}$ or $\mathbb{C}$ then:
(i)  the set $U \subset F$ is open relative to $F$ if and only if to every $x \in U$ there corresponds an open interval $I$ such that $x \in I \cap F \subset U$,

(ii)   if $\{U_\alpha \,|\, \alpha \in \Omega\}$ is any collection of sets $U_\alpha$ open relative to $F$ then $\bigcup_\Omega U_\alpha$ is open relative to $F$,

(iii)   if $\{U_1, \ldots, U_n\}$ is any finite collection of sets $U_i$ open relative to $F$ then $\bigcap_{i=1}^{n} U_i$ is open relative to $F$.

(b)   If $E \subset F$ and $F$ is any subset of either $\mathbb{R}$ or $\mathbb{C}$ then:

(i)   the set $U \subset F$ is open in $E$ relative to $F$ if and only if to every $x \in U \cap E$ there corresponds an open interval $I$ such that $x \in I \cap F \subset U$,

(ii)   if $\{U_\alpha \,|\, \alpha \in \Omega\}$ is any collection of sets $U_\alpha$ open in $E$ relative to $F$ then $\bigcup_\Omega U_\alpha$ is open in $E$ relative to $F$,

(iii)   if $\{U_1, \ldots, U_n\}$ is any finite collection of sets $U_i$ open in $E$ relative to $F$ then $\bigcap_{i=1}^{n} U_i$ is open in $E$ relative to $F$.

*Note*   The results (i) correspond to Theorem 4.1. The results (ii) correspond to Theorem 4.2. The results (iii) correspond to Theorems 4.3 and 4.4. There are results corresponding to the Heine–Borel theorem, but these require that the closed interval $I$ must be a subset of $F$ and in case (b) also $I \subset E$ is needed.

*Proof*   If $E = F$ then '$U$ is open in $E$ relative to $F$' is just the same as '$U$ is open relative to $F$'. The set $G$ is in $F \backslash E$ and so $G = \varnothing$. Hence it is sufficient to establish the result (b) in each case, and the corresponding result (a) will follow on putting $E = F$.

(b)(i)   If $U$ is open in $E$ relative to $F$ then $U = (\hat{U} \cap F) \cup G$ where $\hat{U}$ is an open set and $G \cap E = \varnothing$. Since $\hat{U}$ is open, then to any $x \in \hat{U}$ there corresponds an interval neighbourhood $I$ [open interval containing $x$] such that $x \in I \subset \hat{U}$. Hence if $x \in U \cap E$ we get $x \in I$ and $I \cap F \subset \hat{U} \cap F \subset U$.

For the 'only if' implication, if to each $x \in U \cap E$ there corresponds an interval neighbourhood $I(x)$ such that $x \in I(x)$ and $I(x) \cap F \subset U$, then $\hat{U} = \bigcup_{x \in U \cap E} I(x) \supset U \cap E$ is open. Writing $G = U \backslash (\hat{U} \cap F)$, we get $U = \hat{U} \cup G$ with $\hat{U}$ open and $G \cap E = \varnothing$.

(b)(ii)   Write $U_\alpha = (\hat{U}_\alpha \cap F) \cup G_\alpha$ where $\hat{U}_\alpha$ is open and $G_\alpha \cap E = \varnothing$. Then

$$\bigcup_\Omega U_\alpha = \left[ \left( \bigcup_\Omega \hat{U}_\alpha \right) \cap F \right] \cup \bigcup_\Omega G_\alpha$$

where $\bigcup_\Omega \hat{U}_\alpha$ is open and $\bigcup_\Omega G_\alpha \cap E = \varnothing$. Hence $\bigcup_\Omega U_\alpha$ is open in $E$ relative to $F$.

(b)(iii)   Write $U_i = (\hat{U}_i \cap F) \cup G_i$ where $\hat{U}_i$ is open and $G_i \cap E = \varnothing$. Then

$$\bigcap_{i=1}^{n} U_i = \left[ \left( \bigcap_{i=1}^{n} \hat{U}_i \right) \cap F \right] \cup \bigcap_{i=1}^{n} \hat{G}_i$$

where $\hat{G}_i = G_i \cup (\hat{U}_i \backslash E)$, with $\bigcap_{i=1}^{n} \hat{U}_i$ open and $\bigcap_{i=1}^{n} \hat{G}_i \cap E = \varnothing$. Hence $\bigcap_{i=1}^{n} U_i$ is open in $E$ relative to $F$.

In view of this theorem it will be seen that the proofs of subsequent results are essentially the same as if we were dealing with functions defined and continuous on an open interval. If expressions of the form '*U* is open in *E* relative to *F*' are replaced by corresponding expressions of the form '*U* is open', we get proofs of the results which apply to this special case. So the reader may find it useful to try to construct the proofs of the next few theorems in the form applying only to the special case of functions continuous on their domain which is an open interval. As a further exercise, the reader might construct a proof for the case of functions defined and continuous on a set *F* which need not be open. In this case the form of the expression becomes '*U* is open relative to *F*'.

**Theorem 5.8**   If the real valued function *f* is continuous on a set $E \subset \mathcal{D}f = F$ and $\alpha$ is any real number then the function $\alpha f$ defined by $(\alpha f)(x) = \alpha f(x)$ is continuous on *E*.

*Proof*   If $\alpha = 0$ then for any open interval *I* we have the inverse image

$$(\alpha f)^{-1}(I) = \begin{cases} F & (0 \in I) \\ \varnothing & (0 \notin I) \end{cases}$$

as $\alpha f(x) = 0$ all $x \in F$. Now $F = (\cdot, \cdot) \cap F$ and so is open in *E* relative to *F*, and $\varnothing = \varnothing \cap F$ and is also open in *E* relative to *F*.

If $\alpha \neq 0$ then for any open interval *I* we have

$$(\alpha f)^{-1}(I) = f^{-1}\left(\frac{1}{\alpha} \cdot I\right)$$

Now as *f* is continuous on *E*, $f^{-1}(1/\alpha \cdot I)$ is open in *E* relative to *F*, since $1/\alpha \cdot I$ is an open interval.

**Theorem 5.9**   If the function *f* is continuous on a set *E*, and the function *g* is continuous on a set *H* such that $f(E) \subset H$, then the function $g \circ f$ [that is the function given by $g \circ f(x) = g(f(x))$, Definition 0.12] is continuous on *E*. Provided that the sets $\mathcal{D}g$ and $\mathcal{R}f$ belong to the same field, the domains and ranges of *g* and *f* may be in either $\mathbb{R}$ or $\mathbb{C}$.

*Proof*   Note that $(g \circ f)^{-1} = f^{-1} \circ g^{-1}$. To find $(g \circ f)^{-1}(I)$ for open intervals *I*, notice that $U = g^{-1}(I) \cap H$ is open in *H* relative to $\mathcal{D}g$. Hence $U = (\bigcup_\Omega I_\alpha) \cap H$ where $I_\alpha$ is an open interval for each $\alpha$ in some set $\Omega$. Since *f* is continuous on *E*, for each $\alpha \in \Omega$ we have $f^{-1}(I_\alpha)$ is a set open in *E* relative to $\mathcal{D}f$. Thus $f^{-1}(U) = \bigcup_\Omega f^{-1}(I_\alpha)$ is open in *E* relative to $\mathcal{D}f$, that is $(g \circ f)^{-1}(I) = f^{-1} \circ g^{-1}(I) = f^{-1}(g^{-1}(I)) = f^{-1}(U)$ is open in *E* relative to $\mathcal{D}f$. So $g \circ f$ is continuous on *E*.

In order to prove that the sum of two continuous real valued functions is continuous, we make use of Theorem 5.7(b)(i). We consider each point of the inverse image of an open interval *I* and show that there is an interval neighbourhood of the point whose intersection with $\mathcal{D}f$ is contained in that inverse image. The argument can also be simplified by noting that it is only necessary to consider the case of intervals $I$ which are spherical neighbourhoods, that is only intervals $(y - \varepsilon, y + \varepsilon) = \mathcal{N}(y, \varepsilon)$ where $\varepsilon > 0$ need be considered. If $I = (a, b)$ and $y \in I$ then, on writing $\varepsilon$ for the smaller of $y - a$ and $b - y$, we have $\mathcal{N}(y, \varepsilon) \subset I$. So any open interval $J$ with $x \in J$ and $f(x) = y$ for which $f(J) \subset \mathcal{N}(y, \varepsilon)$ also has $f(J) \subset I$.

**Theorem 5.10**   If $f$ and $g$ are both functions with ranges in $\mathbb{R}$ and with domains which either both lie in $\mathbb{R}$ or both lie in $\mathbb{C}$, and on the set $E \subset F = \mathcal{D}f \cap \mathcal{D}g$ both $f$ and $g$ are continuous, then the function $f + g$ defined by

$$(f + g)(x) = f(x) + g(x)$$

is continuous on $E$.

*Proof*   Take any point $y \in (f + g)(E)$ and any spherical neighbourhood of $y$, $\mathcal{N}(y, \varepsilon) = (y - \varepsilon, y + \varepsilon)$ with $\varepsilon > 0$. Also let $x_0$ be any one of the points in $E$ such that $(f + g)(x_0) = y$. [We don't know how many solutions there may be to the equation $f(x) + g(x) = y$, but the assumption that $y \in (f + g)(E)$ is the same as assuming there is at least one such solution.] Then

$$\left( f(x_0) - \frac{\varepsilon}{2}, f(x_0) + \frac{\varepsilon}{2} \right) + \left( g(x_0) - \frac{\varepsilon}{2}, g(x_0) + \frac{\varepsilon}{2} \right) = (y - \varepsilon, y + \varepsilon)$$

or   $$\mathcal{N}\left( f(x_0), \frac{\varepsilon}{2} \right) + \mathcal{N}\left( g(x_0), \frac{\varepsilon}{2} \right) = \mathcal{N}(y, \varepsilon)$$

To see this, we need only check the end points of the intervals, as we have seen in Theorem 2.5 that the sum of two intervals is an interval. Using the continuity of $f$ on $E$, we have that

$$f^{-1}\left( \mathcal{N}\left( f(x_0), \frac{\varepsilon}{2} \right) \right) = U_f$$

is open on $E$ relative to $\mathcal{D}f$ and so is open on $E$ relative to $F$. As $g$ is also continuous on $E$, the set

$$g^{-1}\left( \mathcal{N}\left( g(x_0), \frac{\varepsilon}{2} \right) \right) = U_g$$

is open on $E$ relative to $F$. Also $x_0 \in U_f$ and $x_0 \in U_g$. So there is an open interval $U$ which is a neighbourhood of $x_0$ with $x_0 \in U \cap F \subset U_f \cap U_g$. Now for $I$ any open interval and $y \in (f + g)(E)$ we can find $\mathcal{N}(y, \varepsilon) \subset I$. Hence for each point $x$ of $(f + g)^{-1}(I) \cap E$ we can find an interval neighbourhood $U(x)$ of $x$ such that $U(x) \cap F \subset (f + g)^{-1}(\mathcal{N}(y, \varepsilon))$. So by Theorem 5.7(b)(i) the set $(f + g)^{-1}(I)$ is open in $E$ relative to $F$ and thus $f + g$ is continuous on $E$.

**Theorem 5.11**   The real functions $f$, $g$ and $h$, given by $f(x) = x$ for $x \in \mathbb{R}$, $g(x) = x^2$ for $x \in \mathbb{R}$, and $h(x) = 1/x$ for $x \in \mathbb{R} \setminus \{0\}$, are continuous on their domains.

*Proof*   [The proof is illustrated by Figs 5.4, 5.5 and 5.6.]

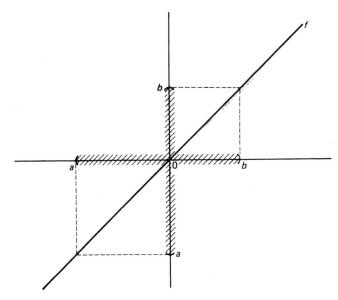

**Fig. 5.4** The function *f*

(i)   *The function f* is continuous because for any set $E$, $f^{-1}(E) = E$ and so in particular for any open interval $I$, $f^{-1}(I) = I$ is an open interval.

(ii)   *The function g* is continuous because, as we have seen, it is sufficient to consider only bounded open intervals $(a, b)$ and

$$g^{-1}((a, b)) = \{x \mid a < x^2 < b\}$$
$$= \begin{cases} (-\sqrt{b}, -\sqrt{a}) \cup (\sqrt{a}, \sqrt{b}) & (0 \leqslant a) \\ (-\sqrt{b}, \sqrt{b}) & (a < 0 \leqslant b) \\ \varnothing & (a \leqslant b < 0) \end{cases}$$

so that in all cases $g^{-1}((a, b))$ is an open set. [It would be quite easy also to check directly the inverse images under $g$ of unbounded open intervals.]

(iii)   *The function h* is continuous because

$$h^{-1}((a, b)) = \begin{cases} (1/b, 1/a) & (a, b > 0) \\ (\cdot, 1/a) \cup (1/b, \cdot) & (a < 0 < b) \\ (1/b, 1/a) & (a, b < 0) \end{cases}$$

which is an open set in each case. Again we note that it is sufficient to consider only bounded open intervals [but there is no difficulty in calculating $h^{-1}(I)$ for unbounded open intervals $I$]. If $a = 0$ then $1/a$ is replaced by $\cdot$, and similarly if $b = 0$.

Combining these results we get the following two constructions for continuous functions.

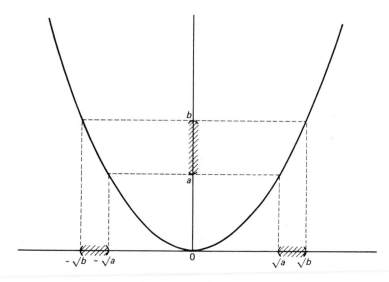

**Fig. 5.5**   The function *g*

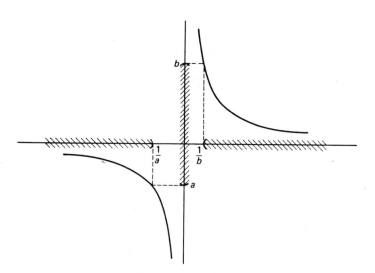

**Fig. 5.6**   The function *h*

**Theorem 5.12**   If the real valued function $f$ is continuous on the set $E$ and $f(x) \neq 0$ for all $x \in E$ then the function $1/f$ defined by $(1/f)(x) = 1/f(x)$ for $x \in E$ is continuous on $E$.

*Proof*   Apply Theorem 5.9 to $h \circ f$, where $h$ is as in Theorem 5.11 but $f$ is the function in the statement of this theorem. The condition $f(x) \neq 0$, $x \in E$, is required in order that $f(E) \subset \mathscr{D}h$.

**Theorem 5.13**  If $f$ and $g$ are both real functions continuous on a set $E \subset \mathscr{D}f \cap \mathscr{D}g$ then the function $f \cdot g$ defined by $f \cdot g(x) = f(x) \cdot g(x)$ is continuous on $E$.

*Proof*  If we use the notation $k^2 = k \cdot k$ for any function $k$ then

$$f \cdot g = \tfrac{1}{2}\{(f+g)^2 - (f^2 + g^2)\}$$

Now $f+g$ is continuous on $E$ by Theorem 5.10 and so $(f+g)^2$, $f^2$ and $g^2$ [using the notation mentioned] are continuous on $E$ by Theorems 5.9 and 5.11. Thus $f^2 + g^2$ is continuous on $E$ by Theorem 5.10 and so $-(f^2 + g^2) = (-1)(f^2 + g^2)$ is continuous on $E$ by Theorem 5.8. Then $(f+g)^2 - (f^2 + g^2) = (f+g)^2 + (-1)(f^2 + g^2)$ is continuous on $E$ by Theorem 5.10. Finally, $f \cdot g$ is continuous on $E$ by Theorem 5.8.

An immediate consequence of this theorem is that the function $f_n$ defined for each $n \in \mathbb{N}'$ by $f_n(x) = x^n$ for $x \in \mathbb{R}$ is continuous on $\mathbb{R}$. The most general form of continuous function which can be built up from $f_1$ by use of these theorems is the rational function $R$ given by

$$R(x) = \frac{a_n x^n + a_{n-1} x^{n-1} + \cdots + a_0}{b_m x^m + b_{m-1} x^{m-1} + \cdots + b_0}$$

where $a_0, \ldots, a_n$ and $b_0, \ldots, b_m$ are constant real numbers. $R$ is the ratio of two polynomials and will be continuous, except for those $x$ for which

$$b_m x^m + b_{m-1} x^{m-1} + \cdots + b_0 = 0$$

Thus, for example, $(x^2 - 1)/(x^2 + 1)$ defines a function continuous on $\mathbb{R}$, and $(x^2 + x + 1)/(x^2 - 1)$ defines a function continuous on $(\cdot, -1) \cup (-1, 1) \cup (1, \cdot)$.

The continuity of $f_n$ given by $f_n(x) = x^n$ also implies the existence of a positive $n$th root for every positive real number $y$. This follows as for every two numbers $n$, $m \in \mathbb{N}'$, we have $m^n \geqslant m$, and also for every $y > 0$ there is an $m \in \mathbb{N}'$ with $m > y$ by Application (i) of Theorem 2.9. Thus there is an $m$ with $f_n(m) = m^n > y$ and so $y \in f_n((0, m))$. Then Theorem 5.1 (the intermediate value theorem) applied to the interval $[0, m]$ shows that there is a number $\xi$ with $0 < \xi < m$ such that $f_n(\xi) = \xi^n = y$.

The functions $f_n$ are monotone increasing over $[0, \cdot)$ and $f_n(0) = 0$. So we can use the Corollary to Theorem 3.3 on the functions $f_n^*$ with $f_n^*(x) = f_n(x)$ for $x \geqslant 0$ and $f_n^*(x) = x$ for $x < 0$. We find $f_n^{-1}$, the inverse of $f_n$ considered as being defined only in $[0, \cdot)$, exists and is continuous. [If we consider $f_n$ as defined only in $[0, \cdot)$ then we only get the continuity of $f_n$ on $(0, \cdot)$.] By the existence of the $n$th roots of positive numbers $\mathscr{D}f_n^{-1} = [0, \cdot)$. In particular, since $|x| = \sqrt{x^2}$, we can now add the modulus to the list of our functions known to be continuous.

The modulus function, mod, is defined by $\mathrm{mod}(x) = |x|$. The same name is used for both the real and the complex functions. The composition $\mathrm{mod} \circ f$ will be written $|f|$.

We can compile a corresponding list of some of the continuous complex functions. Of the Theorems 5.8 to 5.13 used in obtaining the list of real continuous functions, only Theorems 5.8, 5.9 and 5.10 apply to functions which need not have both domain and range in $\mathbb{R}$. Even in Theorem 5.8 the range of the function and the constant $\alpha$ must be real. However, we can use them together with two more simple theorems to prove the continuity of complex rational functions.

**Theorem 5.14**  The complex valued function $f = u + iv$, where $u(z)$, $v(z) \in \mathbb{R}$, is continuous on the set $E$ iff the functions $u$ and $v$ are continuous on the set $E$.

*Proof*  For any open interval $I = I_r \times I_i$ consider $f^{-1}(I) = u^{-1}(I_r) \cap v^{-1}(I_i)$. Since $I_r$ and $I_i$ are both open intervals, if $u$ and $v$ are both continuous on $E$ then $u^{-1}(I_r)$ and $v^{-1}(I_i)$ are open on $E$ relative to $\mathscr{D}f$, and so their intersection $f^{-1}(I)$ is open on $E$ relative to $\mathscr{D}f$. Thus if $u$ and $v$ are continuous on $E$ then so is $f$. [Figure 5.7 illustrates this part of the proof.]

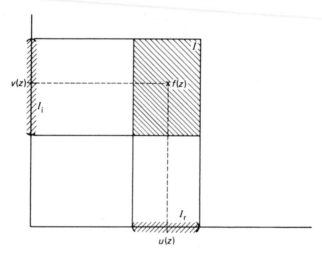

**Fig. 5.7**  $f^{-1}(I) = u^{-1}(I_r) \cap v^{-1}(I_i)$

Conversely, if $f$ is continuous on $E$ consider the open intervals $I = I_r \times (\cdot, \cdot)$ and $J = (\cdot, \cdot) \times I_i$. Then $f^{-1}(I) = u^{-1}(I_r)$ and $f^{-1}(J) = v^{-1}(I_i)$, and for all open intervals $I_r$ and $I_i$ these sets are open on $E$ relative to $\mathscr{D}f$. Thus $u$ and $v$ are continuous on $E$.

**Theorem 5.15**  The real valued functions $u$ and $v$ defined in $\mathbb{C}$ by $u(z) = \text{Re}(z)$ and $v(z) = \text{Im}(z)$ are continuous on $\mathbb{C}$.

*Proof*  Consider any real open interval $I$. Then $u^{-1}(I) = I \times (\cdot, \cdot)$ and $v^{-1}(I) = (\cdot, \cdot) \times I$, and these are open intervals. Thus $u$ and $v$ are both continuous.

Now if $u$ is a continuous real valued function with $\mathscr{D}u \subset \mathbb{C}$, and $g$ is the function given by $g(x) = x^2$ for $x \in \mathbb{R}$, then $g \circ u = u^2$ is continuous. As Theorem 5.10 applies to functions with domain in $\mathbb{C}$ and range in $\mathbb{R}$, exactly the same construction as in Theorem 5.13 shows that $u \cdot v$ is continuous for two such continuous functions $u$ and $v$. Hence the functions $f_2(z) = f_2(x+iy) = z^2 = x^2 - y^2 + i(2xy)$ where $x, y \in \mathbb{R}$ and $f_3(z) = z^3 = x^3 - 3xy^2 + i(3x^2y - y^3)$ where $x, y \in \mathbb{R}$ are continuous. In general, we can construct the real and imaginary parts of $f_n$ given by $f_n(z) = z^n$ as continuous functions to show that $f_n$ is continuous on $\mathbb{C}$ for $n \in \mathbb{N}'$. Since also if $a = \alpha + i\beta$ with $\alpha, \beta \in \mathbb{R}$ then $az^n = \alpha \, \mathrm{Re}(z^n) - \beta \, \mathrm{Im}(z^n) + i[\alpha \, \mathrm{Im}(z^n) + \beta \, \mathrm{Re}(z^n)]$, the functions $af_n$ are continuous on $\mathbb{C}$ for all $a \in \mathbb{C}$. Thus as Theorem 5.10 applies to functions with domains in $\mathbb{C}$, and using again Theorem 5.14, any polynomial $p$ given by

$$p(z) = a_n z^n + a_{n-1} z^{n-1} + \cdots + a_0$$

is continuous on $\mathbb{C}$. Then combining the function $h$ given by $h(x) = 1/x$ for $x \in \mathbb{R} \backslash \{0\}$ with any function $u$ continuous on $E \subset \mathbb{C}$ with $\mathscr{R}u \subset \mathbb{R}$, we get $h \circ u = 1/u$ is continuous on $E$. So that making use of the fact that if $f = u + iv$ with $u, v \in \mathbb{R}$ then $1/f = (u - iv)/(u^2 + v^2)$, we get that if $q$ is continuous on $\mathbb{C}$ then $1/q$ is continuous on $\mathbb{C}$, except for the points at which $q(z) = 0$.

Thus any complex rational function $R$ given by

$$R(z) = \frac{a_n z^n + a_{n-1} z^{n-1} + \cdots + a_0}{b_m z^m + b_{m-1} z^{m-1} + \cdots + b_0}$$

is continuous, except for the points at which $b_m z^m + b_{m-1} z^{m-1} + \cdots + b_0 = 0$. For instance, writing $z = x + iy$ with $x, y \in \mathbb{R}$, the function defined by

$$\frac{z^2 - 1}{z^2 + 1} = \frac{(x^2 + y^2)^2 - 1 + i4xy}{(x^2 + y^2)^2 + 2(x^2 - y^2) + 1}$$

is continuous on $\mathbb{C} \backslash \{i, -i\}$, and the function defined by

$$\frac{z^2 + z + 1}{z^2 - 1} = \frac{(x^2 + y^2)^2 + x(x^2 + y^2) - x - 1 - iy(x^2 + y^2 + 4x + 1)}{(x^2 + y^2)^2 + 2(y^2 - x^2) + 1}$$

is continuous on $\mathbb{C} \backslash \{1, -1\}$.

Also since the real function $\sqrt{\phantom{x}}$ giving positive square roots is continuous on $[0, \cdot)$, and the function given by $z\bar{z} = x^2 + y^2$ for $z \in \mathbb{C}$ is continuous, then the complex function mod which satisfies $\mathrm{mod}(z) = |z| = \sqrt{(z\bar{z})}$ is continuous. Hence also if $f$ is a complex function continuous on $E$ then the function $|f|$ for which $|f|(z) = |f(z)|$ is continuous on $E$.

For both real and complex functions only a few of those functions which can now readily be shown to be continuous have been mentioned. There are, of course, many other continuous functions, some of which we are not even in a position yet to define.

The function $g$ given by $g(z) = z^2$ for $z \in \mathbb{C}$ is continuous on $\mathbb{C}$ but does not map closed intervals to closed intervals. Thus the interval $I = \{z = x + iy \,|\, x, y \in \mathbb{R}, 1 \leqslant x \leqslant 2, -1 \leqslant y \leqslant 1\}$ has $2i, -2i \in g(I)$ and $0 \notin g(I)$. This holds because $1 + i \in I$ and $(1 + i)^2 = 2i$ and also $1 - i \in I$ and $(1 - i)^2 = -2i$. Since $0 \notin I$ but $z^2 = 0$ would imply $z = 0$, we get $0 \notin g(I)$. Thus $g(I)$ is not an interval. The nature of $g(I)$ is shown more fully in Fig. 5.8. So properties of complex functions which correspond to Theorems 5.2, 5.3 and 5.4 are discussed in the next chapter.

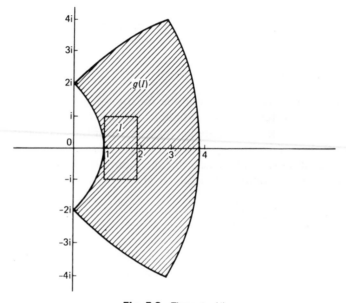

**Fig. 5.8** The set $g(I)$

## Exercises

**1** If the real function $f$ is continuous on $[a, b]$ and $f(x) > 0$ for all $x \in [a, b]$, show that there is a number $\delta > 0$ such that $f(x) > \delta$ all $x \in [a, b]$.

**2** If the real functions $f$ and $g$ are continuous on $(a, b)$ then by considering the inverse images of intervals of the form $(\alpha, \cdot)$ and $(\cdot, \beta)$ prove that the function $m$ defined by $m(x) = \max\{f(x), g(x)\}$ is continuous on $(a, b)$.

**3** Use the considerations of question 2 to show that if $f_1, \ldots, f_n$ are real functions continuous on $(a, b)$ then the function $M$ given by $M(x) = \max\{f_1(x), \ldots, f_n(x)\}$ for $x \in (a, b)$ is continuous on $(a, b)$.

**4** Examine the argument used for question 3 to see if it can be used to prove that if $\{f_1, f_2, \ldots, f_m, \ldots\}$ is a set of real functions each continuous on $(a, b)$ then the function $s$ given by $s(x) = \sup\{f_1(x), \ldots, f_n(x), \ldots\}$ is continuous on $(a, b)$. Assuming that there is a number $N$ such that $f_n(x) \leqslant N$ for all $x \in (a, b)$

and $n \in \mathbb{N}'$, so that $s$ is defined on $(a, b)$, is $s$ necessarily continuous? If so, prove it. If not, give a set of functions $\{f_1, f_2, \ldots, f_n, \ldots\}$ for which $s$ is defined but not continuous on $(a, b)$.

**5**  The real function $f$ is continuous on $[a, b]$ and $f(a) < a$ but $f(b) > b$. Prove that there is a number $x_0 \in (a, b)$ such that $f(x_0) = x_0$.

**6**  Prove that if $a, b, c, d \in \mathbb{R}$ and $a \neq 0$ then there is at least one real number $x$ such that $ax^3 + bx^2 + cx + d = 0$.

**7**  The real functions $f$ and $g$ are continuous on $[a, b]$. Prove that there is a real number $c$ such that $f(x) + c > g(x)$ all $x \in [a, b]$.

**8**  Find the largest sets on which the expressions $\sqrt{x}/(1 + \sqrt{x})$ and $\sqrt{x}/(1 - \sqrt{x})$ define real continuous functions.

**9**  If $f$ is a complex function continuous on $\mathbb{C}$, prove that the function $g$ given by $g(t) = f(t + it)$ for $t \in \mathbb{R}$ is continuous on $\mathbb{R}$.

**10**  Suppose that the two functions $\varphi$ and $\psi$ are either both real or both complex, with $A = \mathcal{D}\varphi$ and $B = \mathcal{D}\psi$, and are such that $\varphi(x) = \psi(x)$ for $x \in A \cap B$. Suppose further that there is a (non-empty) set $C \subset A \cap B$ such that $\varphi$ and $\psi$ are both continuous on $C$. Prove that the function $\theta$ defined by

$$\theta(x) = \begin{cases} \varphi(x) & (x \in A) \\ \psi(x) & (x \in B) \end{cases}$$

is continuous on $C$.

# 6

# Limits and limit points

A question which often arises in connection with a function $f$ defined on a set $F$ is whether or not the definition of $f$ can be extended to include a point $x_0$ in such a way that the extended function is continuous. Stated more precisely, the question is as to whether or not there is a number $l$ such that the function $g$ given by

$$g(x) = \begin{cases} f(x) & (x \in F \backslash \{x_0\}) \\ l & (x = x_0) \end{cases}$$

is continuous on $\{x_0\}$. For example, the function $f$ given by

$$f(x) = \frac{(x-1)^2}{x^2-1}$$

is not defined at $x = 1, -1$, but there is a natural way of extending the definition to include $x = 1$, as for $x^2 \neq 1$ we have

$$\frac{(x-1)^2}{x^2-1} = \frac{(x-1)^2}{(x-1)(x+1)} = \frac{x-1}{x+1}$$

So we can extend the function to be defined at $x = 1$ by putting it equal to $(x-1)/(x+1)$. The resulting function is then continuous on its domain and in particular on $\{1\}$.

However, it may be that a function $f$ arises in such a way that it is defined at $x_0$, but we still need to know how to redefine it in a way that will make $f$ continuous. Hence the $F \backslash \{x_0\}$ in the definition of $g$ can play a role. Thus, for instance, suppose that

$$f_a(x) = \frac{x^2}{x^2+a} \quad \text{for } 0 < a \leq 1$$

Then the functions $f_a$ are defined for all real $x$. Now we can define a function $f$ for all real $x$ by

$$f(x) = \sup\{f_a(x) \mid a \in (0, 1]\}$$

In fact, except for $x = 0$, $f_a(x)$ considered as a function of $a$ for fixed $x$ gives a strictly decreasing function. So if $x \neq 0$ we find $f(x) = x^2/x^2 = 1$. If $x = 0$ then $f_a(x) = 0^2/(0^2 + a) = 0$ for all $a \in (0, 1]$, and so $f(0) = 0$. It is clear that $f$ is not continuous on $\{0\}$. However, defining $g$ as described above by

$$g(x) = \begin{cases} f(x) & (x \neq 0) \\ 1 & (x = 0) \end{cases}$$

we get a function $g$ which is continuous on $\mathbb{R}$, so that $f$ can be made continuous by redefining it at 0.

An important particular case of this problem arises in connection with the definition of powers $a^x$. For any positive real number $a$ and any $n \in \mathbb{N}$ the $n$th power of $a$, $a^n$, is defined by repeated multiplication. We have seen also that for $m \in \mathbb{N}'$, $a$ has a uniquely defined positive root $a^{1/m}$. It is straightforward to check that $(a^{1/m})^n = (a^n)^{1/m}$. Indeed this follows on raising both sides to the power $m$, so that $a^{n/m}$ is defined as the common value of the two expressions. Also we write $a^{-n/m} = (a^{n/m})^{-1}$. Thus we have a function $e$ given by $e(x) = a^x$ for $x \in \mathbb{Q}$. But suppose that we wish to use expressions of the form $a^{\sqrt{2}}$—what is the appropriate meaning for them? If it is possible to find a function $e_1$ defined and continuous on $\mathbb{R}$ with $e_1(x) = e(x)$ for all $x \in \mathbb{Q}$ then this could provide an answer. If moreover we could show that there was only one such function $e_1$ then this condition of continuity of $e_1$ would uniquely define such numbers as $a^{\sqrt{2}}$. The proof that such a function exists is still some way ahead, but Theorem 6.1 will show that there is at most one such function.

Returning to the general case in which $g$ is the function $f$ defined or redefined at $x_0$ in such a way as to be continuous, there is one triviality which must be cleared up. Suppose that $x_0$ is such that there is an interval neighbourhood $(a, b)$ of $x_0$ [that is $x_0 \in (a, b)$] for which $(a, b) \cap F \subset \{x_0\}$, so that $(a, b)$ and $F$ may have $x_0$ in common or no points in common. Then any value $l$ will make $g$ continuous on $\{x_0\}$. This follows by considering open intervals $I \ni l$ which have $g^{-1}(I) \cap \{x_0\} = (a, b) \cap F$, so that $g^{-1}(I)$ is open on $\{x_0\}$ relative to $F$, and open intervals $J$ such that $l \notin J$ which have $g^{-1}(J) \cap \{x_0\} = \varnothing \cap F$, so that $g^{-1}(J)$ is open on $\{x_0\}$ relative to $F$. Figure 6.1 illustrates the situation. The interest is thus centred on points which are not like this.

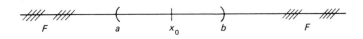

**Fig. 6.1** The relation of $x_0$ to $F$

**Definition 6.1** For subsets of either $\mathbb{R}$ or $\mathbb{C}$, if every open interval $I$ with $x_0 \in I$ meets the set $F$ in at least one point other than $x_0$, that is $I \cap F \backslash \{x_0\} \neq \varnothing$, then $x_0$ is said to be a *limit point* of $F$.

Notice that this definition is not dependent on whether or not $x_0$ belongs to $F$. Limit points will be discussed in more detail later in this chapter. For the moment let us just note that if $U$ is an open set and $x_0 \in U$ then $x_0$ is a limit point of $U$. Thus the situations we are about to discuss do arise in quite simple circumstances.

**Theorem 6.1**    If $x_0$ is a limit point of the subset $F$ of either $\mathbb{R}$ or $\mathbb{C}$ and the real valued functions $f$ and $g$ are defined on $F \cup \{x_0\}$ and are such that $f(x) = g(x)$ for $x \in F \backslash \{x_0\}$, and $f$ and $g$ are both continuous on $\{x_0\}$, then $f(x_0) = g(x_0)$.

*Proof*    Suppose, if possible, that $f(x_0) \neq g(x_0)$ and let $\varepsilon = \frac{1}{2}|f(x_0) - g(x_0)|$, so that $\varepsilon > 0$. Then the open intervals $I = \mathcal{N}(f(x_0), \varepsilon)$ and $J = \mathcal{N}(g(x_0), \varepsilon)$ are disjoint. [The situation is shown in Fig. 6.2.] Now, except at $x_0, f$ and $g$ are the same. Hence $(f^{-1}(I) \backslash \{x_0\}) \cap (g^{-1}(J) \backslash \{x_0\}) = \emptyset$. However, $f^{-1}(I)$ and $g^{-1}(J)$ are open on $\{x_0\}$ relative to $F$, and also $x_0 \in f^{-1}(I) \cap g^{-1}(J)$, so that $f^{-1}(I) \cap g^{-1}(J) \neq \emptyset$. Furthermore, since $f^{-1}(I) \cap g^{-1}(J)$ is open on $\{x_0\}$ relative to $F$, there is an interval neighbourhood $U(x_0)$ of $x_0$ such that $U(x_0) \cap F \subset f^{-1}(I) \cap g^{-1}(J)$. Now $x_0$ is a limit point of $F$ so there are points of $F$ other than $x_0$ in $U(x_0)$. Thus there are points of $F$ other than $x_0$ in $f^{-1}(I) \cap g^{-1}(J)$. This contradicts the fact that $f^{-1}(I) \backslash \{x_0\}$ and $g^{-1}(J) \backslash \{x_0\}$ are disjoint. Hence $f(x_0) = g(x_0)$.

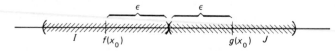

**Fig. 6.2**    $I \cap J = \emptyset$

**Corollary to Theorem 6.1**    If $z_0$ is a limit point of the subset $F$ of $\mathbb{C}$ and the complex functions $f$ and $g$ are defined on $F \cup \{z_0\}$ and are such that $f(z) = g(z)$ for $z \in F \backslash \{z_0\}$, and $f$ and $g$ are both continuous on $\{z_0\}$, then $f(z_0) = g(z_0)$.

*Proof*    Write $f(z) = u(z) + iv(z)$ and $g(z) = u_1(z) + iv_1(z)$ where $u(z), v(z), u_1(z), v_1(z) \in \mathbb{R}$. By Theorem 5.15 the functions $u, v, u_1$ and $v_1$ are continuous on $\{z_0\}$. Now also $u(z) = u_1(z)$ and $v(z) = v_1(z)$ for $z \in F \backslash \{z_0\}$. So by the theorem applied to $u, u_1$ and also to $v, v_1$ we get $u(z_0) = u_1(z_0)$ and $v(z_0) = v_1(z_0)$, that is  $f(z_0) = g(z_0)$.

These results have been formulated in a way which does not state whether $x_0 \in F$ or $x_0 \notin F$. According to which of these holds in any given case, one of $F \cup \{x_0\}$ or $F \backslash \{x_0\}$ could be replaced by $F$.

In view of the theorem and its corollary, we can make the following definition.

**Definition 6.2**    If the function $f$ is defined on a set $F$, and $a$ is a limit point of $F$, then in the case that there is a number $l$ such that the function $g$ defined by

$$g(x) = \begin{cases} f(x) & (x \in F \backslash \{a\}) \\ l & (x = a) \end{cases}$$

is continuous on $a$, we write $l = \lim_{x \to a} f(x)$ and call $l$ the *limit of $f$ as $x$ tends to $a$*, or the *limit of $f$ at $a$*.

This is sometimes written as $f(x) \to l$ as $x \to a$. But it should be noted that the symbols $f(x) \to l$ and the symbols $x \to a$ have not been given any separate meaning. It is only when used together that they are an alternative to writing $l = \lim_{x \to a} f(x)$. Note also that $\lim_{x \to a} f(x)$ is sometimes printed $\lim_{x \to a} f(x)$.

It follows from this definition that if $f$ is continuous on $\{a\}$ then $\lim_{x \to a} f(x) = f(a)$. Thus we can write down immediately such results as $\lim_{x \to a} x^n = a^n$. In the general case it may be that there is no number $l$ which makes $g$ continuous on $\{a\}$. If this holds we say that the limit does not exist, or that the limit is undefined. Thus $\lim_{x \to 0} |x|/x$ does not exist, see Fig. 6.3.

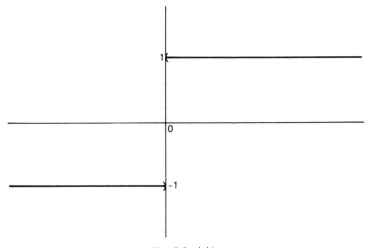

**Fig. 6.3** $|x|/x$

When considering $\lim_{x \to a} f(x)$ we are only concerned with continuity on $\{a\}$. So it is only necessary to test inverse images of intervals which contain $l$. Then if $I$ is an open interval with $l \in I$ we need to see if there is an open interval $J$ such that $f^{-1}(I) \supset J \cap (F \backslash \{a\})$. The interval $J$ should be such that $a \in J$ since it is required that $g^{-1}(I) \cap \{a\} \subset J$. As before, we can confine our attention to spherical neighbourhoods $\mathcal{N}(l, \varepsilon) = (l - \varepsilon, l + \varepsilon)$ and $\mathcal{N}(a, \delta) = (a - \delta, a + \delta)$ where both $\varepsilon, \delta > 0$. This leads to two further ways of characterizing limits.

**Theorem 6.2**  If the real function $f$ is defined on a set $F$, and $a$ is a limit point of $F$, then:

(i)  $\lim_{x \to a} f(x) = l$

iff  (ii)  for every neighbourhood $\mathcal{N}(l, \varepsilon)$ there is an $\mathcal{N}(a, \delta_\varepsilon)$ such that $x \in F \cap \mathcal{N}(a, \delta_\varepsilon)$ and $x \neq a \Rightarrow f(x) \in \mathcal{N}(l, \varepsilon)$

iff  (iii)  for every number $\varepsilon > 0$ there is a number $\delta_\varepsilon > 0$ such that $x \in F$ and $0 < |x - a| < \delta_\varepsilon \Rightarrow |f(x) - l| < \varepsilon$

*Proof*   First note that if the limit $l$ exists then the function $g$ is defined so that $g(a) = l$. So we need only consider what happens on $F\backslash\{a\}$. Now because we are only concerned with open intervals $I$ such that $g(a) \in I$ and also any such $I$ contains a neighbourhood $\mathcal{N}(g(a), \varepsilon) = (g(a) - \varepsilon, g(a) + \varepsilon) = (l - \varepsilon, l + \varepsilon)$ of $g(a)$ with $\varepsilon > 0$, the following equivalences hold.

$$\lim_{x \to a} f(x) = l$$

$\Leftrightarrow$     for every $\mathcal{N}(l, \varepsilon)$ there is some open interval $J$ with $a \in J$ such that $f^{-1}(\mathcal{N}(l, \varepsilon)) \supset J \cap (F\backslash\{a\})$

$\Leftrightarrow$     for every $\mathcal{N}(l, \varepsilon)$ there is some $\mathcal{N}(a, \delta_\varepsilon)$ such that $f^{-1}(\mathcal{N}(l, \varepsilon)) \supset \mathcal{N}(a, \delta_\varepsilon) \cap (F\backslash\{a\})$

This follows as for any $a \in J$ there is always such a neighbourhood $\mathcal{N}(a, \delta) \subset J$. [We write $\delta_\varepsilon$ as $J$ depends on $\varepsilon$.]

Writing this in terms of direct images instead of inverse images, we get:

(i)   $\lim_{x \to a} f(x)$

$\Leftrightarrow$     (ii)   for every $\mathcal{N}(l, \varepsilon)$ there is some $\mathcal{N}(a, \delta_\varepsilon)$ such that if $x \in \mathcal{N}(a, \delta_\varepsilon) \cap F$ and $x \neq a$ then $f(x) \in \mathcal{N}(l, \varepsilon)$

$\Leftrightarrow$     for every $\varepsilon > 0$ there is a $\delta_\varepsilon > 0$ such that if $x \in F\backslash\{a\}$ and $|x - a| < \delta_\varepsilon$ then $|f(x) - l| < \varepsilon$

$\Leftrightarrow$     (iii)   for every $\varepsilon > 0$ there is a $\delta_\varepsilon > 0$ such that if $x \in F$ and $0 < |x - a| < \delta_\varepsilon$ then $|f(x) - l| < \varepsilon$

The theorem also applies to the other types of function which we consider; in particular, to complex functions.

**Corollary to Theorem 6.2**   If the function $f$ is defined on a set $F$, and $a$ is a limit point of $F$, then:

(i)   $\lim_{x \to a} f(x) = l$

iff     (ii)   for every neighbourhood $\mathcal{N}(l, \varepsilon)$ there is an $\mathcal{N}(a, \delta_\varepsilon)$ such that $x \in F \cap \mathcal{N}(a, \delta_\varepsilon)$ and $x \neq a \Rightarrow f(x) \in \mathcal{N}(l, \varepsilon)$

iff     (iii)   for every number $\varepsilon > 0$ there is a number $\delta_\varepsilon > 0$ such that $x \in F$ and $0 < |x - a| < \delta_\varepsilon \Rightarrow |f(x) - l| < \varepsilon$

*Proof*   The argument is exactly the same as for the theorem. However, there are two steps which involve the relation between interval and spherical neighbourhoods, and we have to show that these are valid in $\mathbb{C}$.

In case $\mathcal{R}f$ should be in $\mathbb{C}$ we need to check that for an open interval $I$ of $\mathbb{C}$ with $l \in I$ there is a spherical neighbourhood $\mathcal{N}(l, \varepsilon) \subset I$. Suppose that $I = (\alpha, \beta) \times (\gamma, \delta)$     and     $l = \lambda + i\mu$     where     $\lambda, \ \mu \in \mathbb{R}$.     Then     if     $\varepsilon =$

$\min\{\lambda - \alpha, \beta - \lambda, \mu - \gamma, \delta - \mu\}$ it follows from the fact that for any complex number $z$ both $|\mathrm{Re}(z - l)|, |\mathrm{Im}(z - l)| \leqslant |z - l|$ that $\mathcal{N}(l, \varepsilon) \subset I$. [Figure 6.4 should make this clear.]

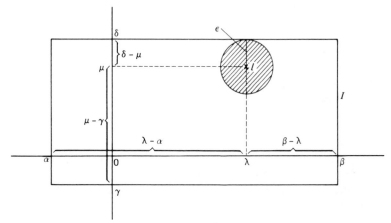

**Fig. 6.4** $\mathcal{N}(\ell, \varepsilon) \subset I$

In case $\mathcal{D}f$ should be in $\mathbb{C}$ we need to check that the openness of $f^{-1}(\mathcal{N}(l, \varepsilon))$ on $\{a\}$ relative to $F$ will follow when it has been shown that $f^{-1}(\mathcal{N}(l, \varepsilon)) \supset \mathcal{N}(a, \delta_{\varepsilon}) \cap (F \backslash \{a\})$. However, any set of the form $\mathcal{N}(a, \delta_{\varepsilon})$ contains the interval neighbourhood $\{z \,|\, |\mathrm{Re}(z - a)| < \delta_{\varepsilon}/\sqrt{2}, |\mathrm{Im}(z - a)| < \delta_{\varepsilon}/\sqrt{2}\}$. [Figure 6.5 illustrates this.]

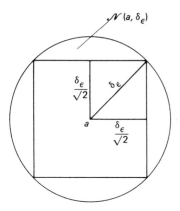

**Fig. 6.5** $\mathcal{N}(a, \delta\varepsilon)$ contains an interval neighbourhood of $a$

So the argument of Theorem 6.2 does apply in cases where $\mathcal{R}f$ or $\mathcal{D}f$ may be in $\mathbb{C}$.

It has been the tradition, and is still largely the custom, to define $\lim_{x \to a} f(x) = l$ by one or other of the two conditions (ii) and (iii). Then continuity is defined in terms of limit. In this approach '$f$ is continuous on $E$' is defined by the condition

$$\lim_{x \to a} f(x) = f(a) \quad \text{for all } a \in E$$

That this definition then leads to the same notion of continuity as we have been using results from the following theorem.

**Theorem 6.3**   If the function $f$ is such that $E \subset \mathcal{D}f$ and for every $a \in E$, $\lim_{x \to a} f(x) = f(a)$ then $f$ is continuous on $E$. [Note that the set $E$ must consist only of limit points of $\mathcal{D}f$.]

*Proof*   We have to show that for every open interval $I$, $f^{-1}(I)$ is open on $E$ relative to $\mathcal{D}f$. If $I \cap f(E) = \emptyset$ then $f^{-1}(I) \cap E = \emptyset$ and so $f^{-1}(I)$ is open on $E$ relative to $\mathcal{D}f$. Suppose that for some $a \in E$, $f(a) \in I$. Then because $\lim_{x \to a} f(x) = f(a)$, $f$ is continuous on $\{a\}$, so that $f^{-1}(I)$ is open on $\{a\}$ relative to $\mathcal{D}f$. Thus there is an open interval $U_a$ with $a \in U_a$ such that $U_a \cap \mathcal{D}f \subset f^{-1}(I)$. This is true for every $a \in f^{-1}(I) \cap E$, and so by Theorem 5.7(b)(i) $f^{-1}(I)$ is open on $E$ relative to $\mathcal{D}f$.

As an immediate result of our theorems on continuity, we can get the following rules for calculating with limits.

**Theorem 6.4**   If $f$ and $g$ are functions such that the point $a$ is a limit point of $\mathcal{D}f \cap \mathcal{D}g$, and $\lim_{x \to a} f(x) = l$ and $\lim_{x \to a} g(x) = k$, then:
(i)   $\lim_{x \to a} [f(x) + g(x)] = l + k,$
(ii)  $\lim_{x \to a} f(x) \cdot g(x) = l \cdot k,$
(iii) if also $k \neq 0$ then $\lim_{x \to a} f(x)/g(x) = l/k.$

*Proof*   As a result of the assumption that the two limits exist, there are functions $\tilde{f}$ and $\tilde{g}$ such that $\tilde{f}(x) = f(x)$ and $\tilde{g}(x) = g(x)$ if $x \neq a$, and $\tilde{f}(a) = l$ and $\tilde{g}(a) = k$. Moreover $\tilde{f}$ and $\tilde{g}$ are continuous on $\{a\}$. It now follows from the results of Chapter 5 that $\tilde{f} + \tilde{g}$, $\tilde{f} \cdot \tilde{g}$ and, if $k \neq 0$, $\tilde{f}/\tilde{g}$ are continuous on $\{a\}$. Hence
(i)   $\lim_{x \to a} [f(x) + g(x)] = \tilde{f}(a) + \tilde{g}(a) = l + k,$
(ii)  $\lim_{x \to a} f(x) \cdot g(x) = \tilde{f}(a) \cdot \tilde{g}(a) = l \cdot k,$
(iii) $\lim_{x \to a} f(x)/g(x) = \tilde{f}(a)/\tilde{g}(a) = l/k$, provided that $k \neq 0$.

Notice particularly that it is necessary to assume that both $f$ and $g$ have limits at $a$. Thus, for example, it is possible for $\lim_{x \to a} g(x)$ not to exist but for $\lim_{x \to a} f(x) \cdot g(x)$ to have a value. This happens in fact if $g$ is given by $g(x) = |x|/x$ and $f$ by $f(x) = x$ and we take $a = 0$. Then $\lim_{x \to 0} x|x|/x = \lim_{x \to 0} |x| = 0$. Notice also that in (iii) the limit can exist even if $k = 0$, but some other method is then needed to calculate it. Thus if both $f$ and $g$ are given by $f(x) = g(x) = x^2$ then $\lim_{x \to 0} f(x) = \lim_{x \to 0} g(x) = 0$ and $\lim_{x \to 0} f(x)/g(x) = \lim_{x \to 0} 1 = 1$. However, if either or both of $f$ or $g$ should fail to have a limit, or in the case of (iii) if $g$ has the limit 0, then the limits of $f(x) + g(x)$,

$f(x) \cdot g(x)$ and $f(x)/g(x)$ may well not exist. Thus if we take $g(x) = 1$ for all $x$ and $f(x) = 1$ for $x > 0$ and 0 for all other real $x$ then none of these combinations of $f$ and $g$ will have a limit at 0.

The next two theorems, although having corresponding results in terms of continuity, are of particular importance in connection with limits. The alternative ways of describing limits given by Theorem 6.2 are used in the proofs which follow.

**Theorem 6.5** If $\lim_{x \to a} f(x) = l$ then there is a neighbourhood $\mathcal{N}(a, \delta)$ such that $f(\mathcal{N}(a, \delta))$ is a bounded set. [This is often expressed by saying that $f$ is bounded on $\mathcal{N}(a, \delta)$.]

*Proof* Let $\mathcal{N}(l, \varepsilon)$ be any spherical neighbourhood of $l$. Then there is a neighbourhood $\mathcal{N}(a, \delta)$ of $a$ such that $f(x) \in \mathcal{N}(l, \varepsilon)$, provided that $x \in \mathcal{N}(a, \delta) \cap \mathcal{D}f$ and $x \neq a$. The set $\mathcal{N}(l, \varepsilon)$ is bounded.

If $f(a)$ is defined but is not in $\mathcal{N}(l, \varepsilon)$ then there is a closed interval containing $f(a)$ and $\mathcal{N}(l, \varepsilon)$. [Thus, for example, if $f$ is a real function then one of $[f(a), l + \varepsilon]$ or $[l - \varepsilon, f(a)]$ will do.] Hence $f(\mathcal{N}(a, \delta))$ is bounded.

**Theorem 6.6  The squeeze theorem** If $f$, $g$ and $h$ are real valued functions and $a$ is a limit point of the set $E = \mathcal{D}f \cap \mathcal{D}g \cap \mathcal{D}h$, and if furthermore $f(x) \geqslant g(x) \geqslant h(x)$ for $x$ in $\mathcal{N}(a, \eta) \cap E$ for some neighbourhood $\mathcal{N}(a, \eta)$ of $a$, then

$$\lim_{x \to a} f(x) = \lim_{x \to a} h(x) = l \Rightarrow \lim_{x \to a} g(x) = l$$

*Proof* If we assume that $\lim_{x \to a} f(x) = \lim_{x \to a} h(x) = l$ then to every number $\varepsilon > 0$ there correspond a $\delta_\varepsilon^f > 0$ and a $\delta_\varepsilon^h > 0$, given by Theorem 6.2(iii) applied to $\lim_{x \to a} f(x)$ and $\lim_{x \to a} h(x)$ respectively, such that, for $\delta_\varepsilon = \min\{\delta_\varepsilon^f, \delta_\varepsilon^h\}$,

$$-\varepsilon < h(x) - l \leqslant g(x) - l \leqslant f(x) - l < \varepsilon$$

provided that $x \in E$ and $0 < |x - a| < \delta_\varepsilon$. Here, in order to get the middle term of the inequalities, $\delta_\varepsilon$ has to be taken so that $\delta_\varepsilon \leqslant \eta$. [There is no difficulty about this, as if we find $\delta_1$ satisfying the rest of the conditions, but $\delta_1 > \eta$, then we can put $\delta_\varepsilon = \eta$.]

Thus we have that to every real number $\varepsilon > 0$ there is a real number $\delta_\varepsilon > 0$ such that $x \in E$ and $0 < |x - a| < \delta_\varepsilon \Rightarrow |g(x) - l| < \varepsilon$. By the Corollary to Theorem 6.2 this is the same as $\lim_{x \to a} g(x) = l$.

Thus, although we have not yet defined the function sin, suppose only that we know that $|\sin t| \leqslant 1$ for $t \in \mathbb{R}$. Then we can find $\lim_{x \to 0} x \sin(1/x)$. We have from $|\sin(1/x)| \leqslant 1$ that

$$-|x| \leqslant x \sin(1/x) \leqslant |x| \quad \text{for all real } x \neq 0$$

but also $\lim_{x \to 0} -|x| = \lim_{x \to 0} |x| = 0$. So Theorem 6.6 shows that $\lim_{x \to 0} x \sin(1/x) = 0$. The calculation is illustrated in Fig. 6.6.

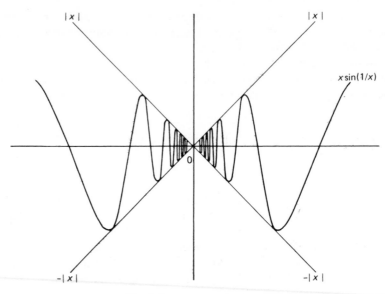

**Fig. 6.6**  $\lim_{x\to 0} x \sin(1/x) = 0$

Another example of how the squeeze theorem can reduce the calculation of the limit of a complicated function to a simple process is provided by the function $f$ defined for all real $x$ by

$$f(x) = \begin{cases} x^2 & (x \text{ rational}) \\ 0 & (x \text{ irrational}) \end{cases}$$

We have $0 \leqslant f(x) \leqslant x^2$ for $x \in \mathbb{R}$ and also $\lim_{x\to 0} 0 = \lim_{x\to 0} x^2 = 0$. Hence $\lim_{x\to 0} f(x) = 0$.

It is now necessary to give some consideration to the concept of limit point. It has so far only been noted that if $U$ is open then any point $x \in U$ is a limit point of $U$, but for closed intervals something stronger is true. Take the case of a real bounded closed interval. If $a < b$ not only is every point $x \in [a, b]$ a limit point of $[a, b]$ but also every limit point of $[a, b]$ is a point of $[a, b]$, that is there are no limit points of $[a, b]$ other than the points of $[a, b]$ themselves. To see this, first notice that if $x \in [a, b]$ and $x \in (\alpha, \beta)$ then $[a, b] \cap (\alpha, \beta)$ is an interval. In fact it is either $[a, b]$ or $(\alpha, \beta)$ or else a half open interval. In all cases it contains points other than $x$. As this holds for all interval neighbourhoods of $x$, $x$ is a limit point of $[a, b]$. Now if $x \notin [a, b]$ let $\varepsilon = \min\{|a - x|, |b - x|\}$. Then the neighbourhood $\mathcal{N}(x, \varepsilon) = (x - \varepsilon, x + \varepsilon)$ does not meet $[a, b]$, see Fig. 6.7 for the two cases $x = x_1 < a$, $\varepsilon_1 = a - x_1$, and $x = x_2 > b$, $\varepsilon_2 = x_2 - b$. Hence such an $x$ is not a limit point of $[a, b]$. It is important to notice that the points $a$ and $b$ are also limit points of the open interval $(a, b)$. In general, the points of a set need not be limit points of that set. Thus if $E = \{1/n \mid n \in \mathbb{N}'\}$ then no member of $E$ is a limit point of $E$. Consider $1/k \in E$ and take $I = (1/(k+1), 1/(k-1))$ if $k \neq 1$ or $I = (\tfrac{1}{2}, \cdot)$ if $k = 1$. Then there is no other point of $E$ in $I$, and so $1/k$ is not a limit point of $E$. However, $E$ does have a limit point, namely 0, but $0 \notin E$. To see that 0 is a limit point of $E$, suppose that $\alpha < 0 < \beta$. Then there is an $n_0 \in \mathbb{N}'$ such that $1/n_0 < \beta$ and hence $1/n_0 \in (\alpha, \beta)$. Thus for every $(\alpha, \beta)$ containing 0, there is a point $1/n_0 \neq 0$ of $E$ in $(\alpha, \beta)$, and 0 is a limit point of $E$.

**Fig. 6.7**  If $x_1 < a$ and $x_2 > b$ then $\mathcal{N}(x_1, \varepsilon_1)$ and $\mathcal{N}(x_2, \varepsilon_2)$ do not meet $[a, b]$

The property of a closed interval that it contains all its limit points is used to give a definition for a general closed set.

**Definition 6.3**    A set $F$ in either $\mathbb{R}$ or $\mathbb{C}$ which is such that if $x$ is a limit point of $F$ then $x \in F$ is called a *closed set*.

We have noted that every closed interval is also a closed set.

The next theorem shows that there is a very close connection between the ideas of open set and closed set.

**Theorem 6.7**    Suppose that $\mathbb{F}$ is one of $\mathbb{R}$ or $\mathbb{C}$. Then for any set $G \subset \mathbb{F}$, $G$ is closed iff $U = \mathbb{F} \setminus G$ is an open set of $\mathbb{F}$.

*Proof*  Suppose first that $U$ is open. [Then all we have to do is to see that no point $x_0 \in U$ is a limit point of $G$.] If $x_0 \in U$ then as $U$ is open there is an open interval $I$ such that $x_0 \in I \subset U$, so that $I \cap G = \varnothing$ and $x_0$ is not a limit point of $G$.

Next suppose instead that $G$ is closed, and let $x_0 \in U$. Since $x_0 \notin G$, then $x_0$ is not a limit point of $G$. Hence there is an open interval $I$ with $x_0 \in I$ and $I \cap G = \varnothing$. Thus $x_0 \in I \subset U$. So by Theorem 4.1 $U$ is open.

We easily get the following results about combinations of closed sets from the results about combinations of open sets.

**Theorem 6.8**
(i)    The union of any finite collection of closed sets is closed.
(ii)   The intersection of any collection of closed sets is closed.

*Proof*  If $\{G_1, \ldots, G_n\}$ is a collection of closed sets in $\mathbb{F}$, where $\mathbb{F}$ is either $\mathbb{R}$ or $\mathbb{C}$, then by Theorem 0.2

$$\mathbb{F} \setminus \bigcup_{k=1}^{n} G_k = \bigcap_{k=1}^{n} (\mathbb{F} \setminus G_k)$$

Each of the sets $\mathbb{F}\backslash G_k$ is open and, there being only a finite number of them, their intersection is open. Hence $\bigcup_{k=1}^{n} G_k = \mathbb{F}\backslash(\mathbb{F}\backslash\bigcup_{k=1}^{n} G_k)$ is a closed set.

Now let $\{G_\alpha | \alpha \in \Omega\}$ be any collection of closed sets in $\mathbb{F}$. Then again by Theorem 0.2

$$\mathbb{F}\backslash \bigcap_{\alpha \in \Omega} G_\alpha = \bigcup_{\alpha \in \Omega} (\mathbb{F}\backslash G_\alpha)$$

For each $\alpha$ the set $\mathbb{F}\backslash G_\alpha$ is open, and so $\bigcup_{\alpha \in \Omega} (\mathbb{F}\backslash G_\alpha)$ is open. As before, $\bigcap_{\alpha \in \Omega} G_\alpha$ is closed.

These results can, of course, be established by a direct use of the properties of limit points. It would be a useful exercise for the reader to do so.

It can be seen directly that the union of a collection of closed sets which contains more than a finite number of sets need not be closed. Thus

$$\bigcup_{n \in \mathbb{N}'} \left[ \frac{1}{n}, 1 - \frac{1}{n} \right] = (0, 1)$$

and the set $(0, 1)$ is not closed because 0 and 1 are limit points which do not belong to $(0, 1)$.

Most sets are neither open nor closed, but it is not customary to have a name for such sets. This sometimes makes people ignore their existence. But there are many simple examples of sets which are neither open nor closed. The half open intervals $(a, b]$ are such. The importance of open and of closed sets is that they are very special types of set with particularly useful properties.

### Examples of closed sets
(i)   Every closed interval is a closed set. This has already been noted, but it was only established for real bounded closed intervals. However, for every closed interval $I$ it is easy to express the difference $\mathbb{R}\backslash I$ or $\mathbb{C}\backslash I$ as a union of a finite number of open intervals. Thus, for example, $(\cdot, b] = (\cdot, \cdot)\backslash(b, \cdot)$ and $I = \{z | |\mathrm{Re}(z)| \leqslant 1, |\mathrm{Im}(z)| \leqslant 2\} = \mathbb{C}\backslash(\{z|\mathrm{Re}(z) > 1\} \cup \{z|\mathrm{Re}(z) < -1\} \cup \{z|\mathrm{Im}(z) > 2\} \cup \{z|\mathrm{Im}(z) < -2\})$. This is shown in Fig. 6.8.
(ii)   In particular, both $\varnothing$ and the whole field $\mathbb{R}$ or $\mathbb{C}$ are closed sets. These are in fact the only sets which are both closed and open.
(iii)   Any set consisting of a finite number of points $E = \{a_1, \ldots, a_n\}$ is a closed set. Indeed such a set $E$ has no limit points, and so no limit point not in $E$. For if $a$ is any point then unless $a_l = a$ for some $l$ we can find corresponding to each $a_l$ an interval neighbourhood $I_l$ of $a$ such that $a_l \notin I$. If for some $l$, $a_l = a$ write $I_l$ equal to the whole field. Then $a \in \bigcap_{l=1}^{n} I_l$ which is an open interval, and contains no point of $E$, except possibly $a$ itself. Hence $a$ is not a limit point of $E$.
(iv)   The circle $\{z | |z| = r, z \in \mathbb{C}\}$ is a closed set. This follows from the continuity of the function mod, which implies that the inverse images of $(\cdot, r)$ and $(r, \cdot)$, namely $\{z | |z| < r\}$ and $\{z | |z| > r\}$, are open. Thus $\{z | |z| \neq r\}$ is an open set, and hence $\{z | |z| = r\}$ is a closed set.

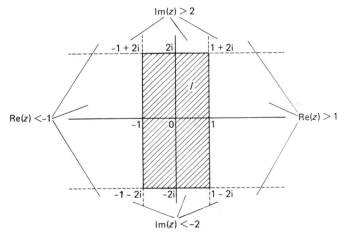

**Fig. 6.8** *I* is a closed set

(v)   The Cantor ternary set $C$, constructed in number (iv) of the Examples of open sets (p. 57), is a closed set as $C = [0, 1]/U$ where $U$ is open and so $C = (\cdot, \cdot)\backslash[U \cup (\cdot, 0) \cup (1, \cdot)]$.

In the last two examples, numbers (iv) and (v), there is no lack of limit points. Indeed every point of each of these sets is a limit point of that set. However, neither set contains any closed interval which does not consist of just one point. Now every set can be written as a union of one-point sets $E = \bigcup_{x \in E} \{x\}$. So there is nothing corresponding to the way in which open sets can be characterized as unions of open intervals.

The following property of real closed sets is an easy consequence of the definition.

**Theorem 6.9**   If $F$ is a non-empty closed set in $\mathbb{R}$ bounded below then inf $F \in F$, or if $F$ is bounded above then sup $F \in F$.

*Proof*   We need only consider the case in which inf $F$ or sup $F$ is not a limit point of $F$. Then there is some open interval $(a, b)$ containing inf $F$ or sup $F$ but no other point of $F$. If inf $F$ or sup $F$ did not belong to $F$ then $F \cap (a, b) = \emptyset$, so that either $b$ is a lower bound for $F$ or $a$ is an upper bound for $F$. This would contradict the definition of inf $F$ or of sup $F$.

The Heine–Borel theorem, Theorems 4.7 and 4.8, can be written in terms of intersections of closed sets. In this form it may be easier to visualize.

**Theorem 6.10**   If $G_n$ is a bounded closed set, in the field $\mathbb{F}$ which is either $\mathbb{R}$ or $\mathbb{C}$, for every $n \in \mathbb{N}$ and $G_{n+1} \subset G_n$ then either $G_m = \emptyset$ for some $m \in \mathbb{N}$ or $\bigcap_{n \in \mathbb{N}} G_n \neq \emptyset$.

*Proof*   First notice that as $G_{n+1} \subset G_n$ all $n \in \mathbb{N}$ then $G_m \subset G_n$ if $m \geq n$ and, in particular, $G_m \subset G_0$ for $m \in \mathbb{N}$. Since $G_0$ is bounded, there is a closed bounded interval $I$ such that $G_0 \subset I$. For every $n \in \mathbb{N}$ let $U_n = \mathbb{F} \backslash G_n$. Then $U_n$ is open by Theorem 6.7. Now if $\bigcap_{n \in \mathbb{N}} G_n = \varnothing$ then $\bigcup_{n \in \mathbb{N}} U_n = \bigcup_{n \in \mathbb{N}} (\mathbb{F} \backslash G_n) = \mathbb{F} \backslash \bigcap_{n \in \mathbb{N}} G_n = \mathbb{F}$, so that $\{U_n\}$ is an open covering of $I$. Hence by the Heine–Borel theorem there is a finite set $\{U_{n_1}, \ldots, U_{n_k}\}$ of the $U_n$ which covers $I$, $\bigcup_{l=1}^{k} U_{n_l} \supset I$. Now the condition that $G_{n+1} \subset G_n$ all $n \in \mathbb{N}$ implies that the intersection $\bigcap_{l=1}^{k} G_{n_l} = G_m$ where $m$ is the largest of the numbers $n_1, \ldots, n_k$, as this is in fact a subset of each of the other $G_{n_l}$. Thus

$$\bigcup_{l=1}^{k} U_{n_l} = \bigcup_{l=1}^{k} (\mathbb{F} \backslash G_{n_l}) = \mathbb{F} \backslash \bigcap_{l=1}^{k} G_{n_l} = \mathbb{F} \backslash G_m$$

However, $G_m \subset I$, so that $\mathbb{F} \backslash G_m \supset I$ implies that $G_m = \varnothing$. Thus if $G_m \neq \varnothing$ for all $m \in \mathbb{N}$ then $\bigcap_{n \in \mathbb{N}} G_n \neq \varnothing$.

If $G_n \neq \varnothing$ for $n \in \mathbb{N}$ then the intersection must contain at least one point. But as the following case shows, it may be that $\bigcap_{n \in \mathbb{N}} G_n$ contains only one point. Let $G_n = [-1/(n+1), 1/(n+1)]$. Then $G_n \neq \varnothing$ for all $n \in \mathbb{N}$ and $\bigcap_{n \in \mathbb{N}} G_n = \{0\}$. Also we can see from particular cases that if either one of the two conditions that the $G_n$ be bounded and that the $G_n$ be closed is dropped then the conclusion need not hold. Thus if we take now $G_n = [n, \cdot)$ then $G_n$ is a closed non-empty set for each $n \in \mathbb{N}$. But $\bigcap_{n \in \mathbb{N}} G_n = \bigcap_{n \in \mathbb{N}} [n, \cdot) = \varnothing$ because for any $x \in \mathbb{R}$ there is an $n \in \mathbb{N}$ with $n > x$. As the third case to illustrate the limitations of this result, suppose now that $G_n = (0, 1/(n+1)]$ for $n \in \mathbb{N}$, so that this time $G_n$ is bounded and non-empty but is not closed. Then $\bigcap_{n \in \mathbb{N}} G_n = \varnothing$, as if $x > 0$ there is an $n \in \mathbb{N}$ with $1/(n+1) < x$.

A similar argument using the Heine–Borel theorem can be used to give a method of proving the existence of a limit point.

### Theorem 6.11   Bolzano–Weierstrass theorem

If the set $E$ of points of the field $\mathbb{F}$ which is either $\mathbb{R}$ or $\mathbb{C}$ is bounded and has infinitely many members then $E$ has at least one limit point.

*Proof*   [We can think of this result in the form: if the bounded set $E$ has no limit points then $E$ has only a finite number of members.] Suppose that $E$ has no limit points. Then for each $x \in E$ there is an interval neighbourhood $I(x)$ such that $x$ is the only point of $E$ in $I(x)$. Also $E$ is closed, there being no limit points of $E$ which could be outside $E$. Let $U = \mathbb{F} \backslash E$, so that $U$ is open. Then $U \cup \bigcup_{x \in E} I(x) = \mathbb{F}$. Now as $E$ is bounded, there is a closed bounded interval $J$ such that $E \subset J$. Thus $J \subset U \cup \bigcup_{x \in E} I(x)$, and the sets $U$ and $I(x)$ are open. So that by the Heine–Borel theorem $J$ is compact and there is a finite collection of the sets $U$ and $I(x)$ whose union contains $J$. Thus $J \subset U \cup \bigcup_{l=1}^{k} I(x_l)$ for some finite set $\{x_1, \ldots, x_k\}$ of points of $E$. However, this implies that $E \subset \bigcup_{l=1}^{k} I(x_l)$, and $I(x_l) \cap E = \{x_l\}$. Thus $E = \{x_1, \ldots, x_k\}$, and so has only a finite number of members.

Once again there need be no more than one limit point. Thus the set $\{1/n \mid n \in \mathbb{N}\}$ is bounded as $0 < 1/n \leqslant 1$, and has as its only limit point 0. Of course, if a set is given explicitly as in this case we can determine its limit points directly, but Theorem 6.11 is of great use where the set $E$ is generated by some theoretical construction. Thus the theorem may be used to derive results corresponding to those of Theorems 5.2, 5.3 and 5.4 for the images of closed sets by real valued functions. By also using the fact that the function mod is continuous, results can be obtained for complex valued functions.

**Theorem 6.12** If the function $f$ is continuous on a closed bounded set $G$ then $f(G)$ is bounded.

*Proof* Consider what would happen if $f(G)$ were not bounded. For any $z \in G$ there would be points of $f(G)$ not in the interval $[-|f(z)|-1, |f(z)|+1]$ for the case of real valued $f$, or not in the interval $[-|f(z)|-1, |f(z)|+1] \times [-|f(z)|-1, |f(z)|+1]$ for the case of complex valued $f$. Thus in either case it would be possible to find a $z_1 \in G$ such that $|f(z_1)| \geqslant |f(z)|+1$. Suppose this to be the case. Then taking any $a_0 \in G$ we can choose an $a_n \in G$ for each $n \in \mathbb{N}'$ such that $|f(a_n)| \geqslant |f(a_{n-1})|+1$. Then $a_n \neq a_m$ if $n \neq m$. The set $\{a_n \mid n \in \mathbb{N}\}$ is thus an infinite set and is bounded as $G$ is bounded. It thus has a limit point $a$ by Theorem 6.11. Furthermore, $G$ is closed, so that $a \in G$.

Consider $f(a)$. As $f$ is continuous on $G$, we have $f(a) = \lim_{z \to a} f(z)$. Hence corresponding to the value $\frac{1}{2}$ for $\varepsilon$ there is a $\delta_{1/2} > 0$ such that if $0 < |z - a| < \delta_{1/2}$ and $f(z)$ is defined then $|f(z) - f(a)| < \frac{1}{2}$. However, since $a$ is a limit point of $\{a_n \mid n \in \mathbb{N}\}$, and the open set $\mathcal{N}(a, \delta_{1/2})$ contains an interval neighbourhood of $a$, there is some $k \in \mathbb{N}$ such that $0 < |a_k - a| < \delta_{1/2}$. Now writing $\delta = |a_k - a| > 0$, we can repeat this argument to find $a_l$ such that $0 < |a_l - a| < \delta < \delta_{1/2}$. Thus $a_l \neq a_k$ and $a_l$, $a_k \in \mathcal{N}(a, \delta_{1/2})$. This gives both $|f(a_l) - f(a)| < \frac{1}{2}$ and $|f(a_k) - f(a)| < \frac{1}{2}$. Thus by the triangle inequality

$$|f(a_l) - f(a_k)| = |f(a_l) - f(a) + f(a) - f(a_k)|$$
$$\leqslant |f(a_l) - f(a)| + |f(a) - f(a_k)|$$
$$< \tfrac{1}{2} + \tfrac{1}{2} = 1$$

However, the $a_n$ were chosen in such a way that $|f(a_l) - f(a_k)| \geqslant 1$, provided only that $a_l \neq a_k$. Hence the assumption that $f(G)$ is not bounded has led to a contradiction, and so $f(G)$ is bounded.

**Theorem 6.13** If $f$ is a real valued function continuous on a non-empty closed bounded set $G$ then there are points $\xi, \eta \in G$ such that

$$f(\xi) = \inf\{f(z) \mid z \in G\}$$

and $\quad f(\eta) = \sup\{f(z) \mid z \in G\}$

*Proof* It is sufficient just to show the existence of $\eta$, as the existence of $\xi$ will then follow by considering $-f$ and $\sup\{-f(z) \mid z \in G\} = -\inf\{f(z) \mid z \in G\}$.

By Theorem 6.12 the set $f(G)$ is bounded, and by the assumption that $G$ is non-empty $f(G)$ is non-empty, so by Theorem 2.15 the number $M = \sup\{f(z)\,|\,z \in G\}$ exists. Then corresponding to every $n \in \mathbb{N}'$ there must be an $a_n \in G$ such that $0 \leqslant M - f(a_n) < 1/n$. Otherwise $M - 1/n$ would be a smaller upper bound for $f(G)$.

Suppose, if possible, that there were no $\eta \in G$ for which $f(\eta) = M$. Then the function $\varphi$ defined on $G$ by $\varphi(x) = 1/[M - f(x)]$ for $x \in G$ would be continuous on $G$ by Theorem 5.12. Thus $\varphi(G)$ would be bounded. However, given any real number $N$, there is some $n \in \mathbb{N}$ such that $n > N$, and so $\varphi(a_n) = 1/[M - f(a_n)] > n > N$. Thus $\varphi(G)$ is not bounded. Hence it must be that there is some $\eta \in G$ with $f(\eta) = M$.

It should be clear that there need not be more than one point $\xi$ or one point $\eta$, but that there can be several points $\eta$ at which $f(\eta) = M$. Indeed there could be infinitely many such points, and similarly for points $\xi$ at which $f(\xi) = \inf\{f(z)\,|\,z \in G\}$.

**Corollary to Theorem 6.13**    If $f$ is a complex valued function and is continuous on the non-empty closed bounded set $G$ then there are points $\xi, \eta \in G$ such that

$$|f(\xi)| = \inf\{|f(z)|\,|\,z \in G\}$$

and    $$|f(\eta)| = \sup\{|f(z)|\,|\,z \in G\}$$

*Proof*    From the continuity of $f$ on $G$, and the continuity of mod on $\mathbb{C}$, it follows by Theorem 5.9 that the function $|f|$ with $|f|(z) = |f(z)|$ is a real valued function continuous on $G$. Thus the existence of $\xi$ and $\eta$ is given by the theorem.

There is no analogue to Theorem 5.1 for general closed sets, even if the function is a real function. Thus if $f$ is the function defined for all $x \in \mathbb{R}$ by $f(x) = x$ and $G = \{1, 2\}$, $f$ is continuous on $\mathbb{R}$ and $G$ is bounded and closed as it has no limit points. But $f(G) = \{1, 2\}$ is not an interval.

**Exercises**

**1**    Using either (ii) or (iii) of Theorem 6.2 as your definition of $\lim_{x \to a} f(x)$, prove directly that if $f$ is a real function with $\lim_{x \to a} f(x) = l_1$ and $\lim_{x \to a} f(x) = l_2$ then $l_1 = l_2$.

**2**    On the same basis as in question 1, give a direct proof that if $f$ and $g$ are real functions and $\lim_{x \to a} f(x) = l$ and $\lim_{x \to a} g(x) = k$ then $\lim_{x \to a} [f(x) + g(x)] = l + k$.

**3**    Find the following limits:

(i)    $\displaystyle\lim_{x \to 1} \frac{x^3 - x^2 + x - 1}{x^3 - x^2 + 2x - 2},$

(ii) $\lim\limits_{x \to 0} \dfrac{x^{-2}-1}{x^{-2}+1}$,

(iii) $\lim\limits_{x \to 2} \sqrt{\left(\dfrac{x^3+1}{x^3-4}\right)}$.

**4** Show that if two real functions $f$ and $g$ are such that both $\lim_{x \to a} f(x) = l$ and $\lim_{x \to a} g(x) = k$ then $\lim_{x \to a} \max\{f(x),\, g(x)\} = \max\{l,\, k\}$.

**5** Show that if $f$ is a complex function, and $\lim_{z \to z_0} f(z) = l$, then writing $\mathrm{Re}(z_0) = x_0$, $\mathrm{Im}(z_0) = y_0$, $\mathrm{Re}(f) = u$ and $\mathrm{Im}(f) = v$, we get $\lim_{x \to x_0} v(x + iy_0) = \mathrm{Im}(l)$ and $\lim_{y \to y_0} u(x_0 + iy) = \mathrm{Re}(l)$.

**6** The sets $U_n$ are open sets on $\mathbb{R}$ and $U_n \subset U_{n+1}$ for $n \in \mathbb{N}'$. Prove that if the closed set $F$ is such that $F \subset \bigcap_{n \in \mathbb{N}'} U_n$ then there is some $k \in \mathbb{N}'$ such that $F \subset U_k$.

**7** If $a \notin E \subset \mathbb{R}$ then the open sets $U_n = \{x \mid |x - a| > 1/n\}$ are such that $\bigcup_{n \in \mathbb{N}'} U_n \supset E$. Hence, or otherwise, deduce that if $E$ is compact then $E$ is closed.

**8** The complex function $f$ has $\mathscr{D}f = \mathbb{C}$ and is continuous. Prove that the set $\{z \mid |f(z)| = 1\}$ is closed.

# 7

# Sequences and countable and uncountable sets

'Sequence' is another name for a function whose domain is a subset of $\mathbb{N}$. We have used some finite sequences, that is sequences with domains which are finite subsets of $\mathbb{N}$, in Chapter 1. Also in the proofs of Theorems 6.12 and 6.13 infinite sequences were used. In all these cases the usual notation has been followed and the value of a sequence at $n$ written as $a_n$ rather than as $a(n)$. Sequences occur in very many computations. For instance, suppose that we are trying to find a solution to an equation of the form $f(x) = x$. Suppose furthermore that $f$ is strictly monotone increasing, that there is a solution $x_0$, so that $f(x_0) = x_0$, and that for $x < y$, $f(y) - f(x) < y - x$. We can set about calculating $x_0$ as follows. Take any number $a_0 > x_0$ with $a_0 \in \mathcal{D}f$. This needs some knowledge of the size of $x_0$, but not very much. Then if $a_1 = f(a_0)$ we have $a_1 - x_0 = f(a_0) - f(x_0) < a_0 - x_0$, so that $x_0 < a_1 < a_0$. Since $a_1 > x_0$, we can apply this construction again to get $a_2 = f(a_1) > x_0$, still closer to $x_0$. Clearly, whenever we have defined $a_n$ by using this construction $n$ times we have $x_0 < a_n$ and can define $a_{n+1} = f(a_n)$ to get $x_0 < a_{n+1} < a_n$. This defines a value $a_n$ corresponding to every $n \in \mathbb{N}$, and so defines a sequence. Now the range of this sequence $\{a_n \mid n \in \mathbb{N}\} \subset (x_0, a_0]$. Also $a_n > a_m$ if $n < m$, so that the range has infinitely many members. Thus by the Bolzano–Weierstrass theorem, Theorem 6.11, there is at least one limit point $a_*$ of $\{a_n \mid n \in \mathbb{N}\}$. Then $a_n > a_*$ for all $n \in \mathbb{N}$, as if $a_k < a_*$ (if $a_k = a_*$ we use $a_{k+1} < a_*$) then $a_n < a_k < a_*$ for $n > k$ and there are at most $k$ numbers $a_0, a_1, \ldots, a_{k-1}$ in $\mathcal{N}(a_*, a_* - a_k)$, and so there is some neighbourhood of $a_*$ not containing any $a_n \neq a_*$, contradicting the fact that $a_*$ is a limit point of $\{a_n \mid n \in \mathbb{N}\}$. Now as $a_*$ is such a limit point, then for every $\varepsilon > 0$ there is some $a_l$ such that $a_* < a_l < a_* + \varepsilon$ and hence $a_n \in [a_*, a_* + \varepsilon)$ for $n > l$. Thus also $f(a_n) \in [a_*, a_* + \varepsilon)$ for $n > l$, so that $|a_n - f(a_n)| < \varepsilon$ if $n > l$. However, $f$ is continuous since $|f(y) - f(x)| < |y - x|$. Hence if we take the limit as $x \to a_*$ we get $|a_* - f(a_*)| \leqslant \varepsilon$ for all $\varepsilon > 0$. Thus $a_* = f(a_*)$. It is not very difficult to show that in fact $a_* = x_0$.

Nowadays the values of square roots of positive numbers are readily available from tables, which have already been calculated. Even so, the above technique is sometimes useful in improving the accuracy of the values available for some square root. Here we consider this calculation mainly because it is a reasonably simple application, and also because we have seen that it can be impossible to find an exact value (i.e. a rational value) for a square root. To find the square root of $s > 0$, consider the function $f$ to be given by

$$f(x) = \tfrac{1}{2}(x + s/x) \quad \text{for } x > \sqrt{s}$$

Then $f(y) - f(x) = \frac{1}{2}(y - x + s/y - s/x) = \frac{1}{2}(1 - s/xy)(y - x)$. If $x, y > \sqrt{s}$ then $s/xy < 1$, so that $f(y) - f(x)$ and $y - x$ are of the same sign. Thus for $y > x$ we get $f(y) - f(x) < y - x$, that is $f$ is a function satisfying the conditions previously considered. So if $a_0 > \sqrt{s}$, e.g. $a_0 > \max\{1, s\}$, then the sequence defined by $a_{n+1} = f(a_n)$ gives a unique limit point which is $\sqrt{s}$.

Apply this to the calculation of $\sqrt{2}$, and let $a_0 = 2$. Since $2^2 = 4 > 2$, we have $a_0 > \sqrt{2}$. Then we get $a_1 = \frac{1}{2}(2 + \frac{2}{2}) = \frac{3}{2}$, $a_2 = \frac{1}{2}(\frac{3}{2} + 2 \times \frac{2}{3}) = \frac{17}{12}$, $a_3 = \frac{1}{2}(\frac{17}{12} + \frac{24}{17}) = \frac{577}{408}$. Correct to seven decimal places, $a_3$ is 1.414 2157 and $\sqrt{2}$ to the same number of places is 1.414 2135, so that already $a_3$ is within $1/500\ 000$ of $\sqrt{2}$. Also $a_4$ will be closer to $\sqrt{2}$ and, as $\sqrt{2}$ is a limit point of $\{a_n \mid n \in \mathbb{N}\}$, we can get $a_n$ as close to $\sqrt{2}$ as is needed for any given purpose.

For the application to finding square roots we needed to consider $a_0 > x_0$, but if $f$ also satisfies the assumptions for $x < x_0$ then an exactly similar discussion shows that we can start with $a_0 < x_0$ and get a limit point $a_* = x_0$. Thus in such a case we can start with any $a_0$. However, then we must notice that if $a_1 = f(a_0) = a_0$ we get $a_n = a_0$ for all $n \in \mathbb{N}$, so there is no limit point. However, we clearly have $a_0$ as the solution.

With these considerations in mind we make the following definitions.

**Definition 7.1** A function $a$ with domain $\mathcal{D}a = \{n \mid n \geq s \text{ and } n \in \mathbb{Z}\}$ is called an *infinite sequence*; the values of the function, called the *terms* of the sequence, are written as $a_n = a(n)$, and the infinite sequence is denoted either by $\{a_n\}$ or by $\{a_n \mid n = s, s+1, \dots\}$ or by $a_s, a_{s+1}, \dots, a_n, \dots$. Where the context should make it clear that an infinite sequence is intended, the word infinite will usually be omitted.

The sequence is said to *start at $s$*, and $s$ is called the *starting point* of the sequence.

If $a$ is a real function then it is called a *real sequence* or a *sequence of real terms*.

If $a$ is a complex valued function then it is called a *complex sequence* or a *sequence of complex terms*.

We shall see that the theory of infinite sequences could be adequately developed using only sequences which start at $s = 0$. Then given a sequence $a_s, a_{s+1}, \dots, a_n, \dots$ for which $s$ was not 0, it could be dealt with by considering instead the sequence $\{b_n\}$ defined by $b_n = a_{n-s}$. In practice it would be too much of a nuisance to have to make such a change every time the form in which a sequence arose naturally started at $s \neq 0$. The most frequent values for $s$ are 0 and 1. Thus, for instance, the sequence $\{1/n\}$ would naturally have starting point $s = 1$, but the sequence $1, \frac{1}{2}, \frac{1}{4}, \dots, 1/2^n, \dots$ would naturally be taken to be $\{1/2^n\}$ with $s = 0$.

**Definition 7.2** If $\{a_n \mid n = r+1, r+2, \dots\}$ is any sequence with starting point $r+1$, let $n = r + 1/x$, and let $\alpha$ be the function defined by $\alpha(x) = a_n$ for $1/x \in \mathbb{N}'$. Then the number $\lim_{x \to 0} \alpha(x)$ if it exists is called the *limit of the sequence* $\{a_n\}$ and is written $\lim_n a_n$, or sometimes $\lim_n a_n$.

Two common notations which will not be used here because they can be misleading at first are: $\lim_{n\to\infty} a_n$, and if $\lim_n a_n = l$ writing $a_n \to l$ as $n \to \infty$.

The domain of a sequence has no limit point, but the transformation given by $x = 1/(n-r)$ maps it onto a set with just one limit point, namely 0. Thus $\mathcal{D}\alpha = \{x \mid 1/x \in \mathbb{N}'\} = \{1/(n-r) \mid n \in \mathbb{Z} \text{ and } n > r\} = \{1/m \mid m \in \mathbb{N}'\}$ has 0 as its only limit point.

Theorem 6.2 can be used to provide an alternative way of stating that $\lim_n a_n = l$.

**Theorem 7.1**   For any sequence $\{a_n \mid n = s, s+1, \ldots\}$

$$\lim_n a_n = l \Leftrightarrow \text{for all } \varepsilon > 0,\ \exists N_\varepsilon \text{ such that } |a_n - l| < \varepsilon \text{ for all } n \in \mathbb{Z} \text{ with}$$
$$n > N_\varepsilon$$

*Proof*   Let $\alpha$ be the function given by $\alpha[1/(n-s+1)] = a_n$ for $n \in \mathbb{Z}$ with $n \geq s$. Then by Definition 7.2 and Theorem 6.2

$$\lim_n a_n = l$$

iff     for every $\varepsilon > 0$, $\exists \delta_\varepsilon > 0$ such that $|\alpha(x) - l| < \varepsilon$ for $0 < |x| < \delta_\varepsilon$, $x \in \mathcal{D}\alpha$

iff     for every $\varepsilon > 0$, $\exists \delta_\varepsilon > 0$ such that $|a_n - l| < \varepsilon$ for $1/(n-s+1) < \delta_\varepsilon$ with $n \in \mathbb{Z}$

iff     for $\varepsilon > 0$, $\exists \delta_\varepsilon > 0$ such that $|a_n - l| < \varepsilon$ for $n > 1/\delta_\varepsilon + s - 1$ with $n \in \mathbb{Z}$

iff     for $\varepsilon > 0$, $\exists N_\varepsilon$ $(= 1/\delta_\varepsilon + s - 1)$ such that $|a_n - l| < \varepsilon$ for $n > N_\varepsilon$ with $n \in \mathbb{Z}$

It is usual to omit specific reference to $\mathbb{Z}$, as we know that if $a_n$ is a term of a sequence then $n \in \mathbb{Z}$.

Notice that $N_\varepsilon$ can be chosen to be an integer, as if there is $N_\varepsilon \notin \mathbb{N}$ for which $|a_n - l| < \varepsilon$ for $n > N_\varepsilon$ then there is some integer $N'_\varepsilon > N_\varepsilon$ and $|a_n - l| < \varepsilon$ for $n > N'_\varepsilon > N_\varepsilon$. Notice also that $N_\varepsilon$ is such that $a_n$ is a term of the sequence for $n > N_\varepsilon$.

In the notation $\lim_n a_n$ the full name of the sequence does not appear. Given a sequence $\{a_n \mid n = s, s+1, \ldots\}$, a second sequence can be formed by taking $r$ to be an integer $r > s$ and writing $\{a_n \mid n = r, r+1, \ldots\}$. It is easy to show that two such sequences have the same limit, or else both fail to have a limit, so the notation $\lim_n a_n$ is not ambiguous.

**Theorem 7.2**   If $r > s$ and either of the two sequences $\{a_n \mid n = s, s+1, \ldots\}$ and $\{a_n \mid n = r, r+1, \ldots\}$ has a limit then so does the other and the two limits are equal.

*Proof*   Suppose that the first sequence has $\lim_n a_n = l$. Then by Theorem 7.1, given $\varepsilon > 0$, there is a number $N_\varepsilon$ such that

$$|a_n - l| < \varepsilon \quad \text{for } n > N_\varepsilon$$

Hence for all $\varepsilon > 0$ there is a number $N_\varepsilon + r - s$ such that

$$|a_n - l| < \varepsilon \quad \text{for } n > N_\varepsilon + r - s$$

For $n > N_\varepsilon + r - s$ the term $a_n$ is a term of the second sequence. Thus we have that $l$ is also the limit of the second sequence.

Now suppose that the second sequence has the limit $l$. Then for all $\varepsilon > 0$ there is a number $N'_\varepsilon$ such that

$$|a_n - l| < \varepsilon \quad \text{for } n > N'_\varepsilon$$

Now $n > N'_\varepsilon$ implies that $a_n$ is a term of the second sequence. Also $r > s$, so that $a_n$ is a term of the first sequence as well. Thus $l$ is the limit of the first sequence also.

As a result of this theorem, when considering $\lim_n a_n$ we can, if necessary, ignore a finite number of terms. What is important is only what happens for *sufficiently large* $n$—that is, what is true of $a_n$ provided that $n > N$ for some fixed $N$. A useful way of writing Theorem 7.2 is as $\lim_n a_n = \lim_n a_{n+k}$ for any fixed $k \in \mathbb{Z}$.

There are some sequences with terms given by simple formulae for which limits can readily be found. Thus the sequence $\{1/n\}$ leads to $\lim_n 1/n = 0$. This follows from Theorem 7.1, as for $\varepsilon > 0$ we can take $N_\varepsilon = 1/\varepsilon$ and then

$$\left| \frac{1}{n} - 0 \right| = \left| \frac{1}{n} \right| < \varepsilon \quad \text{if } n > N_\varepsilon$$

Indeed if $\alpha$ is any positive rational number, so that $1/n^\alpha$ is defined for $n \in \mathbb{N}'$, we get $\lim_n 1/n^\alpha = 0$. Again we can verify this by finding an expression for $N_\varepsilon$. The inequality $1/n^\alpha < \varepsilon$ is equivalent to $1/n < \varepsilon^{1/\alpha}$ and so equivalent to $n > 1/\varepsilon^{1/\alpha}$. Thus to every $\varepsilon > 0$ there corresponds an $N_\varepsilon = 1/\varepsilon^{1/\alpha}$ such that

$$\left| \frac{1}{n^\alpha} - 0 \right| = \left| \frac{1}{n^\alpha} \right| = \frac{1}{n^\alpha} < \varepsilon \quad \text{if } n > N_\varepsilon$$

If we now see what form Theorem 6.4 takes when applied to sequences, we shall be able to deduce the values of the limits of some more complicated sequences.

**Theorem 7.3**   If the two sequences $\{a_n\}$ and $\{b_n\}$ are such that $\lim_n a_n = A$ and $\lim_n b_n = B$ then:
  (i)   $\lim_n (a_n + b_n) = A + B$,
  (ii)   $\lim_n a_n \cdot b_n = A \cdot B$,
  (iii)   if also $B \neq 0$ then $\lim_n a_n / b_n = A / B$.

*Note*   The statement of the theorem is slightly simpler than that of Theorem 6.4. There is a standard domain for the functions in terms of which the limit of a sequence is defined, and so this theorem can be stated without having to make assumptions about that domain.

*Proof*   Write $\alpha$ and $\beta$ for the functions used in the definitions of $\lim_n a_n$ and $\lim_n b_n$. Then $\mathscr{D}\alpha = \mathscr{D}\beta = \{1/n \mid n \in \mathbb{N}'\}$, and so 0 is a limit point of $\mathscr{D}\alpha \cap \mathscr{D}\beta = \mathscr{D}\alpha$. Then:

(i)   the function which is used to define the limit $\lim_n (a_n + b_n)$ is $\alpha + \beta$, and using Theorem 6.4(i)

$$\lim_n (a_n + b_n) = \lim_{x \to 0} (\alpha + \beta)(x) = \lim_{x \to 0} [\alpha(x) + \beta(x)]$$

$$= \lim_{x \to 0} \alpha(x) + \lim_{x \to 0} \beta(x) = \lim_n a_n + \lim_n b_n = A + B$$

(ii)   the function which is used to define the limit $\lim_n a_n \cdot b_n$ is $\alpha \cdot \beta$, and using Theorem 6.4(ii)

$$\lim_n a_n \cdot b_n = \lim_{x \to 0} \alpha \cdot \beta(x) = \lim_{x \to 0} \alpha(x) \cdot \beta(x)$$

$$= \lim_{x \to 0} \alpha(x) \cdot \lim_{x \to 0} \beta(x) = \lim_n a_n \cdot \lim_n b_n = A \cdot B$$

(iii)   the function which is used to define the limit $\lim_n a_n/b_n$ is $\alpha/\beta$ and $\lim_{x \to 0} \beta(x) \neq 0$, so using Theorem 6.4(iii)

$$\lim_n a_n/b_n = \lim_{x \to 0} \alpha(x)/\beta(x) = \lim_{x \to 0} \alpha(x)/\lim_{x \to 0} \beta(x) = \lim_n a_n/\lim_n b_n = A/B$$

By using the rules contained in this theorem we can find, for example,

$$\lim_n \frac{n-1}{n+1} = \lim_n \frac{1 - 1/n}{1 + 1/n} = \frac{\lim_n 1 - \lim_n 1/n}{\lim_n 1 + \lim_n 1/n}$$

provided that the denominator is not 0. Now it is clear that as $|1 - 1| = 0 < \varepsilon$ (if $\varepsilon > 0$) we have $\lim_n 1 = 1$, and so we get

$$\lim_n \frac{n-1}{n+1} = \frac{1-0}{1+0} = 1$$

Now apply a similar calculation to $\lim_n (2n - 1)/(an^2 + n + 1)$, so as to find the limit for all real values of $a$ for which the limit exists. For $a \neq 0$

$$\lim_n \frac{2n-1}{an^2 + n + 1} = \lim_n \frac{2n^{-1} - n^{-2}}{a + n^{-1} + n^{-2}} = \frac{0 - 0}{a + 0 + 0} = 0$$

This does not give a result for $a = 0$, as when we need to use Theorem 7.3(iii) the limit of the denominator is 0. However, the limit does exist in this case since

$$\lim_n \frac{2n-1}{0n^2 + n + 1} = \lim_n \frac{2n-1}{n+1} = \lim_n \frac{2 - n^{-1}}{1 + n^{-1}} = \frac{2-0}{1+0} = 2$$

A similar case, but one in which the limit does not exist for the special value $a = 0$, is $\lim_n (2n - 1)/(an^2 + 1)$. For $a \neq 0$ the limit is 0, the calculation being very much as before. For $a=0$ the limit becomes $\lim_n (2n-1)/1 = \lim_n (2n-1)$. If this existed and was equal to $l$ then for $\varepsilon > 0$ we should have an $N_\varepsilon$ such that $|2n-1-l| < \varepsilon$ for $n > N_\varepsilon$. But if $n > \frac{1}{2}(l+1+\varepsilon)$ then $|2n-1-l| > \varepsilon$. So the limit cannot exist.

The limit $\lim_n [\sqrt{(n+1)} - \sqrt{n}]$ can also be found by this technique, but first notice that neither $\lim_n \sqrt{(n+1)}$ nor $\lim_n \sqrt{n}$ exists. This could be shown by an argument similar to the one used to show that $\lim_n (2n-1)$ does not exist. So we cannot find this limit by writing

$$\lim_n [\sqrt{(n+1)} - \sqrt{n}] = \lim_n \sqrt{(n+1)} - \lim_n \sqrt{n}$$

Instead we write

$$\lim_n [\sqrt{(n+1)} - \sqrt{n}] = \lim_n \frac{[\sqrt{(n+1)} - \sqrt{n}][\sqrt{(n+1)} + \sqrt{n}]}{\sqrt{(n+1)} + \sqrt{n}}$$

$$= \lim_n \frac{n+1-n}{\sqrt{(n+1)} + \sqrt{n}} = \lim_n \frac{1}{\sqrt{(n+1)} + \sqrt{n}}$$

Now $\lim_n \dfrac{1}{\sqrt{n}} = \lim_n \dfrac{1}{\sqrt{(n+1)}} = 0$

and $0 < \dfrac{1}{\sqrt{(n+1)} + \sqrt{n}} < \dfrac{1}{\sqrt{n}}$

so that to $\varepsilon > 0$ there corresponds a number $N_\varepsilon$ such that

$$\left| \frac{1}{\sqrt{(n+1)} + \sqrt{n}} \right| < \frac{1}{\sqrt{n}} < \varepsilon \quad \text{if } n > N_\varepsilon$$

Thus $\lim_n [\sqrt{(n+1)} - \sqrt{n}] = \lim_n \dfrac{1}{\sqrt{(n+1)} + \sqrt{n}} = 0$

The device used to show that $\lim_n (2n-1)$ does not exist, and the way in which the limit of $[\sqrt{(n+1)} - \sqrt{n}]$ was found, are special cases of the theorems which arise from Theorems 6.5 and 6.6.

**Theorem 7.4** If $\lim_n a_n = l$ then the set $\{a_n\}$, the range of the sequence, is bounded. [This is often expressed by saying that the sequence is bounded.]

*Proof* Suppose that $\alpha$ is the function used to define $\lim_n a_n$. Then $\lim_{x\to 0} \alpha(x) = l$. Hence by Theorem 6.5 there is a neighbourhood $\mathcal{N}(0, \delta)$ such that $\alpha(\mathcal{N}(0, \delta))$ is bounded, that is there is an integer $N$ ($\geq 1/\delta + s + 1$) such that the set $\{a_n \,|\, n > N\}$ is bounded. Suppose that $\{\text{Re}(a_n) \,|\, n > N\} \subset [\alpha, \beta]$, and let $\gamma = \max\{|a_s|, |a_{s+1}|, \ldots, |a_N|\}$, so that $\{\text{Re}(a_n) \,|\, n \leq N\} \subset [-\gamma, \gamma]$. Then $\{\text{Re}(a_n) \,|\, n = s, s+1, \ldots\} \subset [\alpha, \beta] \cup [-\gamma, \gamma] \subset [\min\{\alpha, -\gamma\}, \max\{\beta, \gamma\}]$. If the sequence $\{a_n\}$ has complex values then the same inclusions apply to the sets of imaginary parts of the $a_n$. Thus the sequence is bounded.

The fact that the sequence $\{2n-1\}$ has no limit is now a consequence of Theorem 7.4. It is only necessary to note that the set $\{2n-1\}$ is unbounded. Similarly, the sequence $\{(-1)^n \cdot n \mid n \in \mathbb{N}\}$ has no limit since the set of its values is not bounded.

**Theorem 7.5  Comparison test for sequences**  If $\{a_n\}$, $\{b_n\}$ and $\{c_n\}$ are real sequences such that there is a number $N$ for which $a_n \leqslant b_n \leqslant c_n$ for $n > N$ then $\lim_n a_n = \lim_n c_n = l \Rightarrow \lim_n b_n = l$.

*Proof*  If $\alpha$, $\beta$ and $\gamma$ are the functions used to define the limits $\lim_n a_n$, $\lim_n b_n$ and $\lim_n c_n$ then $0$ is a limit point of $\mathscr{D}\alpha \cap \mathscr{D}\beta \cap \mathscr{D}\gamma = \{x \mid 1/x \in \mathbb{N}'\}$. Also for $0 < x < 1/(N-t)$, where if $a_q$, $b_r$ and $c_s$ are the first terms of the sequences $t = \max\{q, r, s\}$, $\alpha(x) \leqslant \beta(x) \leqslant \gamma(x)$. Hence by the squeeze theorem, Theorem 6.6,

$$\lim_{x \to 0} \alpha(x) = l \text{ and } \lim_{x \to 0} \gamma(x) = l \Rightarrow \lim_{x \to 0} \beta(x) = l$$

that is

$$\lim_n a_n = \lim_n c_n = l \Rightarrow \lim_n b_n = l$$

The fact that $\lim_n 1/[\sqrt{(n+1)}+\sqrt{n}] = 0$ can now be seen as a consequence of Theorem 7.5. The sequences are given by $a_n = 0$ for $n \in \mathbb{N}$, $b_n = 1/[\sqrt{(n+1)}+\sqrt{n}]$ for $n \in \mathbb{N}$, and $c_n = 1/\sqrt{n}$ for $n \in \mathbb{N}'$. Since we know that $\lim_n a_n = \lim_n c_n = 0$, the result follows. Notice that in applying this theorem to a sequence of positive terms $b_n \geqslant 0$, we need to consider explicitly only one further sequence $\{c_n\}$ with $c_n \geqslant b_n$ for the case $\lim_n c_n = 0$.

Again if $b_n = (-1)^n/(n+2)^3$ we can take $a_n = -1/(n+2)^3$ and $c_n = 1/(n+2)^3$. Then $a_n \leqslant b_n \leqslant c_n$ and, since $\lim_n 1/n^3 = 0$, we have $\lim_n 1/(n+2)^3 = 0$. Also it follows from Theorem 7.3(ii) that $\lim_n -1/(n+2)^3 = 0$. Thus $\lim_n (-1)^n/(n+2)^3 = 0$. This is a special case of the general result that if $\lim_n |b_n| = 0$ then $\lim_n b_n = 0$. It is worth noting that the corresponding result is true for complex sequences $\{z_n\}$. If $\lim_n |z_n| = 0$ then from $-|z_n| \leqslant \mathrm{Re}(z_n) \leqslant |z_n|$ and $-|z_n| \leqslant \mathrm{Im}(z_n) \leqslant |z_n|$ Theorem 7.5 gives $\lim_n \mathrm{Re}(z_n) = 0$ and $\lim_n \mathrm{Im}(z_n) = 0$. Hence $\lim_n z_n = \lim_n \mathrm{Re}(z_n) + i \cdot \lim_n \mathrm{Im}(z_n) = 0$.

The converse result is a special case of the next theorem.

**Theorem 7.6**  If $\lim_n a_n = l$ where $l$, $a_n \in \mathscr{D}f$ for $n \in \mathbb{N}'$ and the function $f$ is continuous on $\{l\}$ then $\lim_n f(a_n) = f(l)$.

*Proof*  Let $\alpha$ be given by $\alpha(x) = a_{1/x}$ for $1/x \in \mathbb{N}'$. Then $\lim_n f(a_n) = f(l)$ iff $\lim_{x \to 0} f(\alpha(x)) = \lim_{x \to 0} f \circ \alpha(x) = f(l)$. Now $f$ is continuous on $\{l\}$ and the function $\tilde{\alpha}$ given by

$$\tilde{\alpha}(x) = \begin{cases} \alpha(x) & (1/x \in \mathbb{N}') \\ l & (x = 0) \end{cases}$$

is continuous on $\{0\}$, with $\tilde{\alpha}(0) = l$. Thus by Theorem 5.9 $f \circ \tilde{\alpha}$ is continuous on $\{0\}$, which is the same thing as $\lim_{x\to 0} f \circ \tilde{\alpha}(x) = f \circ \tilde{\alpha}(0) = f(l)$. Hence $\lim_n f(a_n) = f(l)$.

Notice that we could not use this theorem to obtain the result that if $\lim_n |a_n| = 0$ then $\lim_n a_n = 0$; the converse of Theorem 7.6 is false. Thus if $a_n = (-1)^n$ then $\lim_n |(-1)^n| = \lim_n 1 = 1$, but $\lim_n (-1)^n$ does not exist. This follows as there is no number $l$ such that both $|-1 - l| < \frac{1}{2}$ and $|1 - l| < \frac{1}{2}$.

We can use the theorem in the following type of calculation. We have first that $\lim_n (8 + 1/n) = 8$. Then as the cube root function is continuous for all real numbers, $\lim_n \sqrt[3]{(8 + 1/n)} = \sqrt[3]{8} = 2$.

It has been shown that if a sequence has a limit then it is bounded, but the sequence $\{(-1)^n\}$ shows that a sequence may well be bounded and still fail to have a limit. However, since a sequence is a function, it makes sense to talk about monotone sequences, and for a monotone sequence boundedness does imply the existence of a limit.

**Theorem 7.7** If the real sequence $\{a_n\}$ is both monotone and bounded then $\lim_n a_n$ exists. In the case of an increasing sequence $\lim_n a_n = \sup_n a_n$. In the case of a decreasing sequence $\lim_n a_n = \inf_n a_n$.

*Note* If $\{a_n\}$ is weakly increasing then it is bounded below by $a_s$, as $a_s \leqslant a_{s+1} \leqslant \cdots \leqslant a_n \leqslant \cdots$ all $n \geqslant s$, so that we have that if $\{a_n\}$ is weakly increasing and bounded above then $\lim_n a_n$ exists. Similarly, if $\{a_n\}$ is weakly decreasing and bounded below then, as $a_s \geqslant a_{s+1} \geqslant \cdots \geqslant a_n \geqslant \cdots$ all $n \geqslant s$, $\{a_n\}$ is bounded and $\lim_n a_n$ exists.

*Proof* Suppose first that the sequence is increasing. The range of the sequence $\{a_n \mid n \geqslant s, n \in \mathbb{Z}\}$ is a bounded non-empty set. Hence $\sup_n a_n$ exists. Then, given $\varepsilon > 0$, there must be an integer $N$ such that $\sup_n a_n - a_N < \varepsilon$, as otherwise $\sup_n a_n - \varepsilon \geqslant a_N$ for all $a_N$ of the sequence and so $\sup_n a_n - \varepsilon$ would be a smaller upper bound than $\sup_n a_n$. As the sequence is increasing, $a_N \leqslant a_r$ for $r \geqslant N$ and so

$$0 \leqslant \sup_n a_n - a_r < \varepsilon \quad \text{if } r \geqslant N$$

Thus $\left| \sup_n a_n - a_r \right| < \varepsilon \quad \text{if } r \geqslant N$

and, since such an $N$ exists for every $\varepsilon > 0$, we get $\lim_n a_n = \sup_n a_n$.

[For the case in which $\{a_n\}$ is decreasing a corresponding argument can be constructed using $\inf_n a_n$ to get $\lim_n a_n = \inf_n a_n$. The reader may find it instructive to do so. However, it is quicker to deduce this case from the previous one.]

Suppose now that $\{a_n\}$ is a decreasing sequence. Then $\{-a_n\}$ is an increasing sequence, and if $\{a_n\}$ is bounded below by $M$ then $\{-a_n\}$ is bounded above by $-M$. Hence by what has already been proved $\lim_n -a_n = \sup_n -a_n$, but $\lim_n -a_n = -\lim_n a_n$ and $\sup_n -a_n = -\inf_n a_n$, so that $\lim_n a_n = \inf_n a_n$.

Theorem 7.7 provides a neat way of finding $\lim_n t^n$ for $t \in [0, 1]$. We first notice that the sequence is decreasing, as on multiplying $0 \leqslant t \leqslant 1$ by $t^n$ we get $0 \leqslant t^{n+1} \leqslant t^n$ for $n \in \mathbb{N}'$. This also shows that the sequence is bounded below by 0. Thus $\lim_n t^n$ exists. Writing its value as $l$, we have

$$\lim_n t^{n+1} = t \cdot \lim_n t^n = t \cdot l$$

But also

$$\lim_n t^n = \lim_n t^{n+1} = l$$

by Theorem 7.2. Thus $t \cdot l = l$, so that either $l = 0$ or $t = 1$. Now if $t = 1$ then $t^n = 1$ all $n \in \mathbb{N}'$, and so $\lim_n t^n = 1$ in that case. Thus

$$\lim_n t^n = \begin{cases} 0 & (0 \leqslant t < 1) \\ 1 & (t = 1) \end{cases}$$

If $-1 < t \leqslant 0$ then $\lim_n |t^n| = 0$ and so $\lim_n t^n = 0$. In the case $t = -1$ the sequence is $\{(-1)^n\}$ which we have already noted has no limit. To see that the sequence also has no limit when $|t| > 1$, we show that in this case it is not bounded. Take some $t$ with $|t| > 1$ and consider any $M > 0$. As $0 < |1/t| < 1$, we can find an $n \in \mathbb{N}'$ such that $|1/t|^n < 1/M$. Then $|t^n| > M$ and, since such an $n$ can be found for each $M > 0$, the sequence is unbounded. We also get $\lim_n z^n = 0$ for complex $z$ with $|z| < 1$ as $\lim_n |z|^n = 0$.

Theorem 7.7 provides a proof that the number $e = \lim_n (1 + 1/n)^n$ exists, although this number cannot be exactly expressed in finite form. First let us establish the equation

$$\frac{1}{2} + \frac{1}{2^2} + \cdots + \frac{1}{2^{n-1}} + \frac{1}{2^n} + \frac{1}{2^n} = 1$$

for $n \in \mathbb{N}'$. We shall need this both here and again later on. We have, writing the sum with a $\sum$,

$$\frac{1}{2^n} + \sum_{l=1}^{n} \frac{1}{2^l} = \frac{1}{2^n} + \frac{1}{2^n} + \sum_{l=1}^{n-1} \frac{1}{2^l} = \frac{1}{2^{n-1}} + \sum_{l=1}^{n-1} \frac{1}{2^l}$$

$$= \frac{1}{2^{n-1}} + \frac{1}{2^{n-1}} + \sum_{l=1}^{n-2} \frac{1}{2^l} = \frac{1}{2^{n-2}} + \sum_{l=1}^{n-2} \frac{1}{2^l}$$

After $n - 1$ such steps we get finally

$$\frac{1}{2^n} + \sum_{l=1}^{n} \frac{1}{2^l} = \tfrac{1}{2} + \tfrac{1}{2} = 1$$

We next see that the sequence is increasing by comparing the expansions of successive terms.

$$\left(1+\frac{1}{n}\right)^n = 1+\frac{n}{n}+\frac{n(n-1)}{2}\frac{1}{n^2}+\cdots$$

$$+\frac{n(n-1)\cdots(n-k)}{(k+1)!}\frac{1}{n^{k+1}}+\cdots+\frac{1}{n^n}$$

$$=1+1+\frac{1}{2}\left(1-\frac{1}{n}\right)+\cdots+\frac{1}{(k+1)!}\left(1-\frac{1}{n}\right)\left(1-\frac{2}{n}\right)\cdots\left(1-\frac{k}{n}\right)$$

$$+\cdots+\frac{1}{n!}\left(1-\frac{1}{n}\right)\left(1-\frac{2}{n}\right)\cdots\left(1-\frac{n-1}{n}\right)$$

$$\left(1+\frac{1}{n+1}\right)^{n+1}=1+1+\frac{1}{2}\left(1-\frac{1}{n+1}\right)+\cdots$$

$$+\frac{1}{(k+1)!}\left(1-\frac{1}{n+1}\right)\left(1-\frac{2}{n+1}\right)\cdots\left(1-\frac{k}{n+1}\right)+\cdots$$

$$+\frac{1}{n!}\left(1-\frac{1}{n+1}\right)\left(1-\frac{2}{n+1}\right)\cdots\left(1-\frac{n}{n+1}\right)$$

Now $\left(1-\frac{1}{n}\right)\leqslant\left(1-\frac{1}{n+1}\right),\left(1-\frac{2}{n}\right)\leqslant\left(1-\frac{2}{n+1}\right),\ldots,\left(1-\frac{k}{n}\right)\leqslant\left(1-\frac{k}{n+1}\right)$

so that

$$\frac{1}{(k+1)!}\left(1-\frac{1}{n}\right)\cdots\left(1-\frac{k}{n}\right)\leqslant\frac{1}{(k+1)!}\left(1-\frac{1}{n+1}\right)\cdots\left(1-\frac{k}{n+1}\right)$$

Also there is one more term in the expansion of $[1+1/(n+1)]^{n+1}$ than in the expansion of $(1+1/n)^n$, and this term $(n+1)^{-(n+1)}$ is positive. Thus

$$\left(1+\frac{1}{n}\right)^n\leqslant\left(1+\frac{1}{n+1}\right)^{n+1}$$

It remains to check that the sequence is bounded above. Using the fact that $(1-k/n)\leqslant 1$, we get

$$\left(1+\frac{1}{n}\right)^n\leqslant 1+1+\frac{1}{2!}+\frac{1}{3!}+\cdots+\frac{1}{k!}+\cdots+\frac{1}{n!}$$

Also $1/k!=1/(1\cdot 2\cdot 3\cdot\ \cdots\ \cdot k)\leqslant 1/2^{k-1}$, as each factor of $k!$ other than 1 is $\geqslant 2$. Thus

$$\left(1+\frac{1}{n}\right)^n\leqslant 1+1+\frac{1}{2}+\cdots+\frac{1}{2^{k-1}}+\cdots+\frac{1}{2^{n-1}}$$

$$=1+2\left(\frac{1}{2}+\cdots+\frac{1}{2^n}\right)=1+2\left(1-\frac{1}{2^n}\right)$$

$$\leqslant 1+2=3$$

The sequence is increasing and bounded above by 3, so that $e = \lim_n (1+1/n)^n$ exists.

The next theorem is not so easy to apply, but actually characterizes sequences which have limits, and holds for both sequences of real terms and sequences of complex terms.

**Theorem 7.8 The general principle of convergence** For any sequence $\{a_n\}$, $\lim_n a_n$ exists $\Leftrightarrow$ for every $\varepsilon > 0$, $\exists N_\varepsilon$ such that $|a_n - a_m| < \varepsilon$ if $n, m > N_\varepsilon$.

*Proof*
(i) [To prove the implication '$\Rightarrow$', we assume that the limit does exist and eliminate specific reference to it.] Suppose that $\lim_n a_n = l$. Then, given $\varepsilon > 0$, $\exists N_\varepsilon$ such that $|a_n - l| < \varepsilon/2$ for $n > N_\varepsilon$. Hence

$$|a_n - a_m| = |(a_n - l) + (l - a_m)| \leqslant |a_n - l| + |l - a_m| < \tfrac{1}{2}\varepsilon + \tfrac{1}{2}\varepsilon = \varepsilon$$

if $n, m > N_\varepsilon$.
(ii) [To prove the converse implication '$\Leftarrow$', we need to find $l$ and then establish that it is in fact the limit of the sequence.] Suppose that for every $\varepsilon > 0$, $\exists N_\varepsilon$ such that $|a_n - a_m| < \varepsilon$ if $n, m > N_\varepsilon$. Then the range of the sequence, the set $\{a_n\}$, is bounded. For taking 1 as a particular value for $\varepsilon$ we get $|a_n - a_m| < 1$ if $n, m > N_1$, that is for a fixed $m$ we have $|a_n - a_m| < 1$ for $n > N_1$. However, there is a bounded interval $I$ containing both $\{x \mid |x - a_m| < 1\}$ and the finite set of points $\{a_n \mid n \leqslant N_1\}$. If the set $\{a_n\}$ is an infinite set then by Theorem 6.11 it has at least one limit point. Let $l$ be such a limit point. If, on the other hand, $\{a_n\}$ is a finite set then there must be at least one number $l$ which is the value of $a_n$ for infinitely many $n$. [In either case it will follow eventually when it has been shown that $l$ is the limit of the sequence that there is only one such number.] Now, given any $\varepsilon > 0$, there is some $N'_\varepsilon$ such that $|a_n - a_m| < \tfrac{1}{2}\varepsilon$ if $n, m > N'_\varepsilon$. Also there is some $m > N'_\varepsilon$ such that $|a_m - l| < \tfrac{1}{2}\varepsilon$, since if $l$ is a limit point of $\{a_n\}$ then it is also a limit point of $\{a_n \mid n > N'_\varepsilon\}$, and otherwise we can choose $m$ so that $a_m = l$. On combining these we get

$$|a_n - l| = |(a_n - a_m) + (a_m - l)| \leqslant |a_n - a_m| + |a_m - l| < \tfrac{1}{2}\varepsilon + \tfrac{1}{2}\varepsilon = \varepsilon$$

if $n > N'_\varepsilon$. Since there is such an $N'_\varepsilon$ corresponding to each $\varepsilon > 0$, this is equivalent to $\lim_n a_n = l$.

If, for example, we were working in the field $\mathbb{Q}$ of rational numbers it would not be true that all sequences for which, given $\varepsilon > 0$, $\exists N_\varepsilon$ such that $|a_n - a_m| < \varepsilon$ for $n, m > N_\varepsilon$, had limits. Sequences satisfying this condition are known as *Cauchy sequences* whether or not they occur in circumstances which ensure that they have limits.

Now let us return to the problem with which we started the chapter. In order to solve the equation $f(x) = x$, a sequence $\{a_n\}$ was constructed such that $a_{n+1} = f(a_n)$. This sequence was decreasing and bounded below. Hence by Theorem 7.7 $\lim_n a_n$ exists. Now by Theorem 7.2 $\lim_n a_{n+1} = \lim_n a_n$, and by Theorem 7.6 $\lim_n f(a_n) = f(\lim_n a_n)$. Thus we say that proceeding to the limit in the equation $a_{n+1} = f(a_n)$ we get $\lim_n a_n = f(\lim_n a_n)$, and so $\lim_n a_n$ is a solution to the equation.

We can use Theorem 7.8 to get a similar solution under different conditions on $f$. Suppose that $f$ is such that $|f(x) - f(y)| < \frac{1}{2}|x - y|$ for all $x, y \in \mathbb{R}$. Then $f$ is continuous by the squeeze theorem, Theorem 6.6. Also if we start with any number $x_0$ and form the sequence $\{x_n\}$ such that $x_{n+1} = f(x_n)$ then $|x_{n+1} - x_1| \le |x_1 - x_0|$. This follows as

$$|x_{n+1} - x_1| = |(x_{n+1} - x_n) + (x_n - x_1)| \le |x_{n+1} - x_n| + |x_n - x_1|$$

$$\le \frac{1}{2}|x_n - x_{n-1}| + |x_n - x_1| \le (\tfrac{1}{2})^2 |x_{n-1} - x_{n-2}| + |x_n - x_1|$$

$$\le \frac{1}{2^n}|x_1 - x_0| + |x_n - x_1| \le \frac{1}{2^n}|x_1 - x_0| + \frac{1}{2^{n-1}}|x_1 - x_0| + |x_{n-1} - x_1|$$

$$\le \left(\frac{1}{2^n} + \frac{1}{2^{n-1}} + \cdots + \frac{1}{2}\right)|x_1 - x_0|$$

for $n \in \mathbb{N}$. However,

$$\left(\frac{1}{2^n} + \frac{1}{2^{n-1}} + \cdots + \frac{1}{2}\right) = 1 - \frac{1}{2^n} < 1$$

all $n \in \mathbb{N}'$. Hence $|x_{n+1} - x_1| \le |x_1 - x_0|$. Now take any number as $a_0$ and form the sequence $\{a_n\}$ such that $f(a_n) = a_{n+1}$ for $n \in \mathbb{N}$. We can show that $\{a_n\}$ is a Cauchy sequence. Since $|f(a_{n+1}) - f(a_n)| \le \frac{1}{2}|a_{n+1} - a_n|$, we get $|f(a_{n+1}) - f(a_n)| \le \frac{1}{4}|a_n - a_{n-1}| \le \cdots \le (1/2^{n+1})|a_1 - a_0|$. Thus $|a_{n+2} - a_{n+1}| \le (1/2^{n+1})|a_1 - a_0|$, and as $\lim_n 1/2^{n+1} = 0$ so also $\lim_n |a_{n+2} - a_{n+1}| = 0$. Hence, given $\varepsilon > 0$, $\exists N_\varepsilon$ such that $|a_{n+1} - a_n| < \varepsilon$ if $n > N_\varepsilon$.

Now for some $n > N_\varepsilon$ take $a_n$ to be $x_0$, so that $|x_{k+1} - x_1| \le |x_1 - x_0|$ for $k \in \mathbb{N}$ becomes $|a_m - a_{n+1}| \le |a_{n+1} - a_n|$ for $m \ge n+1$. Then for $m > n > N_\varepsilon$ we get $|a_m - a_{n+1}| < \varepsilon$. Thus for all $m, n > N_\varepsilon + 1$, $|a_m - a_n| < \varepsilon$. Hence Theorem 7.8 shows that $\lim_n a_n$ exists. Suppose that $\lim_n a_n = l$. Then, as before,

$$l = \lim_n a_{n+1} = f(\lim_n a_n) = f(l) \, .$$

In fact $l$ is the only solution to $x = f(x)$, as if $l_1$ were any other solution we would get $|f(l) - f(l_1)| = |l - l_1| \le \frac{1}{2}|l - l_1|$, so that $l_1 = l$.

Theorem 7.8 will also have important applications in Chapter 9, where a special form of sequence known as a series is considered.

It is very often useful to take a set $E$ and form it into a sequence, that is to find some sequence $\{a_n \mid n \in \mathbb{N}\}$ which is such that the range $\{a_n\}$ of the sequence is $E$. Thus if $E$ is the set of rational numbers in $[0, \cdot)$ we can form the sequence $\{a_n\}$ as follows. If $n \in \mathbb{N}'$, write $n = 2^p(2q-1)$ where $p \in \mathbb{N}$ and $q \in \mathbb{N}'$. There is only one way of writing each $n$ as such a product, as $p$ is the largest integer such that $2^p$ is a factor of $n$, and $q$ is fixed as soon as the odd integer $(2q-1)$ is known. Then write $a_n = p/q$, and $\{a_n \mid n \in \mathbb{N}'\}$ is the required sequence. Any member $x$ of $E$ is given by two integers $p \ge 0$ and $q > 0$, and then for $n = 2^p(2q-1)$ we have $a_n = x$. There are in fact infinitely many terms of the sequence with the value $x$, as $x = (kp)/(kq)$ for $k \in \mathbb{N}'$.

The question arises as to whether this is always possible. By considering sequences whose values are themselves sequences it is not too difficult to show that the answer is no. Thus we can form a sequence $\{A_n\}$ of terms $A_n$, each of which is a sequence with terms either 0 or 1. We might have such a sequence $\{A_n\}$ given by

$$A_0 = 0, 0, 0, \ldots, 0, \ldots$$

$$A_1 = 1, 0, 0, \ldots, 0, \ldots$$

$$A_2 = 1, 1, 0, \ldots, 0, \ldots$$

$$A_3 = 1, 1, 1, 0, \ldots, 0, \ldots$$

. . . . . . . . . . . . . . . . . . . . .

Then $A_n$ would be the sequence the first $n$ terms of which are 1, the rest of the terms being 0. Clearly, the sequence $A$ all of whose terms are 1 is not a value of $\{A_n\}$ for any $n$. If the reader tries to write down a sequence $\{A_n\}$ which includes every sequence of noughts and ones as some $A_n$, he will not be surprised at the next theorem.

**Theorem 7.9**   Let $\mathscr{A}$ be the set of all sequences $\{a_m \,|\, m = 1, 2, \ldots\}$ with values in the set $T = \{a, b\}$ with $a \neq b$. Then there is no sequence $\{A_n \,|\, n = 1, 2, \ldots\}$ with terms $A_n$ which are sequences belonging to $\mathscr{A}$, the range of which $\{A_n\} = \mathscr{A}$.

*Proof*   Suppose that $\{A_n \,|\, n = 1, 2, \ldots\}$ is any sequence of sequences $A_n = \{a_m^{(n)} \,|\, m = 1, 2, \ldots\}$ taking values $a$ and $b$. [Then we have to find a sequence $A \in \mathscr{A}$ such that $A_n \neq A$ all $n \in \mathbb{N}'$.] Define the sequence $A$ by $a_1 \in T$, but $a_1$ is different from the first term of $A_1$, that is

$$a_1 = \begin{cases} a & (a_1^{(1)} = b) \\ b & (a_1^{(1)} = a) \end{cases}$$

and similarly

$$a_n = \left.\begin{cases} a & (a_n^{(n)} = b) \\ b & (a_n^{(n)} = a) \end{cases}\right\} n \in \mathbb{N}'$$

Then $A \neq A_n$ as $a_n \neq a_n^{(n)}$ for all $n \in \mathbb{N}'$, but also $A \in \mathscr{A}$ as $a_n \in T$ for $n \in \mathbb{N}'$. Thus the sequence of sequences $\{A_n \,|\, n = 1, 2, \ldots\}$ does not have $\{A_n\} = \mathscr{A}$.

The method of proof of Theorem 7.9 is known as Cantor's diagonal process. The significance of the word 'diagonal' appears if we think of the terms $a_m^{(n)}$

as being written out in an array:

$$
\begin{array}{ll}
A_1 & \underline{a_1^{(1)}}, a_2^{(1)}, a_3^{(1)}, a_4^{(1)}, \ldots, a_m^{(1)}, \ldots \\[4pt]
A_2 & a_1^{(2)}, \underline{a_2^{(2)}}, a_3^{(2)}, a_4^{(2)}, \ldots, a_m^{(2)}, \ldots \\[4pt]
A_3 & a_1^{(3)}, a_2^{(3)}, \underline{a_3^{(3)}}, a_4^{(3)}, \ldots, a_m^{(3)}, \ldots \\[4pt]
\cdots \\[4pt]
A_n & a_1^{(n)}, a_2^{(n)}, a_3^{(n)}, a_4^{(n)}, \ldots, a_m^{(n)}, \ldots, \underline{a_n^{(n)}}, \ldots \\[4pt]
\cdots
\end{array}
$$

Then the diagonal sequence $\underline{A}$ formed from the terms underlined is

$$a_1^{(1)}, a_2^{(2)}, a_3^{(3)}, \ldots, a_n^{(n)}, \ldots$$

and the sequence $A$ is formed from $\underline{A}$ by making sure that every term of $A$ is different from the corresponding term of $\underline{A}$. Here this has been done by changing every $a$ to $b$ and every $b$ to $a$.

An immediate consequence of Theorem 7.9 is that there are sets of real numbers which cannot be formed into sequences.

**Theorem 7.10**  There is no real sequence $\{a_n \mid n = 1, 2, \ldots\}$ such that the range $\{a_n\}$ of the sequence is the Cantor ternary set $C$.

*Proof*  Let

$$
J(n_1, \ldots, n_k) = \left[ \sum_{l=1}^{k} \frac{n_l}{3^l}, \sum_{l=1}^{k} \frac{n_l}{3^l} + \frac{1}{3^k} \right]
$$

[So that if $I(n_1, \ldots, n_k)$ is one of the open intervals used to define the set $U = [0, 1] \setminus C$, as in number (iv) of the Examples of open sets in Chapter 3 (p. 57), then we have $J(n_1, \ldots, n_k) \supset I(n_1, \ldots, n_k)$. The two intervals have the same mid-point and the closed interval is three times as long as the open interval.] If also for each fixed $k \in \mathbb{N}'$ the set

$$
K_k = \bigcup_{n_1, \ldots, n_k = 0, 2} J(n_1, \ldots, n_k)
$$

then $C = \bigcap_{k \in \mathbb{N}'} K_k$. [Figure 7.1 shows this way of constructing $C$.]

Now $J(n_1, \ldots, n_k) \subset J(m_1, \ldots, m_{k-1})$ precisely if $n_1 = m_1$, $n_2 = m_2, \ldots$, $n_{k-1} = m_{k-1}$. However, a point $x \in \bigcap_{k \in \mathbb{N}'} K_k$ iff there is some sequence of intervals $J(m_1, \ldots, m_k)$, one interval for each $k$, such that $x$ belongs to their intersection. Hence if $x \in C$ there is a sequence $\{m_k \mid k \in \mathbb{N}'\}$ where each $m_k$ is either 0 or 2 such that $x \in \bigcap_{k \in \mathbb{N}'} J(m_1, \ldots, m_k)$.

Thus if $\{a_n \mid n \in \mathbb{N}'\}$ is any sequence of real numbers with $a_n \in C$ all $n \in \mathbb{N}'$, there is a corresponding sequence of sequences $\{A_n \mid n \in \mathbb{N}'\}$ where $A_n = \{m_k^{(n)} \mid k \in \mathbb{N}'\}$ and $m_k^{(n)} \in \{0, 2\}$. Hence by Theorem 7.9 there is some sequence $A = \{m_k \mid k \in \mathbb{N}'\}$ with $m_k \in \{0, 2\}$ which is not the same as any of the $A_n$.

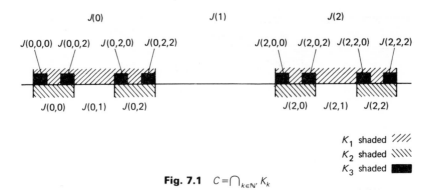

**Fig. 7.1** $C = \bigcap_{k \in \mathbb{N}'} K_k$

However, $\bigcap_{k \in \mathbb{N}'} J(m_1, \ldots, m_k)$ is not empty, by Theorem 6.10, since the $J(m_1, \ldots, m_k)$ are closed intervals and so closed sets. [In fact this intersection contains exactly one point and so every infinite sequence with values 0 and 2 corresponds to one point and just one point of $C$. This follows as the length of $J(m_1, \ldots, m_k)$ is $1/3^k$ and $\lim_k 1/3^k = 0$. So that if $x_1$ and $x_2$ are both in the intersection then $|x_2 - x_1| < 1/3^k$ all $k \in \mathbb{N}$, and this implies that $x_1 = x_2$.] If $x \in \bigcap_{k \in \mathbb{N}'} J(m_1, \ldots, m_k)$ then $x \notin \bigcap_{k \in \mathbb{N}'} J(m_1^{(n)}, \ldots, m_k^{(n)})$ for $n \in \mathbb{N}'$, and so $x \neq a_n$ for $n \in \mathbb{N}'$. Thus $\{a_n\} \neq C$.

In view of this result, the following definitions are made.

**Definition 7.3**   If the set $E$ is not empty and is the range of some sequence then $E$ is said to be *countable*. If also $E$ is infinite then it is said to be *countably infinite* or of *cardinal* $\aleph_0$ or to have $\aleph_0$ members.

If there is no sequence of which the set $E \neq \varnothing$ is the range then $E$ is said to be *uncountable*.

The letter $\aleph$, called aleph, comes from the Hebrew alphabet. The subscript 0 indicates that it is the smallest of a whole system of transfinite cardinal numbers. Here, however, we are only concerned with the distinction between countable and uncountable sets.

We do not have to worry about what value of $s$ gives the first term of the sequence, as clearly, given a sequence $a_s, a_{s+1}, \ldots$, the sequence $b_1, b_2, \ldots$ defined by $b_n = a_{s+n-1}$ for $n \in \mathbb{N}'$ has the same range.

**Theorem 7.11**   If the set $F$ is countable and the set $E$ is not empty, and $E \subset F$, then $E$ is countable.

*Proof*   Let $\{a_n \mid n \in \mathbb{N}\}$ be a sequence such that the set $\{a_n\} = F$, and define a sequence $\{b_n\}$ as follows: take any $e \in E$ and let

$$b_n = \begin{cases} a_n & (a_n \in E) \\ e & (a_n \notin E) \end{cases}$$

Then the range of $\{b_n\}$ is $E$, as if $x \in E$ there is some $m \in \mathbb{N}$ such that $a_m = x$ and by the definition of $b_m$, $b_m = a_m = x$. Thus $E$ is countable.

Since the Cantor set $C \subset [0, 1]$ is uncountable, it now follows that $[0, 1]$ is uncountable. Hence also $\mathbb{R}$ is uncountable, and then also $\mathbb{C}$ is uncountable.

The next few theorems provide some rules enabling us to establish the countability of sets in some cases.

**Theorem 7.12**  If the set $E$ is countable then the set $E \cup \{a\}$ is countable.

*Proof*  Suppose that $E$ is the range of the sequence $a_s, a_{s+1}, \ldots$ Then $E \cup \{a\}$ is the range of the sequence $a, a_s, a_{s+1}, \ldots$

**Theorem 7.13**  If the sets $E$ and $F$ are both countable then the set $E \cup F$ is countable.

*Proof*  Suppose that $E$ is the range of the sequence $a_1, a_2, \ldots, a_n, \ldots$ and that $F$ is the range of the sequence $b_1, b_2, \ldots, b_n, \ldots$. Let $t_n = a_m$ when $n \in \mathbb{N}'$ and $n$ is the odd integer with $n = 2m - 1$, and $t_n = b_m$ when $n \in \mathbb{N}'$ and $n$ is the even integer with $n = 2m$. Then the sequence $\{t_n\}$ can be written $a_1, b_1, a_2, b_2, a_3, b_3, \ldots, a_n, b_n, \ldots$, and clearly has range $E \cup F$.

It follows from this theorem that if $E_1, E_2, \ldots, E_k$ are all countable sets then their union $E_1 \cup E_2 \cup \cdots \cup E_k$ is a countable set. To construct the corresponding sequence explicitly for a few small values of $k$, let us write the sequence $\{t_n\}$ as $\{t_n\} = \{a_n\} * \{b_n\}$. Then

$$\{a_n\} * (\{b_n\} * \{c_n\}) = a_1, b_1, a_2, c_1, a_3, b_2, a_4, c_2, a_5, b_3, a_6, c_3, \ldots$$

$$\{a_n\} * (\{b_n\} * (\{c_n\} * \{d_n\})) = a_1, b_1, a_2, c_1, a_3, b_2, a_4, d_1, a_5, b_3, a_6, c_2, \ldots$$

and $\{a_n\} * (\{b_n\} * (\{c_n\} * (\{d_n\} * \{e_n\}))) = a_1, b_1, a_2, c_1, a_3, b_2, a_4, d_1, a_5, b_3,$

$$a_6, c_2, a_7, b_4, a_8, e_1, \ldots$$

We see that all these sequences start with $a_1$, and that for $k \geqslant 2$ they all start with $a_1, b_1, a_2$, also for $k \geqslant 3$ they all start with $a_1, b_1, a_2, c_1, a_3, b_2, a_4$, and in general for $k \geqslant k_0$ the sequences will all start with the same $2^{k_0} - 1$ terms. Also for $k \geqslant 2$ the place at which every term $a_n$ occurs remains fixed, for $k \geqslant 3$ the place at which every term $b_n$ occurs remains fixed, and in general the places at which the terms of the $k_0$th sequence occur remain fixed for $k \geqslant k_0 + 1$.

In order to be able to write down the terms of the $k$th sequence, we change the notation.

If now we take a countable collection $\{E_k\}$ of countable sets $E_k$, and let $E_k$ be the range of a sequence $\{a_i^{(k)}\}$, we can form the union $\bigcup_k E_k$ into a sequence. For each $k$ we form

$$\{t_i^{(k)}\} = \{a_i^{(1)}\} * (\{a_i^{(2)}\} * ( \cdots * \{a_i^{(k)}\}) \cdots )$$

and define the sequence $\{x_n\}$ by $x_n = t_n^{(n)}$. Since $2^n - 1 \geqslant n$, if $n \in \mathbb{N}'$ this gives $x_n$ the value that $t_n^{(k)}$ takes for all sufficiently large $k$. Then the range of $\{x_n\}$ is $\bigcup_k E_k$. We can picture this by writing down the terms of the sequences $\{a_l^{(k)}\}$ to form an array:

$$a_1^{(1)}, a_2^{(1)}, a_3^{(1)}, a_4^{(1)}, \ldots, a_l^{(1)}, \ldots$$

$$a_1^{(2)}, a_2^{(2)}, a_3^{(2)}, a_4^{(2)}, \ldots, a_l^{(2)}, \ldots$$

$$a_1^{(3)}, a_2^{(3)}, a_3^{(3)}, a_4^{(3)}, \ldots, a_l^{(3)}, \ldots$$

. . . . . . . . . . . . . . . . . . . . . . . . . .

$$a_1^{(k)}, a_2^{(k)}, a_3^{(k)}, a_4^{(k)}, \ldots, a_l^{(k)}, \ldots$$

. . . . . . . . . . . . . . . . . . . . . . . . . .

and then replacing each term $a_l^{(k)}$ by the number $n$ for which $x_n = a_l^{(k)}$. Thus

$$1, \quad 3, \quad 5, \quad 7, \ldots$$

$$2, \quad 6, \quad 10, \quad 14, \ldots$$

$$4, \quad 12, \quad 20, \quad 28, \ldots$$

. . . . . . . . . . . . . . . . . . . . . . . . . .

$$2^{k-1}, 3 \cdot 2^{k-1}, 5 \cdot 2^{k-1}, 7 \cdot 2^{k-1}, \ldots$$

. . . . . . . . . . . . . . . . . . . . . . . . . .

It can be seen that $n = 2^{k-1}(2 \cdot l - 1)$, a relation which we have already discussed with $k - 1 = p$ and $l = q$.

Another way of forming $\bigcup_k E_k$ into a sequence can be represented in the same form as

$$1, 3, 6, 10, 15, \ldots$$

$$2, 5, 9, 14, 20, \ldots$$

$$4, 8, 13, 19, 26, \ldots$$

$$7, 12, 18, 25, 33, \ldots$$

. . . . . . . . . . . . . .

**Theorem 7.14**   If each of the sets $E_k$ for $k \in \mathbb{N}'$ is countable then $\bigcup_{k \in \mathbb{N}'} E_k$ is countable.

*Proof*   Suppose that $E_k$ is the range of the sequence $\{a_l^{(k)} \mid l \in \mathbb{N}'\}$ for $k \in \mathbb{N}'$. Define the sequence $\{x_n \mid n \in \mathbb{N}'\}$ by

$$x_n = a_l^{(k)} \quad \text{for } n = 2^{k-1} \cdot (2l - 1), \ k, \ l \in \mathbb{N}'$$

This is possible, as both $k$ and $l$ are determined by $n$. Then each pair $\langle k, l \rangle$

defines a definite value $n$, and so for each $a_l^{(k)} \in \bigcup_{k \in \mathbb{N}'} E_k$ there is an $x_n = a_l^{(k)}$. Thus the range of $\{x_n \mid n \in \mathbb{N}'\}$ includes $\bigcup_{k \in \mathbb{N}'} E_k$. However, only values belonging to $\bigcup_{k \in \mathbb{N}'} E_k$ are used in defining the values of $x_n$, and so there must be equality between the range and the union.

**Corollary to Theorem 7.14**  The set $\mathbb{Q}$ of rational numbers is countable.

*Proof* Let $a_l^{(k)} = (k-1)/l$ for $k$, $l \in \mathbb{N}'$. Then by Theorem 7.14 the set $\{a_l^{(k)} \mid k, l \in \mathbb{N}'\}$ is countable. Now it is also the set of positive or zero rational numbers. Call this set of rational numbers $r \geqslant 0$, the set $E$, and write $F = \{x \mid -x \in E\}$. Then $E \cup F = \mathbb{Q}$, and both $E$ and $F$ are countable, as the range of $\{x_n\}$ is $E$ so that the range of $\{-x_n\}$ is $F$. Thus by Theorem 7.13 $\mathbb{Q}$ is countable.

Theorem 7.14 is often stated informally as: a countable collection of countable sets is countable. It has a number of very important applications.

**Exercises**

**1**  The sequence $\{\alpha_n\}$ has $\alpha_n = a_n + i b_n$ where $a_n$, $b_n \in \mathbb{R}$ for $n \in \mathbb{N}$. Prove that $\lim_n \alpha_n$ exists if and only if both $\lim_n a_n$ and $\lim_n b_n$ exist. Furthermore, show that if $\lim_n \alpha_n$ does exist then $\lim_n \alpha_n = \lim_n a_n + i \cdot \lim_n b_n$.

**2**  Show that if $\lim_n a_n = l$ and the sequence $\{b_n\}$ is defined by $b_n = a_{2n}$ for $n \in \mathbb{N}$ then $\lim_n b_n = l$.

**3**  Find the values of the following limits:

(i)  $\lim\limits_n \dfrac{n^5 + (x-1)n^4 + 1}{xn^6 + 2n^5 + 1}$ for each $x \in \mathbb{R}$,

(ii)  $\lim\limits_n \sqrt{n} \cdot [\sqrt{(n+1)} - \sqrt{n}]$,

(iii)  $\lim\limits_n \dfrac{x^{2n} - 1}{x^{2n} + 1}$ for each $x \in \mathbb{R}$,

(iv)  $\lim\limits_n \dfrac{[(1+i)n]^3 - 1}{[(1-i)n]^3 + 1}$.

**4**  Show that for each real $t \geqslant 0$ the sequence $\sqrt[n]{t}$ is weakly monotone and bounded. Hence by considering the relation between $\lim_n \sqrt[n]{t}$ and $\lim_n \sqrt[n]{t^2}$ and using the result of question 2, or otherwise, find the value of $\lim_n \sqrt[n]{t}$.

**5**  The real function $f$ has domain $\{x \mid x \geqslant 0\}$ and is continuous and also satisfies $x \geqslant f(x) \geqslant 0$ for all $x$ in the domain of $f$. Show that for any sequence $\{a_n\}$ generated by $a_{n+1} = f(a_n)$, $\lim_n a_n$ exists and is a solution of $f(x) = x$.

**6**  Show that for the function $f$ in question 5 $x = 0$ is always a solution of $f(x) = x$. Construct a function $f$ with a finite number of other solutions to the equation.

**7**  The sequence $\{a_n \mid n \in \mathbb{N}'\}$ has each term different. Construct a sequence $\{x_n\}$ such that for each $k \in \mathbb{N}'$ there are infinitely many $n \in \mathbb{N}'$ such that $x_n = a_k$.

**8**  Suppose that $E$ is an uncountable set of positive real numbers. By considering the sets $\{x \mid x \in E, x > 1/n\}$, or otherwise, show that for any real number $M$ there is a finite set $\{a_1, \ldots, a_k\} \subset E$ such that $M < \sum_{l=1}^{k} a_l$.

**9**  Assuming the result of question 8, show that any real function which is monotone on an interval $[a, b]$ has at most countably many points of discontinuity (points $\alpha$ such that the function is not continuous on $\{\alpha\}$).

# 8

# Differentiation

When performing calculations with functions it is helpful to be able to work in terms of algebraic expressions. In order to facilitate this we consider a stronger condition than continuity, but the definition is still based on the ideas we have been developing. In the first instance we look at the use of linear expressions, that is expressions of the form $ax + b$. When this has been done it will appear that the use of quadratic expressions, and also of higher order polynomials, can be treated by a repeated application of the theory.

One very basic and important case in the realm of physics of the process to be considered is the definition of velocity. For something which moves at a constant speed it is very easy to find the velocity. All that is needed is to take two points along the path at a measured distance $d$ from each other and find the times $t_1$ and $t_2$ at which each point is passed. Then the velocity is $d/(t_2 - t_1)$. This method would be used, for instance, in an attempt at a land speed record in a car. The car is run over a carefully measured track and timed at each end of the measured distance. It may well be that the car does not travel at a constant speed: it might perhaps be still accelerating when it reaches the start of the measured distance. In this case the measurement yields some sort of average velocity. We are used to being able to tell the speed of a car at any instant by reading the speedometer, but the design of a speedometer depends on having available a physical theory in which velocity at an instant is a well-defined quantity. In any case the accuracy of the instrument may well be dependent on measurements of the sort that have been described.

Suppose now that the object whose speed is to be measured is a ball which is dropped and allowed to fall freely. We might be interested in finding out how the speed of the ball increases with the distance it has fallen. If the time from the moment at which the ball is released is $t$ (measured, say, in seconds) then there will be a function $f$ which gives the distance through which the ball has fallen as $f(t)$. Then if we manage to find the position of the ball at two different times $t_0$ and $t_1$ the average velocity in the interval $[t_0, t_1]$ will be $[f(t_1) - f(t_0)]/(t_1 - t_0)$. The same expression results when $t_1 < t_0$ for the velocity over the time interval $[t_1, t_0]$. Now to try to find the velocity at time $t_0$, it is natural to take $t_1$ as near to $t_0$ as possible. In fact if we write

$$f(t) = f(t_0) + (t - t_0)\varphi(t)$$

then for any $t_1 \neq t_0$, $\varphi(t_1)$ is the average velocity over the interval between $t_0$ and $t_1$. The velocity at $t_0$ should be the value $\varphi(t_0)$ of $\varphi$ at $t_0$. As we have seen, in order that $\varphi(t_0)$ can be found from measurements of $\varphi(t_1)$ with $t_1 \neq t_0$ we must have $\varphi$ continuous on $\{t_0\}$. In geometrical terms $\varphi(t_1)$ is the slope of the chord of the graph of $f$ between $\langle t_0, f(t_0) \rangle$ and $\langle t_1, f(t_1) \rangle$. When $\varphi$ is continuous $\varphi(t_0)$ is the slope of the tangent to the graph at $\langle t_0, f(t_0) \rangle$. This is illustrated in Fig. 8.1. A further way of looking at this is by writing

$$f(t) = f(t_0) + (t - t_0)\varphi(t_0) + (t - t_0)[\varphi(t) - \varphi(t_0)]$$

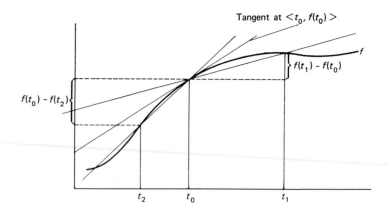

**Fig. 8.1**   The tangent to the graph of $f$

When $\varphi$ is continuous on $\{t_0\}$ we have $\lim_{t \to t_0} [\varphi(t) - \varphi(t_0)] = 0$. Thus for any given degree of accuracy there is a neighbourhood of $t_0$ in which the term $(t - t_0)[\varphi(t) - \varphi(t_0)]$ is sufficiently small compared with $(t - t_0)\varphi(t_0)$ for $f(t)$ to be represented by the linear expression $f(t_0) + (t - t_0)\varphi(t_0)$.

In view of the preceding considerations, the following definitions are made.

**Definition 8.1**   If the function $f$ and the point $x_0 \in \mathscr{D}f$ are such that $x_0$ is a limit point of $\mathscr{D}f$ and there is a function $\varphi$ continuous on the set $\{x_0\}$ for which

$$f(x) = f(x_0) + (x - x_0)\varphi(x) \quad \text{for all } x \in \mathscr{D}f$$

then $f$ is said to be *differentiable at a*.

If the function $f$ is differentiable at each point $a \in E$ then $f$ is said to be *differentiable on E*.

It follows immediately that differentiability is a stronger condition than continuity.

**Theorem 8.1**   If the function $f$ is differentiable on $E$ then $f$ is continuous on $E$.

*Proof*   For any point $x_0 \in E$ there is a function $\varphi$ continuous on $\{x_0\}$ such that

$$f(x) = f(x_0) + (x - x_0)\varphi(x) \quad \text{for } x \in \mathscr{D}f$$

Or if $\alpha$ is the function which has the constant value $f(x_0)$ and $\beta$ is the function defined by $\beta(x) = x - x_0$ for all $x$ then $f = \alpha + \beta \cdot \varphi$. Since $\alpha$, $\beta$ and $\varphi$ are all continuous on $\{x_0\}$, so is $f$. Now because this is true for all $x_0 \in E$, $f$ is continuous on $E$.

## Examples of differentiable and non-differentiable functions

(i)   $f$ given by $f(z) = az + b$ all $z \in \mathbb{C}$ is differentiable on $\mathbb{C}$, as for any $x_0 \in \mathbb{C}$

$$f(z) = ax_0 + b + a(z - x_0) = f(x_0) + (z - x_0)\varphi(z)$$

provided that $\varphi(z) = a$ for all $z \in \mathbb{C}$.

(ii)   $g$ given by $g(x) = x^2$ for $x \in \mathbb{R}$ is differentiable on $\mathbb{R}$, as

$$g(x) = x^2 = x_0^2 + (x - x_0)(x + x_0) = g(x_0) + (x - x_0)\varphi(x)$$

provided that $\varphi(x) = x + x_0$ for all $x \in \mathbb{R}$.

(iii)   $h$ given by $h(x) = |x|$ for $x \in \mathbb{R}$ is not differentiable on $\{0\}$, as if

$$h(x) = h(0) + (x - 0)\varphi(x) = 0 + x \varphi(x) \quad \text{for } x \neq 0$$

then

$$\varphi(x) = \frac{|x|}{x} = \begin{cases} 1 & (x > 0) \\ -1 & (x < 0) \end{cases}$$

so that, however $\varphi(0)$ is defined, $\varphi$ is not continuous on $\{0\}$.

Notice that $h$ is continuous on $\{0\}$, so that there is no converse to Theorem 8.1, and it is clear that the class of all functions differentiable on a set $E$ is a proper subset of the class of all functions continuous on $E$.

(iv)   $k$ given by

$$k(x) = \begin{cases} x^2 & (x > 0) \\ 0 & (x \leq 0) \end{cases}$$

is differentiable on $\mathbb{R}$. For any point $x_0 > 0$ the argument is as for the function $g$, and for any point $x_0 < 0$ the argument is as for the function $f$. To see that $k$ is differentiable on $\{0\}$, write

$$k(x) = k(0) + x \varphi(x) = 0 + x \varphi(x)$$

Then

$$\varphi(x) = \begin{cases} x & (x > 0) \\ 0 & (x \leq 0) \end{cases}$$

and it is easy to check that this function is continuous on $\{0\}$.

(v)   $l$ given by

$$l(x) = \begin{cases} x^2 & (x \text{ rational}) \\ -x^2 & (x \text{ irrational}) \end{cases} x \in \mathbb{R}$$

is differentiable at 0 but nowhere else. As $l$ is not continuous outside $\{0\}$, this is the only place where it can be differentiable, and if

$$l(x) = l(0) + (x - 0)\varphi(x) = x\,\varphi(x)$$

then

$$\varphi(x) = \begin{cases} x & (x \text{ rational}) \\ -x & (x \text{ irrational}) \end{cases}$$

provided that we define $\varphi(0) = 0$. The fact that $\varphi$ is continuous on $\{0\}$ follows from the squeeze theorem, Theorem 6.6.

Notice that starting from Example (iii) it is possible to construct continuous functions which fail to be differentiable at a number of points. We add together several functions of the form $a|x - b|$. It is even possible, by using a countable set of such functions, to construct a function which is continuous on $\mathbb{R}$ but differentiable nowhere. Unfortunately, the verification that it has these properties is too deep for inclusion at this point.

**Definition 8.2**   If the function $f$ is differentiable at $x_0 \in \mathcal{D}f$ and

$$f(x) = f(x_0) + (x - x_0)\varphi(x) \quad \text{for } x \in \mathcal{D}f$$

with $\varphi$ continuous on $\{x_0\}$, then the number $\varphi(x_0)$ is called the *derivative of $f$ at $x_0$* and written $f'(x_0)$.

The function $f'$ which takes the value $f'(x_0)$ for each $x_0$ at which $f$ is differentiable is called the *derivative of $f$* or the *derived function of $f$.*

It is clear that the definition implies that

$$f'(x_0) = \lim_{x \to x_0} \frac{f(x) - f(x_0)}{x - x_0}$$

so that both $\varphi$ and $f'(x_0)$, if they exist, are uniquely determined. The function $\varphi$ will be called the *difference quotient function for $f$ at $x_0$*, or simply the *difference quotient function.* This terminology will be used even when $f$ is not differentiable at $x_0$, in which case $\mathcal{D}\varphi = \mathcal{D}f \setminus \{x_0\}$. If more than one function is being discussed the notation used will be $\varphi_f$.

Applying Definition 8.2 to the functions of our examples above we get the following.

(1)   $f'(z) = a$ for all $z \in \mathbb{C}$. In particular, we see on taking $a = 0$ that the derivative of any constant real or complex function is always 0, and that the derivative of the identity function whose value at $x$ is $x$, or whose value at $z$ is $z$, is always 1.

(2)   $g'(x_0) = x_0 + x_0 = 2x_0$.

(3)   $\mathcal{D}h' = (\cdot, 0) \cup (0, \cdot)$ and proceeding on the basis of (1) with $a = 1$, $b = 0$ for $x_0 > 0$ and $a = -1$, $b = 0$ for $x_0 < 0$ we can see that $h'(x_0) = |x_0|/x_0$ for $x_0 \neq 0$.

(4)   For $x_0 > 0$ we have as for $g$ that $k'(x_0) = 2x_0$, for $x_0 < 0$ we have as for $f$ when $a = b = 0$ that $k'(x_0) = 0$, and for $x_0 = 0$ the function $\varphi$ has been found

and $\varphi(0) = 0$. So in all

$$k'(x_0) = \begin{cases} 2x_0 & (x_0 > 0) \\ 0 & (x_0 \le 0) \end{cases}$$

(5) $\mathcal{D}l' = \{0\}$, as $l$ is not continuous outside this set. We have seen that $\varphi(0) = 0$, so that $l'(0) = 0$.

We see that for each point $x_0 \in \mathcal{D}f'$ there is a function $\varphi$, so that really we have a function of the pair of variables $\langle x, x_0 \rangle$. However, since we have not investigated functions of two variables, it has been more convenient to fix $x_0$ and have $\varphi$ a function of only one variable.

There are two other important notations for the derivative in use. The first is simply to write $f'$ as $Df$. This, for instance, enables us to write $D(f \circ g) = (f \circ g)'$ and this may be found easier to read. The second is for use where we have a function expressed by a formula. If the formula gives the value of the function at $x$ then $d/dx$ is placed in front of the formula to denote the corresponding formula for the derived function. If instead the formula gave the value of the function at $y$ then $d/dy$ would be used. Thus

$$\frac{d}{dx}\left(\frac{x^2+1}{x^2-1}\right) = \theta'$$

where $\theta$ is the function given by $\theta(x) = (x^2+1)/(x^2-1)$, and

$$\frac{dy^3}{dy} = \psi'$$

where $\psi$ is the function given by $\psi(y) = y^3$. This notation is very useful when performing computations, but becomes more clumsy when values of the functions at particular points need to be considered. When it is desired to use the notation in this book the following device will be used. Let $x$ be the function such that for every real number $x$ we have $x(x) = x$. Similarly, let $z$ be the function defined by $z(z) = z$ for all $z \in \mathbb{C}$. Also denote by $c, 0, 1, 2, \ldots$ the constant functions taking the values $c, 0, 1, 2, \ldots$, the functions to be real or complex according to context. Thus we will write

$$\theta' = D\left(\frac{x^2+1}{x^2-1}\right)$$

and $\quad \psi' = Dx^3$

In this notation we have established the following derivatives

$\quad Dc = 0 \qquad$ (in both the real and complex cases)

$\quad Dc \cdot z = c \qquad$ (and hence also $Dc \cdot x = c$)

$\quad Dx^2 = 2 \cdot x \qquad$ (and by a similar argument $Dz^2 = 2z$)

With the aid of these results, and the rules for calculation contained in the following theorem, we can find the derivatives of a wide class of functions.

**Theorem 8.2**   If the functions $f$, $g$ and $h$ are either all real or all complex functions and $f$ and $g$ are both differentiable at $a$, $h$ is differentiable at $b = f(a)$, and $a$ is a limit point of $\mathscr{D}(h \circ f)$ and also of $\mathscr{D}f \cap \mathscr{D}g$, then:

(i)   $(f+g)'(a) = f'(a) + g'(a)$,

(ii)   $(f \cdot g)'(a) = f'(a) \cdot g(a) + f(a) \cdot g'(a)$,

(iii)   $(h \circ f)'(a) = h'(b) \cdot f'(a) = h' \circ f(a) \cdot f'(a)$.

If also $f$ is monotone on some interval $(\alpha, \beta) \ni a$ then:

(iv)   $(f^{-1})'(b) = \dfrac{1}{f' \circ f^{-1}(b)}$   unless $f'(f^{-1}(b)) = 0$.

*Note*   Any reader who is the least bit worried or confused by the condition that $a$ is a limit point of $\mathscr{D}(h \circ f)$, or the implied conditions that $a$ is a limit point of $\mathscr{D}f$, and of $\mathscr{D}g$, should assume that there is an open interval $(\alpha, \beta) \subset \mathscr{D}f \cap \mathscr{D}g$ with $a \in (\alpha, \beta)$ and an interval $(\alpha', \beta')$ with $b \in (\alpha', \beta') \subset \mathscr{D}h$. In practice, the only sort of case which we may need to discuss where these assumptions do not hold is that of a function whose domain is a closed interval and where the derivative is required at an end point of the interval.

*Proof*

(i)   $\varphi_{f+g}(x) = \dfrac{f(x) + g(x) - f(a) - g(a)}{x - a}$   (for $x \neq a$)

$= \dfrac{f(x) - f(a)}{x - a} + \dfrac{g(x) - g(a)}{x - a}$   (for $x \neq a$)

$= \varphi_f(x) + \varphi_g(x)$   (for $x \neq a$)

By Theorem 5.10 this is continuous on $\{a\}$ and so

$$(f+g)'(a) = \varphi_{f+g}(a) = \varphi_f(a) + \varphi_g(a) = f'(a) + g'(a)$$

(ii)   $f \cdot g(x) - f \cdot g(a) = (x - a)\varphi_{fg}(x)$

$= [f(x) - f(a)]g(x) + [g(x) - g(a)]f(a)$

$= (x - a)\varphi_f(x) \cdot g(x) + (x - a)\varphi_g(x) \cdot f(a)$

$= (x - a)[\varphi_f(x) \cdot g(x) + \varphi_g(x) \cdot f(a)]$

Now $\varphi_f \cdot g + \varphi_g \cdot f$ is continuous by Theorems 5.10 and 5.13. Hence by the uniqueness of the value at $a$

$$\varphi_{fg}(x) = \varphi_f(x) \cdot g(x) + \varphi_g(x) \cdot f(a)$$

and   $(f \cdot g)'(a) = \varphi_{fg}(a) = \varphi_f(a) \cdot g(a) + \varphi_g(a) \cdot f(a)$

$= f'(a) \cdot g(a) + g'(a) \cdot f(a)$

(iii)   By the definition of $\varphi_h$

$$h(y) - h(b) = (y - b)\varphi_h(y)$$

Now we put $y = f(x)$ and use $b = f(a)$ to get

$$h(f(x)) - h(f(a)) = h \circ f(x) - h \circ f(a) = [f(x) - f(a)]\varphi_h(f(x))$$
$$= (x - a)\varphi_f(x) \cdot \varphi_h(f(x))$$

by the definition of $\varphi_f$. Also

$$h \circ f(x) - h \circ f(a) = (x - a)\varphi_{h \circ f}(x)$$

By Theorems 5.13 and 5.9 $\varphi_f \cdot \varphi_h \circ f$ is continuous on $\{a\}$ and so, again using the uniqueness of $\varphi_{h \circ f}(a)$, we get

$$\varphi_{h \circ f}(x) = \varphi_f(x) \cdot \varphi_h(f(x))$$
$$(h \circ f)'(a) = \varphi_{h \circ f}(a) = \varphi_f(a) \cdot \varphi_h \circ f(a) = f'(a)h' \circ f(a) = f'(a)h'(b)$$

(iv)   We have on writing $y = f(x)$

$$(x - a) = f^{-1}(y) - f^{-1}(b) = (y - b)\varphi_{f^{-1}}(y) = (x - a)\varphi_f(x) \cdot \varphi_{f^{-1}}(y)$$

so that if $x \neq a$

$$1 = \varphi_f(x) \cdot \varphi_{f^{-1}}(y) = \varphi_f(x) \cdot \varphi_{f^{-1}}(f(x))$$

However, we are assuming that $f'(f^{-1}(b)) = f'(a) = \varphi_f(a) \neq 0$, so that

$$\frac{1}{\varphi_f} = \varphi_{f^{-1}} \circ f$$

is continuous on $\{a\}$. Also by Theorem 5.12 $f^{-1}$ is continuous on $\{b\}$ and so

$$\varphi_{f^{-1}} = \varphi_{f^{-1}} \circ f \circ f^{-1} = \frac{1}{\varphi_f \circ f^{-1}}$$

is continuous on $\{b\}$. Hence

$$(f^{-1})'(b) = \frac{1}{\varphi_f(f^{-1}(b))} = \frac{1}{f'(f^{-1}(b))}$$

**Corollary to Theorem 8.2**   If the function $f$ is either a real or a complex function, and is differentiable at a point $a \in \mathscr{D}f$ with $f'(a) \neq 0$, then $1/f$ is differentiable at $a$ and

$$\left(\frac{1}{f}\right)'(a) = \left(-\frac{f'}{f^2}\right)(a)$$

If furthermore $g$ is a function of the same kind as $f$ and is also differentiable at $a$ which is a limit point of $\mathscr{D}f \cap \mathscr{D}g$ then the function $g/f$ is differentiable at $a$ and

$$\left(\frac{g}{f}\right)'(a) = \frac{g(a)f'(a) - f(a)g'(a)}{[f(a)]^2}$$

*Proof*  We note that

$$D\frac{1}{x} = -\frac{1}{x^2}$$

This follows as writing $h = 1/x$ we get

$$h'(x_0) = \lim_{k \to 0} \frac{h(x_0 + k) - h(x_0)}{k} = \lim_{k \to 0} \frac{1}{k}\left(\frac{1}{x_0 + k} - \frac{1}{x_0}\right)$$

$$= \lim_{k \to 0} \frac{x_0 - (x_0 + k)}{k(x_0 + k)x_0} = \lim_{k \to 0} \frac{-1}{(x_0 + k)x_0} = -\frac{1}{x_0^2}$$

provided that $x_0 \neq 0$. The first result now follows on using Theorem 8.2(iii) to calculate $(h \circ f)'(a)$. The second result follows on using this and Theorem 8.2(ii) to calculate $(g \cdot (h \circ f))'(a)$.

We can now show that for any rational number $r \in \mathbb{Q}$

$$Dx^r = rx^{r-1}$$

Firstly, for $n \in \mathbb{N}$ we have already seen that $Dx^0 = 0$ ($x^0$ is the constant $\mathbf{1}$) and $Dx^1 = \mathbf{1}$, and so if for some particular value of $n$ it is true that

$$Dx^n = nx^{n-1}$$

then by Theorem 8.2(ii)

$$Dx^{n+1} = D(x \cdot x^n) = (Dx) \cdot x^n + x \cdot Dx^n$$

$$= \mathbf{1} \cdot x^n + x \cdot nx^{n-1} = (n+1)x^n$$

Thus the formula must hold for all $n \in \mathbb{N}$. Next $x^m$ is monotone if $m \in \mathbb{N}$ is odd, and monotone on $(0, \cdot)$ if $m \in \mathbb{N}'$ is even. Thus by Theorem 8.2(iv)

$$Dx^{1/m} = \frac{1}{(Dx^m) \circ x^{1/m}} = \frac{1}{mx^{m-1} \circ x^{1/m}} = \frac{1}{m}x^{(1/m)-1}$$

as $(x^{1/m})^{m-1} = x^{(m-1)/m} = x^{1-(1/m)}$. Now using Theorem 8.2(ii) for $r \geq 0$, $r \in \mathbb{Q}$ we can write $r = n/m$ with $m \in \mathbb{N}'$ and $n \in \mathbb{N}$ to get

$$Dx^r = Dx^{n/m} = D(x^n \circ x^{1/m}) = nx^{n-1} \circ x^{1/m} \cdot \left(\frac{1}{m}x^{(1/m)-1}\right)$$

$$= nx^{(n-1)/m} \cdot \frac{1}{m}x^{(1/m)-1} = \frac{n}{m}x^{n/m-1} = rx^{r-1}$$

Finally, if $r < 0$, $r \in \mathbb{Q}$ we can use the corollary to get

$$\mathbf{D}x^r = \mathbf{D}\frac{1}{x^{-r}} = \frac{-\mathbf{D}x^{-r}}{x^{-2r}} = \frac{-(-rx^{-r-1})}{x^{-2r}} \qquad (\text{as } -r > 0)$$

$$= rx^{-r-1+2r} = rx^{r-1}$$

We do not have a definition of $x^\alpha$ for irrational $\alpha$ as yet, so this is the most general form of the result we can hope for. Nor do we have a function $z^{1/n}$, as there is no easy way of ensuring uniqueness for the complex $n$th root. However, it is clear that precisely the same arguments do lead to

$$\mathbf{D}z^n = nz^{n-1}$$

for $n \in \mathbb{Z}$.

We can now differentiate many functions. By combining the result $\mathbf{D}x^n = nx^{n-1}$ for various $n$, using the rules in the theorem and its corollary, we can find the derivative of a rational function. Also we can differentiate expressions involving square roots, cube roots, etc.

It should be noticed in connection with Theorem 8.2 that the differentiability of $f$, $g$ and $h$ at $a$ is fundamental to all that follows. If $f$, $g$ and $h$ should fail to be differentiable then the theorem does not tell us whether or not the various functions constructed from them are differentiable. Thus $x$ is differentiable at 0 but $|x|$ is not; however, $x|x|$ is differentiable at 0. This follows from $\lim_{k \to 0} (k|k| - 0)/k = \lim_{k \to 0} |k| = 0$. Here $k = x(k)$ and $0 = x(0)$ have been used. However, $|x|$ multiplied by a constant function other than $\mathbf{0}$ is not differentiable at 0, but it is possible on the basis of the theorem to get some information about such cases.

Turning now to properties of the derivative, we look at the relation between a function and its derived function. In particular, we shall see how to get certain sorts of information about a function from its derivative. For this purpose it is convenient to restrict our attention to real functions. This will enable us to think of the results in geometric terms.

**Theorem 8.3   Rolle's theorem**   If the real function $f$ is continuous on $[a, b]$ with $a < b$ and is differentiable on $(a, b)$, and if moreover $f(a) = f(b)$, then there is at least one point $\xi \in (a, b)$ at which $f'(\xi) = 0$ [that is, there is a point at which the tangent to the graph of $f$ is horizontal and so parallel to the chord from $\langle a, f(a) \rangle$ to $\langle b, f(b) \rangle$].

*Proof*   [We first deal with a very easy special case for which the main proof fails.] If $f(x) = f(a)$ for all $x \in [a, b]$ then we have seen that $f'(x) = 0$ in $[a, b]$, and any $\xi \in (a, b)$ can be chosen.

If there is some $x_1 \in (a, b)$ with $f(x_1) > f(a)$ then there is some point $\xi \in (a, b)$ at which $f(x)$ attains its supremum. This follows from Theorem 5.4 and the fact that $f(a) = f(b)$ is less than the supremum. Then

$$\varphi(x) = \frac{f(x) - f(\xi)}{x - \xi} \geq 0 \quad \text{for } x \in [a, \xi)$$

as $f(x) \leq f(\xi)$, and

$$\varphi(x) = \frac{f(x) - f(\xi)}{x - \xi} \leq 0 \quad \text{for } x \in (\xi, b]$$

as $f(x) \leq f(\xi)$. Now $\varphi(\xi) = \lim_{x \to \xi} \varphi(x)$, as $f$ is differentiable at $\xi$, and so we have for every $\varepsilon > 0$ values $x'$ for which $|\varphi(x') - \varphi(\xi)| < \varepsilon$ and $\varphi(x') \geq 0$, and also values $x''$ for which $|\varphi(x'') - \varphi(\xi)| < \varepsilon$ and $\varphi(x'') \leq 0$. Hence $\varphi(x'') - \varphi(\xi) > -\varepsilon$ and $\varphi(x') - \varphi(\xi) < \varepsilon$ give $|\varphi(\xi)| < \varepsilon$. Thus $f'(\xi) = \varphi(\xi) = 0$. [This case is illustrated in Fig. 8.2.]

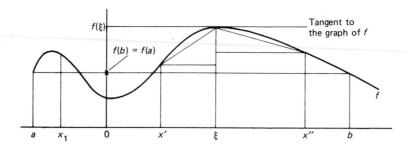

**Fig. 8.2**  The case $f(x_1) > f(a)$

The remaining possibility is that there is no $x_1 \in (a, b)$ with $f(x_1) > f(a)$, but that there is an $x_2 \in (a, b)$ with $f(x_2) < f(a)$. In this case by Theorem 5.4 there is a point which will again be called $\xi$ such that $f(\xi) = \inf\{f(x) \mid x \in [a, b]\}$ and $\xi \in (a, b)$. This time

$$\varphi(x) \leq 0 \quad \text{for } x \in [a, \xi)$$

as $f(x) \geq f(\xi)$, and

$$\varphi(x) \geq 0 \quad \text{for } x \in (\xi, b]$$

as $f(x) \geq f(\xi)$. As before, this implies that $f'(\xi) = 0$. [This case is illustrated in Fig. 8.3.]

The most important use of Rolle's theorem is to prove the mean value theorem. This is a generalization of Theorem 8.3 in which we do not need $f(b)$ to be the same as $f(a)$, but we still get a point at which the tangent to the graph is parallel to the chord.

**Theorem 8.4  The mean value theorem**  If $f$ is a real function continuous on $[a, b]$ and differentiable on $(a, b)$, with $a < b$, then there is at least one point

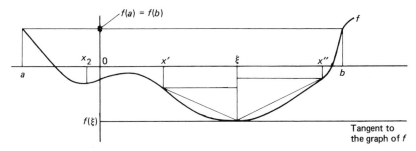

**Fig. 8.3** The case $f(x_2) < f(a)$

$\xi \in (a, b)$ such that

$$f'(\xi) = \frac{f(b) - f(a)}{b - a}$$

*Proof* Apply Rolle's theorem to the function $F$ defined by

$$F(x) = f(x) - f(a) - \frac{x - a}{b - a}[f(b) - f(a)]$$

$\left[\text{The function } f(a) + (x - a)\dfrac{f(b) - f(a)}{b - a} \text{ has the chord from } \langle a, f(a) \rangle \text{ to } \langle b, f(b) \rangle\right.$

as its graph, so that $F(x)$ is the length marked in Fig. 8.4.$\Big]$ We have

$$F(a) = f(a) - f(a) - \frac{a - a}{b - a}[f(b) - f(a)] = 0$$

and $\quad F(b) = f(b) - f(a) - \dfrac{b - a}{b - a}[f(b) - f(a)] = 0$

so that $F(a) = F(b)$. Also $F$ is formed from $f, x$ and constants by addition and multiplication, so that $F$ is continuous on $[a, b]$. Furthermore, by

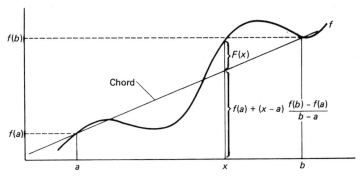

**Fig. 8.4** The function $F$

Theorem 8.2

$$F'(x) = f'(x) - \frac{f(b) - f(a)}{b - a}$$

Thus by Theorem 8.3 there is some $\xi \in (a, b)$ such that

$$F'(\xi) = f'(\xi) - \frac{f(b) - f(a)}{b - a} = 0$$

Whence

$$f'(\xi) = \frac{f(b) - f(a)}{b - a}$$

The only points of $[a, b]$ where the condition that $f$ be differentiable is not needed are $a$ and $b$. Thus $|x|$ fails to be differentiable at 0 and in $[-1, 1]$ there is no point at which $D|x|$ is 0.

Although it is nice to have equality between $f'(\xi)$ and the difference quotient function for the point $a$ evaluated at $b$, it is difficult to use in this form. Since we do not know the exact value of $\xi$ (except in some cases where we do not need the mean value theorem to show that $\xi$ exists), we can only use properties of $f'(\xi)$ which belong to $f'$ in the whole of $(a, b)$. The mean value theorem is most often used in the following form.

**Corollary to Theorem 8.4**   If $f$ is a real function continuous on $[a, b]$ and differentiable on $(a, b)$ then:
  (i)   if $f'(x) \leqslant M$ all $x \in (a, b)$ then $f(b) - f(a) \leqslant M(b - a)$,
  (ii)  if $m \leqslant f'(x)$ all $x \in (a, b)$ then $f(b) - f(a) \geqslant m(b - a)$,
  (iii) if $f'(x) < M$ all $x \in (a, b)$ then $f(b) - f(a) < M(b - a)$,
  (iv)  if $m < f'(x)$ all $x \in (a, b)$ then $f(b) - f(a) > m(b - a)$.

*Proof*   In each case as the inequality for $f'(x)$ applies for all $x \in (a, b)$, it applies in particular to $f'(\xi)$ where $\xi$ is as given by the theorem. Use this inequality in the conclusion of the theorem and multiply through by $(b - a)$.

By far the most important application of this corollary is the following characterization of the constant functions in terms of their derivatives.

**Theorem 8.5**   If $I$ is any real interval with $A$ its set of end points and $f$ is a real function continuous on $I$ and differentiable on $I \backslash A$, then $f$ is a constant on $I$ iff $f'(x) = 0$ for $x \in I \backslash A$.

*Note*   $I$ may be any real interval, $A$ could be empty, or have one member or two members, and the end points in $A$ may or may not belong to $I$.

*Proof*   If $f'(x) = 0$ in $I \backslash A$ then for any $a, b \in I$ with $a < b$ we have the conditions of the Corollary to Theorem 8.4 for both parts (i) and (ii) with

$m = M = 0$. Hence

$$0 \leqslant f(b) - f(a) \leqslant 0$$

Thus $f(a) = f(b)$ for all $a, b \in I$, that is $f$ is constant on $I$.

We have already seen that for any constant function $c$ we get $Dc = \mathbf{0}$. Then, clearly, restricting the domain of $c$ to $I$ only results in the restriction of the domain of $\mathbf{0}$ to $I$ as well.

Theorem 8.5 is really just a combination of parts (i) and (ii) of the following theorem.

**Theorem 8.6** If $I$ is any real interval and $A$ the set of its end points then for any real function $f$ continuous on $I$ and differentiable on $I \backslash A$:
  (i) if $f'(x) \geqslant 0$ for $x \in I \backslash A$ then $f$ is weakly increasing on $I$,
  (ii) if $f'(x) \leqslant 0$ for $x \in I \backslash A$ then $f$ is weakly decreasing on $I$,
  (iii) if $f'(x) > 0$ for $x \in I \backslash A$ then $f$ is strictly increasing on $I$,
  (iv) if $f'(x) < 0$ for $x \in I \backslash A$ then $f$ is strictly decreasing on $I$.

*Proof* For any two numbers $a, b \in I$ with $a < b$ the appropriate case of the Corollary to Theorem 8.4 with $M$ or $m$ put equal to $0$ gives an inequality between $f(a)$ and $f(b)$. The fact that this inequality holds for all $a, b \in I$ with $a < b$ when stated in words is the conclusion in each case.

It is important to realize that for both the Corollary to Theorem 8.4 and Theorem 8.6 the condition on $f'$ must hold in some open interval. If, for instance, $f'(x_0) = 0$ there need not be any interval about $x_0$ in which $f$ is constant. Thus $Dx^2 = 2x$ and $x(0) = 0$, but $x^2(t) > 0$ for all other $t$. We can construct a counter example to (iii) from the function $h$, number (i) of the Examples of continuous and discontinuous functions given in Chapter 3 (p. 59). Firstly, because $h$ is not everywhere differentiable, we consider $\theta = (1 - h^2)^2$. It is not too difficult to check that $\mathcal{D}\theta' = (\cdot, \cdot)$ and that $0 \leqslant \theta(x) \leqslant 1$ all $x$. Then $f$ defined by

$$f(x) = \begin{cases} x + x^2 \, \theta(1/x) & (x \neq 0) \\ 0 & (x = 0) \end{cases}$$

has $f'(0) = 1 > 0$, as $Dx^2 \, \theta(1/x)$ is $0$ at $0$ by an argument similar to that used to show that $l'(0) = 0$, where $l$ was the function given as number (v) of the Examples of differentiable and non-differentiable functions (p. 127). However, there is no neighbourhood of $0$ on which $f$ is monotone. When the function sin has been defined it will be seen that the function $g$ given by

$$g(x) = \begin{cases} x + x^2 \sin(1/x) & (x \neq 0) \\ 0 & (x = 0) \end{cases}$$

has similar properties. Figure 8.5 illustrates this type of behaviour.

One often used application of Theorem 8.6 is as a method of locating the maximum or minimum value of a function defined on an interval. Suppose

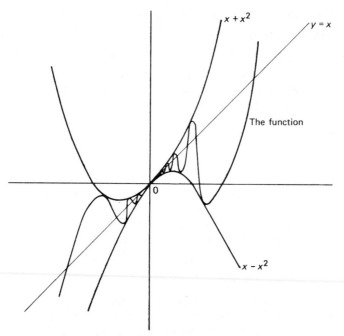

**Fig. 8.5**   A function like $f$ or $g$

that $f$ is differentiable on $[a, b]$, and that on looking at $f'$ we find that $[a, b]$ can be divided up into intervals on each of which $f'$ has constant sign. More precisely, suppose that there is a finite set of points $\{x_i | i = 0, \ldots, n\}$ such that $a = x_0 < x_1 < \cdots < x_{n-1} < x_n = b$, and each interval $[x_{i-1}, x_i]$ is such that either $f'(x) > 0$ all $x \in (x_{i-1}, x_i)$ or $f'(x) < 0$ all $x \in (x_{i-1}, x_i)$. Then on each such interval $f$ is strictly monotone and takes its greatest value at one end point and its least value at the other. Moreover if $f$ is increasing in $(x_{i-1}, x_i)$ and decreasing in $(x_i, x_{i+1})$ then $f(x)$ is greatest in $(x_{i-1}, x_{i+1})$ for $x = x_i$. Thus the greatest value of $f(x)$ is either at one such $x_i$ or, if $f$ is decreasing on $(a, x_1)$, it could be at $a$ or, if $f$ is increasing on $(x_{n-1}, b)$, it could be at $b$. The points $x_i$ such that there is a neighbourhood $(x_{i-1}, x_{i+1})$ in which $f(x) < f(x_i)$ if $x \neq x_i$ are called *local maxima*, or such a point may just be called a *maximum*. Similarly, defining a *minimum* of $f$ to be a point $x'$ with a neighbourhood in which $f(x) > f(x')$ for $x \neq x'$, the points $x_i$ for which $f$ is decreasing on $(x_{i-1}, x_i)$ and increasing on $(x_i, x_{i+1})$ are minima of $f$. Then the least value of $f(x)$ is taken at one of the $x_i$ which is a minimum or at $a$ or $b$. Furthermore, $a$ can only give the least value to $f(x)$ if $f$ is increasing on $(a, x_1)$ and $b$ can only give the least value to $f(x)$ if $f$ is decreasing on $(x_{n-1}, b)$.

Let us apply this to the function $(x^2 - 1)^2/(x^4 + 1)$ over $[-2, 2]$. We get

$$D[(x^2 - 1)^2/(x^4 + 1)] = [4x(x^2 - 1)(x^4 + 1) - 4x^3(x^2 - 1)^2](x^4 + 1)^{-2}$$

$$= 4x \cdot (x^4 - 1) \cdot (x^4 + 1)^{-2}$$

Since this is continuous on $(\cdot, \cdot)$, we can find the numbers $x_i$ as the points which make this derivative 0. Take $x_0 = -2$, $x_1 = -1$, $x_2 = 0$, $x_3 = 1$ and $x_4 = 2$. Then as $(x^4 + 1)^{-2} > 0$ all $x$ and $x^4 - 1 < 0$ for $x < -1$ and $x > 1$ but $x^4 - 1 < 0$ in $(-1, 0)$ and $(0, 1)$, we have that the derivative is

$$\text{negative} \quad \text{in} \quad (-2, -1)$$

$$\text{positive} \quad \text{in} \quad (-1, 0)$$

$$\text{negative} \quad \text{in} \quad (0, 1)$$

and $\quad$ positive $\quad$ in $\quad (1, 2)$

This is shown in Fig. 8.6. Thus 1 and −1 are minima of the function and 0 is a maximum. Hence the smallest value of the function in $[-2, 2]$ is 0, taken at −1 and 1. Also since the value of the function at both 2 and −2 is $\frac{9}{17}$ and at 0 it is 1, the greatest value of the function in $[-2, 2]$ is 1, taken at 0.

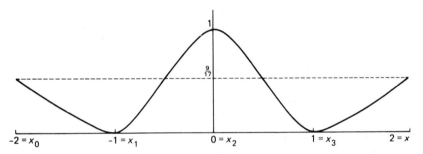

**Fig. 8.6** $(x^2 - 1)^2 / (x^2 + 1)$

If we do the corresponding calculation for the function $(x^2 - \frac{1}{4})^2 / (x^4 + 1)$ we find $x_0 = -2$, $x_1 = -\frac{1}{2}$, $x_2 = 0$, $x_3 = \frac{1}{2}$ and $x_4 = 2$. As before, the derivative is negative in $(x_0, x_1)$ and $(x_2, x_3)$ and positive in $(x_1, x_2) \cup (x_3, x_4)$. The smallest value of the function is again 0, taken at the minima $\frac{1}{2}$ and $-\frac{1}{2}$. However, the largest value is $\frac{225}{272}$, taken at both −2 and 2, since this is greater than the value $\frac{1}{16}$, taken at the maximum 0. This is shown in Fig. 8.7.

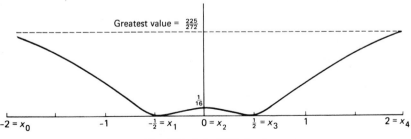

**Fig. 8.7** $(x^2 - \frac{1}{4})^2 / (x^4 + 1)$

The function $(x^4+1)^{-1}$, taken over the same interval $[-2, 2]$, is readily seen to take its greatest value at a maximum at 0 and its least value of $\frac{1}{17}$ at each of the points $-2$ and 2.

The function $x^3$ has $Dx^3 = 3x^2$. So considered over some interval, say $[-1, 3]$, has $x_0 = -1$, $x_1 = 0$ and $x_2 = 3$. However, $3x^2 > 0$ for all $x \neq 0$, and so $x^3$ is monotone increasing over $[-1, 3]$ and 0 is neither a maximum nor a minimum.

The method of differentiating and equating the derivative to zero, to find the points which may be maxima or minima, seems to depend on two things.

Firstly, there must be only a finite number of points at which the derivative is zero. However, if we consider a function like $x^2 \theta(1/x)$ we can see that the same considerations can be made to apply, although the arguments do become more complicated.

Secondly, we need to know that the derivative is continuous on the interval. In fact we need the intermediate value theorem to tell us that if $f'(x)$ is of one sign in $(x_{i-1}, x_i)$ and of the opposite sign in $(x_i, x_{i+1})$ then $f'(x_i) = 0$. However, although derivatives need not be continuous, $Dx^2 \theta(1/x)$ is a case in point, they can be shown to have the intermediate value property.

**Theorem 8.7    The intermediate value theorem for derivatives**    If $f$ is a real function, $f'(x)$ exists on $[a, b]$, and $\mu$ is any number between $f'(a)$ and $f'(b)$, then $\exists \xi \in (a, b)$ such that $f'(\xi) = \mu$.

*Proof*  [It is more convenient to look for a point at which a derivative is zero.] Let $g$ be the function given by $g(x) = f(x) - \mu x$ if $f'(a) < f'(b)$, and the function given by $g(x) = \mu x - f(x)$ if $f'(a) > f'(b)$. Then either $g'(x) = f'(x) - \mu$ for all $x \in [a, b]$, or $g'(x) = \mu - f'(x)$ for all $x \in [a, b]$. Since $\mu$ is between $f'(a)$ and $f'(b)$, this gives $g'(a) < 0$ and $g'(b) > 0$. Now as $g$ is differentiable on $[a, b]$, by Theorem 8.1 $g$ is continuous on $[a, b]$, and hence there is some $\xi \in [a, b]$ such that

$$g(\xi) = \inf\{g(x) \mid x \in [a, b]\}$$

Now as $g'(a) < 0$, there must be some $x_1 \in (a, b)$ such that

$$\frac{g(x_1) - g(a)}{x_1 - a} < 0$$

Since $x_1 - a > 0$, this is $g(x_1) - g(a) < 0$. Thus $g(x_1) < g(a)$ and we have $\xi \neq a$. Also $g'(b) > 0$, so that for some $x_2 \in (a, b)$ we can get

$$\frac{g(x_2) - g(b)}{x_2 - b} > 0$$

Now $x_2 - b < 0$, so that $g(x_2) - g(b) < 0$ and again $\xi \neq b$. Thus $\xi \in (a, b)$ and, as $g(\xi) \leq g(x)$ all $x \in (a, b)$, every neighbourhood of $\xi$ contains both points

$x_3 \in (a, \xi) \subset (a, b)$ for which

$$\frac{g(x_3) - g(\xi)}{x_3 - \xi} \leqslant 0$$

and $x_4 \in (\xi, b) \subset (a, b)$ for which

$$\frac{g(x_4) - g(\xi)}{x_4 - \xi} \geqslant 0$$

Just as in the proof of Rolle's theorem, Theorem 8.3, these two inequalities imply that

$$g'(\xi) = \lim_{h \to 0} \frac{g(\xi + h) - g(\xi)}{h} = 0$$

From the definition of g this gives $f'(\xi) = \mu$. [The proof is illustrated in Fig. 8.8.]

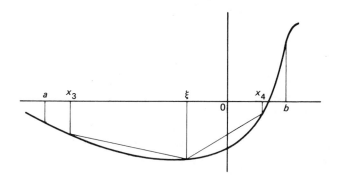

**Fig. 8.8** The function g

From this theorem it follows that although a derivative need not be continuous, nevertheless it is true that if $f$ is differentiable on $[a, b]$ then its greatest and least values are each taken at a point of the set $\{a, b\} \cup \{x \mid x \in (a, b), f'(x) = 0\}$.

It is sometimes helpful in determining whether a point at which $f'(x) = 0$ is a maximum or a minimum, and if so which, to look at the derivative of $f'$. Suppose that $f'(x_i) = 0$, and $f'$ is differentiable at $x_i$ with $f''(x_i) > 0$. (Here $f''$ is the derivative of $f'$.) Then as $\varphi_{f'}(x_i) > 0$ there is some neighbourhood $N$ of $x_i$ such that $\varphi_{f'}(x) > 0$ in $N$. Thus $f'(x) - 0$ is of the same sign in $N$ as $x - x_i$, so that $f'(x) < 0$ in $(x_{i-1}, x_i) \cap N$ and $f'(x) > 0$ in $(x_i, x_{i+1}) \cap N$. Clearly, $x_i$ is then a minimum. In the same way, if $f''(x_i) < 0$ then the point $x_i$ is a maximum. However, even if $f'$ is differentiable on the whole of $[a, b]$, the question may still not have been settled for all $x_i$. It can happen that $f''(x_i) = 0$, in which case $x_i$ may be a maximum or a minimum or neither. Thus on $[-1, 1]$, $x^3$ has $Dx^3 = 3x^2$ and $D(Dx^3) = 6x$ both zero at 0 and, since $x^3$ is of the same sign as $x$, it is clear that 0 is neither a maximum nor a minimum of $x^3$. Again $x^4$ has

$Dx^4 = 4x^3$ and $D(Dx^4) = 12x^2$ both zero at 0, but $12x^2 > 0$ unless $x = 0$, and so 0 is the smallest value of $x^4$ and is a minimum. Similarly, $-x^4$ has a maximum at 0. If $f''$ is differentiable at $x_i$ it may be possible to determine which of these possibilities holds from the value of $f'''(x_i)$. The discussion of this problem in general is helped by the generalizations of Rolle's theorem and the mean value theorem which follow.

We must first take a look at the notation for differentiation. Since $f'$ is a function, its derivative must be $f''$, and the derivative of $f''$ must be $f'''$. Clearly, some modification of this notation is needed if the process is to be continued much further. One device which is used is to replace the dashes or primes with a roman numeral, so that $f''''''$ becomes $f^v$. A more powerful modification of the notation is to put a number or a letter standing for a number in brackets in place of the dashes. Thus $f''''' = f^{(5)}$. This has the advantage of allowing us to write $f^{(n)}$ where $n \in \mathbb{N}'$. It is convenient to extend this to the case $n = 0$ by $f^{(0)} = f$. The definition of the derived function has been so chosen that $f^{(n)}$ always exists; however, it could be that for some $n$, $\mathscr{D}f^{(n)} = \emptyset$. In any case we get

$$\mathscr{D}f \supset \mathscr{D}f^{(1)} \supset \mathscr{D}f^{(2)} \supset \cdots \supset \mathscr{D}f^{(n)} \supset \mathscr{D}f^{(n+1)} \supset \cdots$$

When the derivative is written $Df$ it is natural to write $DDf$ as $D^2f$, $DDDf$ as $D^3f$, and in general $D^nf$ for $f^{(n)}$. We take this to include $D^0f = f$. In the notation $(d/dx)f(x)$ the second derivative is written $(d^2/dx^2)f(x)$, and in general $f^{(n)}$ is written $(d^n/dx^n)f(x)$.

**Theorem 8.8    Generalized Rolle's theorem**    If $f$ is a real function with $f(a) = f(b)$ for some $a$ and $b$ with $a < b$, $f^{(n-1)}$ exists and is continuous on $[a, b]$ and differentiable on $(a, b)$, and either

$$f'(a) = f''(a) = \cdots = f^{(n-1)}(a) = 0$$

or    $$f'(b) = f''(b) = \cdots = f^{(n-1)}(b) = 0$$

then $\exists \xi \in (a, b)$ for which

$$f^{(n)}(\xi) = 0$$

*Proof*    Since $f$ is differentiable on $[a, b]$ and $f(a) = f(b)$, Rolle's theorem gives $\xi_1 \in (a, b)$ such that $f'(\xi_1) = 0$. Now by assumption this is equal to one of $f'(a)$ or $f'(b)$. As $f'$ is differentiable on $[a, b]$, we can use Rolle's theorem again to get $\xi_2 \in (a, \xi_1) \subset (a, b)$ if $f'(a) = 0$ or $\xi_2 \in (\xi_1, b) \subset (a, b)$ if $f'(b) = 0$. Suppose that for $r \leq n - 1$ we have found $\xi_r \in (a, b)$ with $f^{(r)}(\xi_r) = 0$. Then Rolle's theorem gives us $\xi_{r+1} \in (a, \xi_r)$ if $f^{(r)}(a) = 0$ or $\xi_{r+1} \in (\xi_r, b)$ if $f^{(r)}(b) = 0$ such that $f^{(r+1)}(\xi_{r+1}) = 0$. Hence on writing $\xi = \xi_n$ we have the result. [The process is illustrated by Fig. 8.9.]

The generalization of the mean value theorem is usually called Taylor's theorem and gives an approximation to the function by an $n$th degree polynomial.

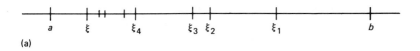

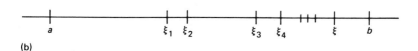

(a)

(b)

**Fig. 8.9** Generalized Rolle's theorem

**Theorem 8.9 Taylor's theorem** If $f$ is a real function with $f^{(n-1)}$ continuous on a closed interval $I$, and $I$ is such that $x_0$ and $x_0+h$ are its end points, $h$ not being 0, and if also $f^{(n-1)}$ is differentiable on $I\backslash\{x_0, x_0+h\}$, then $\exists\xi\in I\backslash\{x_0, x_0+h\}$ such that

$$f(x_0+h)=f(x_0)+f'(x_0)h+f''(x_0)\frac{h^2}{2!}+\cdots+f^{(n-1)}(x_0)\frac{h^{n-1}}{(n-1)!}$$

$$+f^{(n)}(\xi)\frac{h^n}{n!}$$

$$=\sum_{r=0}^{n-1}f^{(r)}(x_0)\frac{h^r}{r!}+f^{(n)}(\xi)\frac{h^n}{n!}$$

*Note* Here $r!$ is a notation for $\prod_{k=1}^{r}k$, with $0!=1$. Thus, for example, $7!=1\cdot2\cdot3\cdot4\cdot5\cdot6\cdot7=5040$.

*Proof* Since $h\neq0$, we can define a number $\alpha$ by the equation

$$f(x_0+h)=\sum_{r=0}^{n-1}f^{(r)}(x_0)\frac{h^r}{r!}+\alpha\frac{h^n}{n!}$$

[We now have to show that $\alpha$ can be put in the form $f^n(\xi)$.] Write $g$ for the function defined by

$$g(t)=\sum_{r=0}^{n-1}f^{(r)}(x_0)\frac{t^r}{r!}+\alpha\frac{t^n}{n!}$$

and define $F$ by $F(t)=f(x_0+t)-g(t)$. Then $g$ is a polynomial and so is differentiable on $\mathbb{R}$, and the function of $t$ given by $f(x_0+t)$ is differentiable on the open interval with end points 0 and $h$ and continuous on the closed interval with the same end points. Thus $F$ is differentiable and continuous on these same intervals. Now $g(0)=f(x_0)$ and $g(h)=f(x_0+h)$, so that $F$ satisfies

$F(0) = F(h) = 0$. Also by direct calculation we have that

$$D^l x^r = \begin{cases} r(r-1)(r-2) \cdots (r-l+1)x^{r-l} & (l < r) \\ r! & (l = r) \\ 0 & (l > r) \end{cases}$$

and $x^{r-l}$ evaluated at 0 is 0, provided that $l < r$. We see that $g^{(l)}(0) = f^{(l)}(x_0)$ for $l = 1, \ldots, n-1$. From this it follows that $F^{(l)}(0) = 0$ for $l = 1, \ldots, n-1$. [It is this calculation of the derivatives which determines the form of the *Taylor polynomial* used to express $f(x_0 + h)$.] Thus the function $F$ satisfies the hypotheses of Theorem 8.8 on $I - \{x_0\}$, and so there is a number $\xi \in I \backslash \{x_0, x_0 + h\}$ such that $F^{(n)}(\xi - x_0) = 0$. However, $F^{(n)}(t) = f^{(n)}(x_0 + t) - g^{(n)}(t) = f^{(n)}(x_0 + t) - \alpha$. Hence $f^{(n)}(\xi) = \alpha$.

The term $f^{(n)}(\xi) h^n / n!$ is known as the *remainder term*. It is obvious that the usefulness of the Taylor polynomial $\sum_{r=0}^{n-1} f^{(r)}(x_0) h^r / r!$ in representing $f$ depends on being able to show that the remainder term is sufficiently small; because of this several forms of the remainder have been developed. The one given is known after Lagrange. As with the mean value theorem, the simplest way of dealing with the fact that we have no information as to the exact position of $\xi$ is to use bounds for $f^{(n)}(t)$ in $I$. Thus suppose that $x_0 \in (a, b)$ gives $f'(x_0) = 0$, and we wish to find out if $x_0$ is a maximum or a minimum or neither. If $f^{(l)}(x_0) = 0$ for $l = 1, \ldots, n-2$ and $f^{(n)}$ is continuous on $[a, b]$ we have by Theorem 8.9

$$f(x_0 + h) = f(x_0) + f^{(n-1)}(x_0) \frac{h^{n-1}}{(n-1)!} + \frac{f^{(n)}(\xi)}{n!} h^n$$

where $\xi$ is between $x_0$ and $x_0 + h$. We must have, of course, $h$ sufficiently small for $x_0 + h \in [a, b]$. If $f^{(n-1)}(x_0) \neq 0$ then, writing $M \geq |f^{(n)}(t)|$ for $t \in [a, b]$, we can find a $\delta > 0$ such that $|h f^{(n)}(t)| < n|f^{(n-1)}(x_0)|$ for $|h| < \delta$. We take $\delta < n|f^{(n-1)}(x_0)| / M$, and also small enough to make $x_0 + h \in [a, b]$. Since this holds for $t = \xi$, we have that

$$\left| \frac{f^{(n)}(\xi)}{n!} h^n \right| < \left| \frac{f^{(n-1)}(x_0)}{(n-1)!} h^{n-1} \right|$$

Thus $f(x_0 + h)$ is greater than or less than $f(x_0)$ according as $h^{n-1} f^{(n-1)}(x_0) / (n-1)!$ is positive or negative. If $n-1$ is even then $h^{n-1} > 0$, and so $f(x_0 + h) > f(x_0)$ for $0 < |h| < \delta$ if $f^{(n-1)}(x_0) > 0$, that is $x_0$ is a minimum. If $f^{(n-1)}(x_0) < 0$ we get similarly that $x_0$ is a maximum. Finally, if $n-1$ is odd then the sign of $h^{n-1}$ is the same as the sign of $h$. Thus in this case there are always two numbers $h_1$ and $h_2$ with $0 < |h_1|, |h_2| < \delta$ such that $f(x_0 + h_1) < f(x_0) < f(x_0 + h_2)$, and so $x_0$ is neither a maximum nor a minimum.

In this argument we have $\xi$ between $x_0$ and $x_0 + h$, and so $\xi$ depends on $h$. However, $h$ has to be chosen after the choice of $\delta$ has been made. This is why we have to use $M$ and not $|f(\xi)|$ in making the choice of $\delta$.

If $f$ has derivatives of all orders in $[a, b]$, it can still happen that we cannot determine whether $x_0$ is a maximum or a minimum in this way. If $k$ is the smallest positive integer such that $f^{(k)}(x_0) \neq 0$, and $k$ is odd, then $x_0$ is neither a maximum nor a minimum. If $k$ is even then $x_0$ is a maximum if $f^{(k)}(x_0) < 0$ and a minimum if $f^{(k)}(x_0) > 0$. Suppose, however, that $f^{(n)}(x_0) = 0$ for $n \in \mathbb{N}'$. It can be shown that there are real functions satisfying this condition which are not constants and which can have $x_0$ a maximum, a minimum, or neither.

To see how close the approximation to a function by a Taylor polynomial can be, let us consider $\sqrt{(1+x)}$ in $[-\frac{1}{10}, \frac{1}{10}]$. If we take $x_0 = 0$ then for the case $n = 3$ we require the values of

$$D\sqrt{(1+x)} = \tfrac{1}{2}(1+x)^{-1/2}$$

and $\quad D^2\sqrt{(1+x)} = -\tfrac{1}{4}(1+x)^{-3/2}$

at 0. These are $\frac{1}{2}$ and $-\frac{1}{4}$ respectively. For the remainder term we require a bound for

$$D^3\sqrt{(1+x)} = \tfrac{3}{8}(1+x)^{-5/2}$$

on $[-\frac{1}{10}, \frac{1}{10}]$. It is clear that the largest value of this function is $\frac{3}{8}(\frac{10}{9})^{5/2}$ and is taken at $-\frac{1}{10}$. Now by direct calculation we get $(\frac{10}{9})^{5/2} < 2$. So if $h \in [-\frac{1}{10}, \frac{1}{10}]$ the remainder term satisfies

$$\left| \frac{h^3}{3!} \frac{3}{8} [1+x(\xi)]^{-5/2} \right| < \left( \frac{1}{10} \right)^3 \frac{3 \cdot 2}{3! \cdot 8} = \frac{1}{8000}$$

as also $\xi \in [-\frac{1}{10}, \frac{1}{10}]$.

Thus in $[-\frac{1}{10}, \frac{1}{10}]$ the quadratic expression

$$1 + \tfrac{1}{2}h + (-\tfrac{1}{4}) \frac{h^2}{2!} = 1 + \tfrac{1}{2}h - \tfrac{1}{8}h^2$$

differs from $\sqrt{(1+h)}$ by less than $\frac{1}{8000}$. Or perhaps it is more natural to write the variable as $x$ and say

$$|\sqrt{(1+x)} - (1 + \tfrac{1}{2}x - \tfrac{1}{8}x^2)| < \tfrac{1}{8000}$$

We see that the quadratic is quite close enough to $\sqrt{(1+x)}$ for many numerical calculations, but this has been achieved by keeping the interval small. However, we can expect to get a better approximation by taking a larger value of $n$. For example, taking $n = 5$ we require

$$D^4\sqrt{(1+x)} = D\tfrac{3}{8}(1+x)^{-5/2} = -\tfrac{15}{16}(1+x)^{-7/2}$$

and $\quad D^5\sqrt{(1+x)} = D - \tfrac{15}{16}(1+x)^{-7/2} = \frac{3 \cdot 5 \cdot 7}{32}(1+x)^{-9/2}$

so that, using $x$ as the variable, the Taylor polynomial now is

$$1+\frac{1}{2}x-\frac{1}{8}x^2+\frac{3}{8}\frac{x^3}{3!}-\frac{15}{16}\frac{x^4}{4!}=1+\frac{1}{2}x-\frac{1}{8}x^2+\frac{1}{16}x^3-\frac{5}{128}x^4$$

Now if we take the interval in which $x$ lies to be $[-\frac{1}{3},\frac{1}{3}]$ we get something approaching the same degree of accuracy. The function $(1+x)^{-9/2}$ takes its largest value in the interval at $-\frac{1}{3}$, and the value is $(1-\frac{1}{3})^{-9/2}=(\frac{2}{3})^{-9/2}=(\frac{3}{2})^{9/2}$. Thus we get for the remainder term

$$\left|\frac{x^5}{5!}\frac{3\cdot5\cdot7}{32}[1+x(\xi)]^{-9/2}\right|<\left(\frac{1}{3}\right)^5\cdot\frac{1}{5!}\cdot\frac{3\cdot5\cdot7}{2^5}\cdot\frac{3^{9/2}}{2^{9/2}}$$

$$=\frac{7}{2^3}\cdot\frac{1}{2^5}\cdot\frac{1}{2^4\cdot2^{1/2}\cdot3^{1/2}}=\frac{7}{2^{12}\sqrt6}$$

$$=\frac{7}{4096\sqrt6}<\frac{7}{8192}$$

as $\sqrt6>2$.

It is even more useful to be able to calculate values of a function from its Taylor polynomials when that function is not given by an algebraic expression. All the particular functions we have considered so far are constructed by the use of algebraic expressions. Now let us assume that there is some function e such that $e'=e$ in an interval $[-a,a]$. Since e is continuous on the closed interval, there will be some number $M$ such that $|e(x)|\leqslant M$ for $x\in[-a,a]$. It is easy to calculate the Taylor polynomials of e since $e''=e'=e$, and so $e'''=e''=e$, and in general $e^{(n)}=e$. Hence also $|e^{(n)}(x)|\leqslant M$ for $x\in[-a,a]$. Thus the Taylor polynomial of degree $n$ is

$$\sum_{r=0}^{n}\frac{e^{(r)}(0)}{r!}x^r=e(0)\sum_{r=0}^{n}\frac{x^r}{r!}$$

The remainder is

$$\frac{e^{(n+1)}(\xi)}{(n+1)!}x^{n+1}$$

and so

$$\left|e(x)-e(0)\sum_{r=0}^{n}\frac{x^r}{r!}\right|\leqslant\frac{M}{(n+1)!}a^{n+1}$$

Thus if we take both $a$ and $M$ to be 1 we can calculate e in $[-1,1]$, always supposing that there is such a function, to within $1/5040$ by using

$$e(0)\left(1+x+\frac{1}{2!}x^2+\frac{1}{3!}x^3+\frac{1}{4!}x^4+\frac{1}{5!}x^5+\frac{1}{6!}x^6\right)$$

However, whatever values of $a$ and $M$ we take the accuracy improves as $n$ increases. Let $R_n$ be the remainder term, for convenience, that is to be $R_n = (x^n/n!)f^{(n)}(\xi_n)$ where $\xi_n$ is used to make it clear that the $\xi$ of Theorem 8.9 depends on $n$. Then $|R_n| \le (M/n!)a^n$. Now suppose that $n_0$ is such that $a/n < \frac{1}{2}$ when $n \ge n_0$. Then

$$\frac{a^n}{n!} = \frac{a}{1} \cdot \frac{a}{2} \cdot \frac{a}{3} \cdots \cdot \frac{a}{n} = \frac{a^{n_0}}{n_0!} \cdot \frac{a}{n_0+1} \cdots \cdot \frac{a}{n} \qquad \text{(for } n \ge n_0)$$

$$\le \frac{a^{n_0}}{n_0!} \left(\frac{1}{2}\right)^{n-n_0}$$

Now $\lim_n (\frac{1}{2})^n = 0$, so that $\lim_n a^{n_0}/(n_0! \, 2^{n_0})(\frac{1}{2})^n = 0$. Hence $\lim_n R_n = 0$ and so

$$e(x) = \lim_n e(0) \sum_{r=0}^{n} \frac{x^r}{r!}$$

so that in the sense of a limit we have an expression for e, provided that such a function exists.

This particular sort of limit is of great importance, especially in representing functions. There are very many functions for which the remainder term $R_n$ in the Taylor expansion has the limit 0. For instance, we get $\lim_n R_n = 0$ for certain values of $x$ in the case of $\sqrt{(1+x)}$. The limits of sums, taken as the number of terms increases, are called series, and are the object of study in the next chapter.

It will have been noticed that once the rules for calculating derivatives had been established, attention was restricted to real functions. Part of the reason for this is that as $\mathbb{C}$ is not an ordered field we cannot discuss maxima and minima of complex functions. More fundamentally, however, it is because there is a big difference between the effect of assuming a real function to be differentiable on an open interval and the effect of the same assumption on a complex function. For a complex function the assumption is much more restrictive. Thus, for example, it implies the existence of derivatives of all orders throughout the interval. For these reasons the study of differentiable complex functions is a special topic of its own. It is treated in books such as *Complex Analysis* by L V Ahlfors.

### Exercises

**1** Find the following derivatives stating their domains in each case:
  (i) $\mathrm{D}x$,
  (ii) $\mathrm{D}x^{1+r}$ for $r \in \mathbb{Q}, r > 0$,
  (iii) $f', f'', f'''$ for $f$ given by $f(x) = \begin{cases} x^2 & (x \ge 0) \\ x^3 & (x < 0) \end{cases}$,
  (iv) $\mathrm{D}1/(1+x^2)$.

**2** Suppose that the real function $f$ is defined and continuous on the interval $[0, b]$ and is differentiable on $(0, b)$. By applying Rolle's theorem to a suitable function show that if $f(0) = 0$ then $\exists \xi \in (0, b)$ such that $f'(\xi)f(\xi) = [f(b)]^2/(2b)$.

**3**  If $g$ is a real function differentiable on $(\cdot, \cdot)$, except possibly at $a$, but continuous on the whole of $(\cdot, \cdot)$, and the function $h$ given by

$$h(x) = \begin{cases} g'(x) & (x \neq a) \\ c & (x = a) \end{cases}$$

is continuous on $\{a\}$, then by using the mean value theorem or its corollary show that $g'(a) = c$.

**4**  The function $f$ is such that $f'''(x)$ exists at all $x \in [a, b]$ where $a < b$, and $f(a) = f(b) = f'(a) = f'(b) = f''(a) = f''(b) = 0$. Show that there are at least three distinct points $x_1$, $x_2$ and $x_3$ such that $x_i \in (a, b)$ and $f'''(x_i) = 0$ for $i = 1, 2, 3$.

**5**  The function $k$ is differentiable on $[a, b]$ and $k'(a), k'(b) \in (a, b)$. Use the fact that $2x = Dx^2$ to show that there is a point $\xi \in (a, b)$ such that $k'(\xi) = \xi$.

**6**  The real function $f$ is continuous on $[a, b]$ with $f(a) = f(b) = 0$ and $f'$ is differentiable on $(a, b)$. Also there is some $c \in (a, b)$ such that $f(c) > 0$. Prove that there is some point $\xi \in (a, b)$ such that $f''(\xi) < 0$.

**7**  Find the greatest and least values of the expression $x^2/(x^4 + x^2 + 16)$.

**8**  Find the maxima and also the minima of the function $(1 - x)^3$. You should state clearly how you determine which are maxima and which are minima.

**9**  Let $f$ be the real function given by $f(x) = 1/(1 + x)$. Find the Taylor polynomials for $f$ of degree 3 in $x$ and in $(x - 1)$. Give an estimate for the difference between $f(x)$ and the polynomial in each case, considering the intervals $[-\frac{1}{2}, \frac{1}{2}]$ and $[\frac{1}{2}, 1]$ respectively. Calculate the values of the two polynomials at $x = \frac{1}{2}$ and the value of $f(\frac{1}{2})$.

**10**  Suppose that there is a real function $f$ defined in $[-1, 1]$ such that $f'' = f$. Prove that

$$f(x) = \lim_n \left[ f(0) \sum_{r=0}^{n} \frac{x^{2r}}{(2r)!} + f'(0) \sum_{r=0}^{n} \frac{x^{2r+1}}{(2r+1)!} \right]$$

for $x \in [-1, 1]$.

# 9

# Series and sequences and series of functions

In Chapter 8 we were led to consider limits of the form $\lim_N s_N$ where $s_N = \sum_{n=0}^N a_n x^n$ for $N \in \mathbb{N}$. Here it will be assumed that $x$ is a fixed number. So let us simplify the notation by writing $b_n = a_n x^n$ for $n \in \mathbb{N}$, and so $s_N = \sum_{n=0}^N b_n$. We see that what is being discussed is a sequence $\{s_N\}$ which is defined by means of sums of the terms of a second sequence $\{b_n\}$. It is also possible, given any sequence $\{s_N\}$, to write it in this form. Clearly, $b_N = \sum_{n=0}^N b_n - \sum_{n=0}^{N-1} b_n$ if $N - 1 \in \mathbb{N}$, and $b_0 = \sum_{n=0}^0 b_n$, so that $b_N = s_N - s_{N-1}$ for $N \in \mathbb{N}$ if we define $s_{-1}$ to be 0.

**Definition 9.1** If the two sequences, the sequence of *partial sums* $s = \{s_N \mid N = k_0 + k, k_0 + k + 1, \ldots\}$ and the sequence of *terms* $b = \{b_n \mid n = k, k+1, \ldots\}$, are related by

$$s_{N+k_0} = \sum_{n=k}^N b_n$$

the pair $\langle s, b \rangle$ of sequences is called a *series* or an *infinite series*.

Almost always, and unless otherwise indicated, $k_0$ will be taken to be 0 [indeed $k_0$ is included here for the convenience of notation on a very few occasions]. The series $\langle s, b \rangle$ with $k_0 = 0$ is written $\sum_{n=k} b_n$, or when the value of $k$ is not important simply as $\sum b_n$.

The object of studying series is to try to relate the properties of the two sequences $s$ and $b$. In particular, we would like to know when $\lim_N s_N$ exists, and to be able to determine this from the sequence $\{b_n\}$. It is therefore necessary always to be clear which properties of a series relate to $s$ and which relate to $b$. Particular notice should be taken in all definitions relating to series, so as to be quite clear to which sequence the definition refers.

**Definition 9.2** If the series $\sum_{n=k} b_n$ is such that the limit $\lim_N s_N = \lim_N \sum_{n=k}^N b_n$ exists then the series is said to be *convergent* and the value of the limit is written $\sum_{n=k}^{\infty} b_n = \lim_N \sum_{n=k}^N b_n$, and called the *sum* of the series. If the limit $\lim_N s_N$ fails to exist then the series $\sum_{n=k} b_n$ is said to be *divergent*.

We notice first that the convergence or divergence of a series is a property of the sequence $s$ of partial sums. Secondly, we see that the process of calculating the sum of a convergent series from its sequence $\{b_n\}$ of terms consists of two stages. The partial sums are calculated $s_N = \sum_{n=k}^{N} b_n$, and then the limit is taken. In the notation this second process of taking the limit is denoted by the symbol $\infty$. This can make the two formulae $\sum_{n=k}^{N} b_n$ and $\sum_{n=k}^{\infty} b_n$ look to be more similar than the concepts represented are. In the first case $N$ is just a number, but in the second $\infty$ is an abbreviation indicating the presence of $\lim_N$. It may be found useful to write the sums of series out in the form $\lim_N \sum_{n=k}^{N} b_n$ until the basic properties of series are familiar.

An alternative notation which is sometimes used, either for a series or for its sum, is to write

$$\sum_{n=k}^{\infty} b_n = b_k + b_{k+1} + \cdots + b_n + \cdots$$

Although strictly speaking this is a notation for the sum of a series, it is clear that if we say $b_k + b_{k+1} + \cdots + b_n + \cdots$ is convergent we mean that $\sum_{n=k} b_n$ is convergent.

There are some series which can easily be summed. A very important case is that of the geometric series $\sum_{n=0} z^n$. The partial sum $s_N$ is

$$s_N = \sum_{n=0}^{N} z^n = \begin{cases} \dfrac{1-z^{N+1}}{1-z} & (z \neq 1) \\[2mm] N+1 & (z=1) \end{cases}$$

This follows as

$$(1-z) \sum_{n=0}^{N} z^n = \sum_{n=0}^{N} z^n - \sum_{n=0}^{N} z^{n+1} = 1 + \sum_{n=1}^{N} z^n - \sum_{n=0}^{N-1} z^{n+1} - z^{N+1}$$

$$= 1 - z^{N+1}$$

In the case $z=1$ we cannot get the answer by dividing by $1-z$, but then $z^n = 1$ for all $n$ and so the sum is equal to the number of terms. Now we have seen that $\lim_N z^N = 0$, provided that $|z| < 1$, so in this case we get

$$\sum_{n=0}^{\infty} z^n = \lim_N s_N = \lim_N \frac{1-z^{N+1}}{1-z} = \frac{1}{1-z}$$

The series is convergent with sum $1/(1-z)$. It will be a consequence of our first theorem on series, Theorem 9.1, that $\sum_{n=0} z^n$ diverges for $|z| \geq 1$. However, it is clear from the values of $s_N$ that the series diverges if $|z| > 1$. This follows as then $\{|z|^{N+1}\}$ is unbounded and so $\{z^{N+1}\}$ cannot have a limit. Similarly, $\lim_N (N+1)$ does not exist, so the geometric series diverges if $z=1$.

**Theorem 9.1**　If the series $\sum b_n$ is convergent then $\lim_n b_n = 0$.

*Proof*　By the general principle of convergence, Theorem 7.8, we have that the sequence $s$ of partial sums satisfies the condition: for $\varepsilon > 0$ there is a number $N_\varepsilon$ such that $|s_n - s_m| < \varepsilon$ if $n, m > N_\varepsilon$. In particular, putting $m = n-1$,

$$|s_n - s_{n-1}| = \left| \sum_{r=k}^{n} b_r - \sum_{r=k}^{n-1} b_r \right| = |b_n| < \varepsilon \quad \text{if } n > N_\varepsilon + 1$$

Thus $\lim_n b_n = 0$.

This result supplies us with a test for divergence. Thus $\sum z^n$ diverges if $|z| \geqslant 1$, as then $|z|^n \geqslant 1$, so that we cannot have $\lim_n z^n = 0$. However, it cannot be used to prove the convergence of a series. We shall see as a consequence of Theorem 9.7 that the series $\sum 1/n$ is divergent although $\lim_n 1/n = 0$. In using the special case of $|s_n - s_m| < \varepsilon$ in which $n = m - 1$, and only this case, something has been lost.

It will have become apparent that the convergence of a series depends only on $|s_n - s_m|$ for both $n$ and $m$ sufficiently large, so that the first few terms can never make any difference as to whether or not a series converges. In fact we get the following result.

**Theorem 9.2** If $l > m$ and $l, m \in \mathbb{N}$ then either both the series $\sum_{n=m} b_n$ and $\sum_{n=l} b_n$ diverge or else they both converge and

$$\sum_{n=m}^{\infty} b_n = \sum_{n=m}^{l-1} b_n + \sum_{n=l}^{\infty} b_n$$

*Proof* Let $c = \sum_{n=m}^{l-1} b_n$ and $s_N = \sum_{n=m}^{N} b_n$, while $\sigma_N = \sum_{n=l}^{N} b_n$ for $N \geqslant l$, $N \in \mathbb{N}'$. Then $s_N = c + \sigma_N$. Hence by Theorem 7.3 if $\lim_N s_N$ exists then so does $\lim_N \sigma_N$ and $\lim_N s_N = c + \lim_N \sigma_N$. Also if $\lim_N \sigma_N$ exists then so does $\lim_N s_N$ and $\lim_N s_N = c + \lim_N \sigma_N$. That is either both these limits exist and the equation holds or else neither limit exists.

Theorem 7.3 gives some other immediate results for series.

**Theorem 9.3** If the two series $\sum b_n$ and $\sum c_n$ are convergent then $\sum (b_n + c_n)$ is convergent and

$$\sum_{n=k}^{\infty} (b_n + c_n) = \sum_{n=k}^{\infty} b_n + \sum_{n=k}^{\infty} c_n$$

where $k$ is any integer large enough for both $b_n$ and $c_n$ to be defined for $n \geqslant k$.

*Proof* By Theorem 9.2 we need only consider $\sum_{n=k} b_n$, $\sum_{n=k} c_n$ and $\sum_{n=k} (b_n + c_n)$ for some fixed $k$. Let $\{s_N\}$, $\{\sigma_N\}$ and $\{\tau_N\}$ be the sequences of partial sums of these three series respectively. Then $\tau_N = s_N + \sigma_N$ and so by Theorem 7.3

$$\lim_N \tau_N = \lim_N s_N + \lim_N \sigma_N$$

or $\quad \displaystyle\sum_{n=k}^{\infty} (b_n + c_n) = \sum_{n=k}^{\infty} b_n + \sum_{n=k}^{\infty} c_n$

**Theorem 9.4**   For any number $c \neq 0$ the series $\sum_{n=k} cb_n$ is convergent if and only if the series $\sum_{n=k} b_n$ is convergent, and $\sum_{n=k}^{\infty} cb_n = c \sum_{n=k}^{\infty} b_n$.

*Note*   In the case $c = 0$ the series $\sum cb_n = \sum 0$ is convergent whatever the sequence $\{b_n\}$, as all the partial sums are 0.

*Proof*   If $\{s_N\}$ is the sequence of partial sums of $\sum_{n=k} b_n$ then $\{cs_N\}$ is the sequence of partial sums of $\sum_{n=k} cb_n$. By Theorem 7.3(ii) applied to the sequence $\{s_N\}$ and the sequence each of whose terms is $c$ we get that if $\sum_{n=k} b_n$ is convergent then $\lim_N cs_N = c \lim_N s_N$. Now by this result, if $\sum_{n=k} cb_n$ is convergent then $\sum_{n=k} b_n$ is convergent and

$$\sum_{n=k}^{\infty} b_n = \sum_{n=k}^{\infty} c^{-1} cb_n = c^{-1} \sum_{n=k}^{\infty} cb_n$$

Thus we can show that $\sum_{n=0}^{\infty} (3^n + 2^{n+1})/6^n = 5$. Since $0 < \frac{1}{3} < \frac{1}{2} < 1$, we get that the geometric series $\sum_{n=0}^{\infty} (\frac{1}{2})^n = 2$ and $\sum_{n=0}^{\infty} 1/3^n = \frac{3}{2}$. Hence by Theorem 9.4 $\sum_{n=0}^{\infty} 2/3^n = 3$. Then by Theorem 9.3

$$5 = \sum_{n=0}^{\infty} (\tfrac{1}{2})^n + \sum_{n=0}^{\infty} 2/3^n = \sum_{n=0}^{\infty} [(\tfrac{1}{2})^n + 2/3^n] = \sum_{n=0}^{\infty} \frac{3^n + 2^{n+1}}{6^n}$$

We do not get simple corresponding results for products and quotients of series. If $\{s_N\}$ and $\{\sigma_N\}$ are sequences of partial sums of two series then the terms of the series with sequences of partial sums $\{s_N \sigma_N\}$ and $\{s_N/\sigma_N\}$ are not simply enough related to the terms of the given series. It can be seen that the partial sums $\{s_N \sigma_N\}$ lead to a general term of the form $(b_0 + b_1 + \cdots + b_{N-1})c_N + (c_0 + c_1 + \cdots + c_{N-1})b_N + b_N c_N$. For this reason there are theorems giving special cases where there is a simpler series with the same sum.

An operation on series which suggests itself when they are written in the form $b_0 + b_1 + \cdots + b_n + \cdots$ is the insertion or removal of brackets from a series. It is easy to see that we cannot in general remove brackets from a series. Thus $\sum (1-1) = (1-1) + (1-1) + \cdots + (1-1) + \cdots$ is convergent with sum 0, as every term and so every partial sum is 0, whereas $\sum_{n=0} (-1)^n = 1 - 1 + 1 - \cdots + (-1)^n + \cdots$ is divergent in view of Theorem 9.1, as $\{(-1)^n\}$ does not have a limit, and so does not have the limit 0. On the other hand, insertion of brackets is always possible. Consider

$$(b_0 + b_1 + \cdots + b_{n_1}) + (b_{n_1+1} + \cdots + b_{n_2}) + \cdots + (b_{n_m+1} + \cdots + b_{n_{m+1}}) + \cdots$$

and write $c_{m+1} = b_{n_m+1} + \cdots + b_{n_{m+1}}$. Then if $\{s_N\}$ is the sequence of partial sums of $\sum_{n=0} b_n$ and $\{\sigma_M\}$ is the sequence of partial sums of $\sum_{m=0} c_m$ we get

$$\sigma_M = s_{n_M}$$

Thus if $\sum_{n=0}^{\infty} b_n = B$, and hence, given $\varepsilon > 0$, there is an $N_\varepsilon$ such that $|s_N - B| < \varepsilon$

for $N > N_\varepsilon$, we have $|\sigma_M - B| = |s_{n_M} - B| < \varepsilon$ for $n_M > N_\varepsilon$. Since $n_M \geqslant M$, the inequality certainly holds for $M > N_\varepsilon$. Thus $\sum_{m=0}^{\infty} c_m = B$.

In order to set up some tests to establish convergence of series, we consider first a special case. If the terms $b_n$ of a series are all positive or zero, $b_n \geqslant 0$ for $n \in \mathbb{N}'$, then the sequence $\{s_N\}$ of partial sums is increasing. We have $s_N - s_M = b_{M+1} + b_{M+2} + \cdots + b_N$ for $N > M$, and so $s_N - s_M \geqslant 0$. So we get the following theorem.

**Theorem 9.5** If the real numbers $b_n \geqslant 0$ form a series $\sum b_n$ then either $\sum b_n$ is convergent or else for every $M \in \mathbb{R}$ there is some integer $N$ such that $\sum_{n=k}^{N} b_n > M$.

*Proof* Write $\{s_N\}$ for the sequence of partial sums. Then $\{s_N\}$ is an increasing sequence. Thus by Theorem 7.7, if there is an $M$ such that $s_N \leqslant M$ all $N$ then $\lim_N s_N$ exists, that is the series converges.

We can use this result to establish convergence if we can find an $M$ for which $s_N \leqslant M$ all $N$. The other case is sometimes written as $\sum_{n=k}^{\infty} b_n = \infty$. This indicates not only that the series diverges, but also that it diverges in this special way. There has to be an $N_0$ corresponding to any given $M$ such that $s_N > M$ for all $N \geqslant N_0$. The use of this notation is avoided in this book, as it makes it appear that there is a number $\infty$, and neither $\mathbb{R}$ nor $\mathbb{C}$ has such a number.

One of the most frequently used ways of employing Theorem 9.5 is in the form of the comparison test.

**Theorem 9.6 The comparison test** If $\{b_n\}$ and $\{c_n\}$ are two real sequences such that $0 \leqslant b_n \leqslant c_n$ all $n$ then

$$\sum c_n \text{ convergent} \Rightarrow \sum b_n \text{ convergent}$$

and $\quad \sum b_n \text{ divergent} \Rightarrow \sum c_n \text{ divergent}$

*Proof* The second implication follows from the first. [To prove the first implication] assume $\sum c_n$ to be convergent and let $\sigma_N$ be its sequence of partial sums, so that $\lim_N \sigma_N$ exists. If $\{s_N\}$ is the sequence of partial sums of $\sum b_n$ then $s_N \leqslant \sigma_N$ all $N$, so that as $\{\sigma_N\}$ is increasing, $s_N \leqslant \sup_N \sigma_N = \lim_N \sigma_N$. Hence $s_N = \sum_{n=k}^{N} b_n \leqslant \lim_N \sigma_N = \sum_{n=k}^{\infty} c_n$ for all $N$. Thus $\sum b_n$ is convergent.

Thus, for instance, $\sum x^n$ is convergent if $0 \leqslant x < 1$ and also $0 \leqslant x^n/n \leqslant x^n$ for these $x$. Hence $\sum x^n/n$ is convergent if $x \in [0, 1)$. On the other hand, we shall see that $\sum 1/n$ is divergent and $0 < 1/n < (1 + 1/n)/n = (n+1)/n^2$. Thus the comparison test will show that $\sum (n+1)/n^2$ is divergent.

Another way of using Theorem 9.5 is given by the next theorem.

**Theorem 9.7　The condensation test**　If the real sequence $\{b_n\}$ is decreasing and $b_n \geq 0$ for all $n$ then either both the series $\sum b_n$ and $\sum 2^n \cdot b_{2^n}$ converge, or else they both diverge.

*Proof*　[We first show that if $\sum 2^n \cdot b_{2^n}$ is convergent then so is $\sum b_n$.] Since $\{b_n\}$ is decreasing, $b_0 \leq b_0$, $b_1 \leq 1 \cdot b_1$, $b_2 + b_3 \leq 2 \cdot b_2$, $b_4 + b_5 + b_6 + b_7 \leq 4 \cdot b_4, \ldots, b_{2^n} + b_{2^n+1} + \cdots + b_{2^{n+1}} \leq 2^n \cdot b_{2^n}, \ldots$. Thus provided we take $M$ so large that $2^M > N$ we have for any partial sum of $\sum b_n$ that

$$\sum_{n=0}^{N} b_n \leq \sum_{n=0}^{M} 2^M \cdot b_{2^M}$$

Then by Theorem 9.5, if $\sum_{n=0} 2^M \cdot b_{2^M}$ is convergent then $\sum_{n=0} b_n$ is convergent.

Again using the fact that $\{b_n\}$ is decreasing, $b_0 \geq b_0$, $b_1 \geq b_1$, $b_2 \geq 1 \cdot b_2$, $b_3 + b_4 \geq 2 \cdot b_4$, $b_5 + b_6 + b_7 + b_8 \geq 4 \cdot b_8, \ldots, b_{2^{n-1}+1} + \cdots + b_{2^n} \geq 2^{n-1} \cdot b_{2^n}, \ldots$. Then if $\sum b_n$ is convergent the bracketed series $\sum (b_{2^{n-1}+1} + \cdots + b_{2^n})$ is convergent. As $2^{n-1} \cdot b_{2^n} \geq 0$, the comparison test shows that $\sum 2^{n-1} \cdot b_{2^n}$ is convergent. Thus if either of the series $\sum b_n$ and $\sum 2^n \cdot b_{2^n}$ is convergent then so is the other.

Although the condition that $\{b_n\}$ be monotone is restrictive, this test can be used to give some very important standard results. These will then be available for use in the comparison test.

Consider $\sum 1/n$. We have $0 < 1/(n+1) < 1/n$, and so the conditions of the test apply. Now $\sum 2^n \cdot (\frac{1}{2})^n = \sum 1$, which is clearly divergent. Thus we have at last established that $\sum 1/n$ diverges. Again $0 < 1/(n+1)^2 < 1/n^2$, and so the conditions of the test apply to $\sum 1/n^2$. This time, however, $\sum 2^n \cdot [(\frac{1}{2})^n]^2 = \sum 2^{-n}$ is convergent, as it is a geometric series with terms $(\frac{1}{2})^n$ and $0 < \frac{1}{2} < 1$. Thus $\sum 1/n^2$ converges.

In fact the test will settle what happens for all series of the form $\sum 1/n^\alpha$ for $\alpha > 0$. For $\alpha \leq 0$ the terms do not have a limit 0 and so the series cannot converge. For $\alpha > 0$ we have

$$0 < \frac{1}{(n+1)^\alpha} < \frac{1}{n^\alpha} \quad \text{for } n \in \mathbb{N}'$$

Hence $\sum n^{-\alpha}$ converges or diverges according to whether $\sum 2^n (2^n)^{-\alpha}$ converges or diverges. Now $2^n \cdot (2^n)^{-\alpha} = 2^{n(1-\alpha)} = (2^{(1-\alpha)})^n$, so that we have a geometric series and will get convergence precisely if $2^{1-\alpha} < 1$, that is if $1 - \alpha < 0$. Hence we have

$$\sum \frac{1}{n^\alpha} \text{ converges if } \alpha > 1 \text{ and diverges if } \alpha \leq 1$$

However, even for series with positive decreasing sequences of terms, this is very far from being the whole story. Define a function $l$ in such a way that $l(2^n) = n$, and that $l$ is monotone increasing. There are many ways of doing this. We might, for instance, give $l(x)$ a value by first finding $n$ such that $2^n \leq x < 2^{n+1}$ and then writing $l(x) = n + 2^{-n}(x - 2^n)$. However, any reader who has already met the logarithmic functions may prefer to think of $l$ as being given by $l(x) = \log_2 x = \log x / \log 2$. We shall only use the values of $l$ for positive integer values of $x$. With such a function $l$ we have

$$0 < \frac{1}{(n+1)l(n+1)} < \frac{1}{nl(n)} \quad \text{for } n - 1 \in \mathbb{N}'$$

(as $l(1) = 0$, we must have $n \neq 1$). Then $\sum 1/[nl(n)]$ is convergent only if $\sum 2^n/[2^n l(2^n)]$ is convergent. However, $2^n/[2^n l(2^n)] = 1/n$ and $\sum 1/n$ is divergent. Thus $\sum 1/[nl(n)]$ is divergent. Similar considerations show that $\sum 1/(n[l(n)]^2)$ is convergent, as $\sum 2^n/(2^n n^2) = \sum 1/n^2$ is convergent.

Because of the importance of series arising from Taylor's theorem, there are special tests for comparing series of positive terms with geometric series $\sum x^n$.

**Theorem 9.8   D'Alembert's ratio test**   If all the terms $b_n$ of the sequence $\{b_n\}$ satisfy $b_n > 0$, and for some number $l$, $b_{n+1}/b_n \leqslant l < 1$, then the series $\sum b_n$ is convergent. If $b_{n+1}/b_n \geqslant 1$ then $\sum b_n$ is divergent.

*Proof*   If there is a number $l$ for which $b_{n+1}/b_n \leqslant l$ for $n = k, k+1, \ldots$ then $b_n \leqslant l b_{n-1} \leqslant l^2 b_{n-2} \leqslant \cdots \leqslant l^{n-k} b_k$. If $0 \leqslant l < 1$ the series $\sum l^n$, and hence also the series $\sum l^n (b_k/l^k)$, is convergent. Thus by the comparison test $\sum b_n$ is convergent.

If, on the other hand, $b_{n+1}/b_n \geqslant 1$ for $n = k, k+1, \ldots$ then $b_n \geqslant b_{n-1} \geqslant \cdots \geqslant b_k$ for $n > k$. Since $b_k > 0$, we cannot have $\lim_n b_n = 0$. Thus in this case $\sum b_n$ is divergent.

Of course, it may well be that $b_{n+1}/b_n > 1$ for some arbitrarily large $n$ and also $b_{n+1}/b_n < l < 1$ for other arbitrarily large $n$. In such a case the test is not applicable.

Notice particularly that the condition $b_{n+1}/b_n \leqslant l < 1$ is not the same as $b_{n+1}/b_n < 1$. If there were only one value of $n$ involved they would be the same, but the condition must hold for all $n \in \mathbb{N}$, $n \geqslant k$, and $l$ is to be fixed. Thus if $b_n = 1/n$ we have $b_{n+1}/b_n = n/(n+1) < 1$ for $n \in \mathbb{N}'$. Now $\lim_n n/(n+1) = 1$, so that there can be no $l < 1$ with $n/(n+1) \leqslant l$ all $n \in \mathbb{N}'$. As we have seen, $\sum 1/n$ is divergent.

In cases where this limit $\lim_n b_{n+1}/b_n$ exists, the test can be put in a particularly neat form.

**Theorem 9.9   Limit form of ratio test**   If all the terms of the sequence $\{b_n\}$ satisfy $b_n > 0$ then the series $\sum b_n$ is convergent if $\lim_n b_{n+1}/b_n < 1$ and divergent if $\lim_n b_{n+1}/b_n > 1$.

*Proof*   Suppose that $\lim_n b_{n+1}/b_n = B$. If $B < 1$ then we can find an integer $N_\varepsilon$ corresponding to $\varepsilon = \frac{1}{2}(1 - B)$ such that $b_{n+1}/b_n - B < \frac{1}{2}(1 - B)$ for $n \geqslant N_\varepsilon$.

Hence $b_{n+1}/b_n < \frac{1}{2}(1+B) < 1$ for $n \geq N_\varepsilon$, and so by Theorem 9.8 $\sum_{n=N_\varepsilon} b_n$ is convergent. Thus $\sum b_n$ is convergent.

If $B > 1$ then we can find an integer $N'_\varepsilon$ corresponding to $\varepsilon = \frac{1}{2}(B-1)$ such that $-b_{n+1}/b_n + B < \frac{1}{2}(B-1)$ for $n \geq N'_\varepsilon$. Hence $b_{n+1}/b_n > \frac{1}{2}(1+B) > 1$ for $n \geq N'_\varepsilon$, and so by Theorem 9.8 $\sum b_n$ is divergent.

If the limit $\lim_n b_{n+1}/b_n$ exists and $b_n > 0$ for all sufficiently large $n$ (we have seen that we can neglect a finite number of terms) there is still one case not settled by this test. If $\lim_n b_{n+1}/b_n = 1$ the series may diverge, as shown by $\sum 1/n$, but also it may converge. Thus $\sum 1/n^2$ is convergent, $1/n^2 > 0$, and $\lim_n [1/(n+1)]^2/(1/n)^2 = \lim_n [n/(n+1)]^2 = 1$. So that if $\lim_n b_{n+1}/b_n = 1$ this test gives no answer. It can be thought of like an answer to an opinion poll. If the limit is less than one, the answer is 'yes we have convergence'. If the limit is greater than one, the answer is 'no we do not have convergence'. If the limit is one then the answer is 'don't know'.

Some cases in which the limit is one can be settled by use of Theorem 9.8. Thus $\sum n/(n+1)$ is divergent by Theorem 9.8, as

$$\frac{n+1}{(n+1)+1} \cdot \frac{n+1}{n} = \frac{n^2+2n+1}{n^2+2n} > 1$$

for all $n \in \mathbb{N}'$. However,

$$\lim_n \frac{n^2+2n+1}{n^2+2n} = 1$$

so that Theorem 9.9 gives no answer.

**Theorem 9.10   Cauchy's $n$th root test**   If the terms of the sequence $\{b_n\}$ satisfy $b_n \geq 0$ and $\sqrt[n]{b_n} \leq l < 1$ then the series $\sum b_n$ is convergent. If the terms satisfy $b_n \geq 0$ and $\sqrt[n]{b_n} \geq 1$ then the series $\sum b_n$ is divergent.

*Proof*   In the case $\sqrt[n]{b_n} \leq l < 1$ we have $0 \leq b_n \leq l^n$ and $0 \leq l < 1$, and so the convergence of $\sum l^n$ yields the convergence of $\sum b_n$ by the comparison test.

In the case $\sqrt[n]{b_n} \geq 1$ we have $b_n \geq 1$ and so the terms $b_n$ do not have the limit 0, and $\sum b_n$ is divergent.

As in the case of the ratio test, we must have the number $l$ not depending on $n$; it is not good enough just to have $\sqrt[n]{b_n} < 1$. Also we get a form of this test applicable to the case in which $\lim_n \sqrt[n]{b_n}$ exists.

**Theorem 9.11   Limit form of $n$th root test**   If the terms of the sequence $\{b_n\}$ satisfy $b_n \geq 0$ and $\lim_n \sqrt[n]{b_n} < 1$ then the series $\sum b_n$ is convergent. If $b_n \geq 0$ and $\lim_n \sqrt[n]{b_n} > 1$ then the series $\sum b_n$ is divergent.

*Proof*   Let $B = \lim_n \sqrt[n]{b_n}$. Then if $B < 1$, corresponding to $\varepsilon = \frac{1}{2}(1-B)$ there is an integer $N_\varepsilon$ such that $\sqrt[n]{b_n} - B < \frac{1}{2}(1-B)$ or $\sqrt[n]{b_n} < \frac{1}{2}(1+B) < 1$ for $n \geq N_\varepsilon$. Thus by Theorem 9.10 $\sum_{n=N_r} b_n$ converges and hence $\sum b_n$ converges.

If $B > 1$ we take $\varepsilon = \frac{1}{2}(B-1)$ and find $N'_\varepsilon$ such that $-\sqrt[n]{b_n} + B < \frac{1}{2}(B-1)$ or $\sqrt[n]{b_n} > \frac{1}{2}(B+1) > 1$ for $n \geqslant N'_\varepsilon$. Hence $\sum_{n=N'_\varepsilon} b_n$ is divergent by Theorem 9.10 and so also $\sum b_n$ is divergent.

Once again if the limit is 1, $\lim_n \sqrt[n]{b_n} = 1$, the test does not tell us whether $\sum b_n$ converges or diverges. Indeed if we assume that $\lim_n n^{1/n}$ exists we can find it from our knowledge that $\sum 1/n$ diverges and $\sum 1/n^2$ converges. The terms $1/n$ and $1/n^2$ are positive, so that as $\sum 1/n$ does not converge, $\lim_n \sqrt[n]{(1/n)}$ is not less than 1. However, $\lim_n \sqrt[n]{(1/n^2)} = [\lim_n \sqrt[n]{(1/n)}]^2$, so that the convergence of $\sum 1/n^2$ gives us that $[\lim_n \sqrt[n]{(1/n)}]^2$ is not greater than 1. If we write $l = \lim_n n^{1/n}$, and note that $l > 0$ as $n^{1/n} \geqslant 1$, we get $1/l \geqslant 1$ and $1/l^2 \leqslant 1$. Hence $1 \leqslant l^{-1} \leqslant l^{-2} \leqslant 1$, and so $l = 1$. The existence of $\lim_n n^{1/n}$ can be shown by proving that $\{n^{1/n}\}$ is decreasing. However, the result $\lim_n n^{1/n} = 1$ is derived in a different way in Chapter 11.

Using the ratio test we can prove that $\sum x^n/n!$ is convergent for all $x \geqslant 0$ and that $\sum n! x^n$ is divergent for all $x > 0$. Thus $x^n/n! \geqslant 0$ if $x \geqslant 0$ and $x^{n+1}/(n+1)! \cdot n!/x^n = x/(n+1)$ and $\lim_n x/(n+1) = 0 < 1$. Hence $\sum x^n/n!$ is convergent if $x \geqslant 0$. Now $n! x^n \geqslant 0$ if $x \geqslant 0$ and $[(n+1)! x^{n+1}]/(n! x^n) = (n+1)x \geqslant 1$ for $n \geqslant 1 + 1/x$. Hence $\sum n! x^n$ is divergent for $x > 0$. Similarly, the $n$th root test shows that $\sum x^n/n^n$ is convergent for all $x \geqslant 0$ but that $\sum n^n x^n$ is divergent for $x > 0$. The terms of the series are positive or 0 in both cases, and $\sqrt[n]{(x^n/n^n)} = x/n$. Now $\lim_n x/n = 0 < 1$, so that $\sum x^n/n^n$ is convergent for $x \geqslant 0$. Again $\sqrt[n]{(n^n x^n)} = nx \geqslant 1$ if $n \geqslant 1/x$ for $x > 0$, so that $\sum n^n x^n$ is divergent for all $x > 0$.

Consider now the series $\sum n^3 x^n$ for positive $x$. Apply the ratio test:

$$\lim_n \frac{(n+1)^3 x^{n+1}}{n^3 x^n} = \lim_n \left(\frac{n+1}{n}\right)^3 x = x$$

as $\lim_n (n+1)/n = \lim_n (1+1/n) = 1$. Hence $\sum n^3 x^n$ is convergent if $0 \leqslant x < 1$ and divergent if $x > 1$. This test does not tell us what happens if $x = 1$. However, the series is then $\sum n^3$ and, as $n^3 \geqslant 1$ for $n \in \mathbb{N}'$, it is clear that $\{n^3\}$ cannot have limit 0. Thus in this case $\sum n^3 x^n$ is divergent. The series $\sum x^n/n^3$ is also convergent for $0 \leqslant x < 1$ and divergent for $x > 1$, since the limit of the ratio is

$$\lim_n \left(\frac{x^{n+1}}{(n+1)^3} \cdot \frac{n^3}{x^n}\right) = \lim_n x \left(\frac{n}{n+1}\right)^3 = x$$

In this case the series has been shown to be convergent when $x = 1$, as it is $\sum 1/n^3$.

If we use the result $\lim_n n^{1/n} = 1$ we could get the same results from the $n$th root test. The limits this time would be $\lim_n \sqrt[n]{(n^3 x^n)} = \lim_n x(n^{1/n})^3 = x$ and $\lim_n \sqrt[n]{(x^n/n^3)} = x/(\lim_n n^{1/n})^3 = x$. The choice as to whether to use the $n$th root test or the ratio test can usually be made on which limit is easier to calculate. In cases where the terms contain factorials, $n!$, $(2n)!$, etc., or similar expressions, e.g. $1 \cdot 3 \cdot 5 \cdots (2n-1)$, this will usually be the limit of the ratio. Theoretically, that is if we suppose that the limits can be found, the $n$th root test is more powerful than the ratio test. Thus suppose that $b_0 = 1$ for simplicity and that the ratio test's conditions for convergence hold. Then $b_{n+1}/b_n \leqslant l$ for $n \in \mathbb{N}$ gives $b_n \leqslant l^n$ for $n \in \mathbb{N}$, so that $\sqrt[n]{b_n} \leqslant l$ for $n \in \mathbb{N}$ and, as $l < 1$, the conditions for the $n$th root test to give convergence hold.

So far we only have tests for convergence of series with real positive terms. The next theorem enables these to be applied in many cases where the terms may have a mixture of signs or may be complex.

**Theorem 9.12**    If the series $\sum |b_n|$ is convergent then the series $\sum b_n$ is convergent.

*Proof*    Let $\{s_N\}$ be the sequence of partial sums of $\sum b_n$ and $\{\sigma_N\}$ be the sequence of partial sums of $\sum |b_n|$. If $\sum |b_n|$ is convergent, that is $\lim_N \sigma_N$ exists, then by the general principle of convergence, Theorem 7.8, for every $\varepsilon > 0$, $\exists N_\varepsilon$ such that $|\sigma_N - \sigma_M| < \varepsilon$ if $N \geqslant M > N_\varepsilon$. However,

$$|s_N - s_M| = |b_{M+1} + \cdots + b_N| \leqslant |b_{M+1}| + \cdots + |b_N|$$

by repeated applications of the triangle inequality and, as this is positive or zero,

$$|s_N - s_M| \leqslant ||b_{M+1}| + \cdots + |b_N|| = |\sigma_N - \sigma_M| < \varepsilon$$

if $N \geqslant M > N_\varepsilon$. So, using the general principle of convergence as a sufficient condition for the existence of a limit, $\lim_N s_N$ exists. Thus $\sum b_n$ is convergent.

Any test for the convergence of a series of positive terms can now be applied to $\sum |b_n|$ in order to prove the convergence of $\sum b_n$. This does not supply tests for divergence. Thus we have immediately that for any $z$ with $|z| = 1$, $\sum z^n/n^2$ is convergent, as $\sum |z^n|/n^2 = \sum 1/n^2$ is convergent, but we shall see that $\sum z^n/n$ need not be divergent for $|z| = 1$, although $\sum 1/n$ is divergent.

However, in the cases of both the ratio test and the $n$th root test we do get tests for divergence. This is because these tests, when they establish divergence, do so by proving that the sequence of terms does not have the limit 0. Now if $\lim_n b_n = 0$ then $\lim_n |b_n| = 0$, so if we prove that $\lim_n |b_n| = 0$ is not true then it also is not true that $\lim_n b_n = 0$, and so $\sum b_n$ is divergent.

We have then that $\sum z^n/n!$ is convergent for all $z$ but that $\sum n! z^n$ diverges if $z \neq 0$. If $z = 0$ then every term $n! z^n$ for $n \in \mathbb{N}$ is 0, and so the series converges. Again both the series $\sum z^n/n$ and $\sum z^n/n^2$ are convergent for all $z$ with $|z| < 1$ and divergent for $z$ with $|z| > 1$.

In light of Theorem 9.12 we make the following definition.

**Definition 9.3**    If the series $\sum |b_n|$ is convergent then the series $\sum b_n$ is said to be *absolutely convergent*.

This definition is not directly about either of the two sequences which form $\sum b_n$. It is about the sequence of partial sums of the new series formed of the moduli or absolute values of the $b_n$.

Theorem 9.12 can now be stated as follows: an absolutely convergent series is convergent.

The limit forms of the ratio test and the $n$th root test can now be stated as follows: if either $\lim_n |b_{n+1}/b_n| < 1$ or $\lim_n \sqrt[n]{|b_n|} < 1$ then $\sum b_n$ is absolutely convergent; if either $\lim_n |b_{n+1}/b_n| > 1$ or $\lim_n \sqrt[n]{|b_n|} > 1$ then $\sum b_n$ is divergent.

Absolute convergence can also be looked at in the following way. Write $\sum b_n$ as a sum of 4 series, one consisting of the positive real parts, one of the negative real parts, one of the positive imaginary parts, and one of the negative imaginary parts of $b_n$. Thus let $c_n = \frac{1}{2}(\mathrm{Re}(b_n) + |\mathrm{Re}(b_n)|)$, $d_n = \frac{1}{2}(\mathrm{Re}(b_n) - |\mathrm{Re}(b_n)|)$, $e_n = \frac{1}{2}(\mathrm{Im}(b_n) + |\mathrm{Im}(b_n)|)$ and $f_n = \frac{1}{2}(\mathrm{Im}(b_n) - |\mathrm{Im}(b_n)|)$. Thus if $b_n = i^n$ the sequences would be: 0, 0, 0, 1, 0, 0, 0, 1, ... ; 0, 1, 0, 0, 0, 1, 0, 0, ... ; 1, 0, 0, 0, 1, 0, 0, 0, ... ; 0, 0, 1, 0, 0, 0, 1, 0, ... ; if we start the sequences at $b_1$. Then $0 \leqslant c_n \leqslant |b_n|$, $0 \leqslant -d_n \leqslant |b_n|$, $0 \leqslant e_n \leqslant |b_n|$ and $0 \leqslant -f_n \leqslant |b_n|$, so that $\sum c_n$, $\sum d_n$, $\sum e_n$ and $\sum f_n$ all converge when $\sum b_n$ is absolutely convergent. On the other hand, $|b_n|^2 = (c_n + d_n)^2 + (e_n + f_n)^2 = c_n^2 + d_n^2 + e_n^2 + f_n^2$ (as one of $c_n$ or $d_n$ is zero and one of $e_n$ or $f_n$ is zero), so that $|b_n| \leqslant |c_n| + |d_n| + |e_n| + |f_n|$. Thus if each of the series $\sum c_n$, $\sum d_n$, $\sum e_n$ and $\sum f_n$ converges then $\sum b_n$ converges absolutely.

To complete the justification of Definition 9.3, we have to see that there are some series which are convergent but not absolutely convergent.

**Theorem 9.13   Leibnitz test**   If the series $\sum b_n$ can be written as $\sum (-1)^n B_n$ where $B_n \geqslant 0$ for all $n$, $\{B_n\}$ is decreasing, and $\lim_n B_n = 0$, then $\sum b_n$ is convergent.

*Proof*   Let $\{s_N\}$ be the sequence of partial sums, and since this can be done without loss of generality suppose that the first term of the series is $b_0$. [Now if we consider the way in which $s_N$ oscillates we see that it goes one way and then comes back again, but not beyond the place where it was. This is shown in Fig. 9.1. Therefore we consider successive pairs of terms.] Write

$$s_N = \sum_{n=0}^{N} (-1)^n B_n$$

$$= (B_0 - B_1) + 0(B_1 - B_2) + (B_2 - B_3) + \cdots + \delta_{N-1}(B_{N-1} - B_N) + \delta_N B_N$$

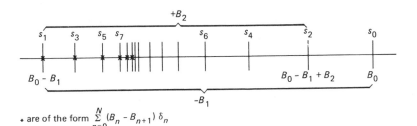

* are of the form $\sum_{n=0}^{N} (B_n - B_{n+1}) \delta_n$

**Fig. 9.1**   The partial sums of $\sum (-1)^n B_n$

where

$$\delta_n = \begin{cases} 1 & (n \text{ even}) \\ 0 & (n \text{ odd}) \end{cases}$$

Thus $s_N = \sum\limits_{n=0}^{N-1} (B_n - B_{n+1})\delta_n + \delta_N B_N$

Now $0 \le \delta_N B_N \le B_N$ and $\lim_N B_N = 0$, so that $\lim_N \delta_N B_N = 0$. Also

$$\sum_{n=0}^{N-1} (B_n - B_{n+1})\delta_n \le \sum_{n=0}^{N-1} (B_n - B_{n+1}) = B_0 - B_1 + B_1 - B_2 + \cdots - B_N$$

$$= B_0 - B_N \le B_0$$

as $\delta_n \le 1$. However, $\{B_n\}$ is decreasing, so that $B_n - B_{n+1} \ge 0$ all $n$ and hence $\sum (B_n - B_{n+1})\delta_n$ is a series of positive terms with every partial sum $\le B_0$. Hence by Theorem 9.5 $\sum (B_n - B_{n+1})\delta_n$ is convergent. Thus

$$\lim_N s_N = \sum_{n=0}^{\infty} (B_n - B_{n+1})\delta_n + \lim_N \delta_N B_N = \sum_{n=0}^{\infty} (B_n - B_{n+1})\delta_n$$

so that $\sum b_n = \sum (-1)^n B_n$ is convergent.

Thus series such as $\sum (-1)^n/n$ and $\sum (-1)^n/\sqrt{n}$ are convergent, but as $\sum 1/n$ and $\sum 1/\sqrt{n}$ are divergent, these series are not absolutely convergent.

**Definition 9.4**    If the series $\sum b_n$ is convergent, but is not absolutely convergent, then it is said to be *conditionally convergent*.

The difference between absolutely convergent and conditionally convergent series is not merely in the method by which convergence can be proved. There are some ways in which we need to be more careful in handling conditionally convergent series.

Consider the sequence defined by putting $b_0 = 1$, $b_1 = b_2 = \frac{1}{2}$, the next 4 terms equal to $\frac{1}{4}$, the next 8 terms equal to $\frac{1}{8}$, and so on. Then $b_n > 0$, $n \in \mathbb{N}$, $\{b_n\}$ is decreasing, and $\lim_n b_n = 0$. Thus $\sum (-1)^n b_n$ is convergent. In fact $\sum_{n=0}^{\infty} (-1)^n b_n = 1 = s_0 = 1 - \frac{1}{2} + \frac{1}{2} = s_2 = 1 - \frac{1}{2} + \frac{1}{2} - \frac{1}{4} + \frac{1}{4} = s_4 = 1 - \frac{1}{2} + \frac{1}{2} - \frac{1}{4} + \frac{1}{4} - \frac{1}{4} + \frac{1}{4} = s_6 = \cdots$. However, $\sum (-1)^n b_n$ is not absolutely convergent, as $b_0 = 1$, $b_0 + b_1 + b_2 = 2$, $\sum_{n=0}^{6} b_n = 3$, $\sum_{n=0}^{14} b_n = 4$, and it is clear that it is possible to find a partial sum equal to any $n \in \mathbb{N}'$. Now it is apparent that the insertion or deletion of zero terms in a series cannot change its convergence or its sum. Thus

$$1 - \tfrac{1}{2} + \tfrac{1}{2} - \tfrac{1}{4} + \tfrac{1}{4} - \tfrac{1}{4} + \tfrac{1}{4} - \tfrac{1}{8} + \cdots + (-1)^n b_n + \cdots \qquad = 1$$

$$0 + \tfrac{1}{2} + 0 - \tfrac{1}{4} + 0 + \tfrac{1}{4} + 0 - \tfrac{1}{8} + \cdots + 0 + (-1)^n b_n/2 + \cdots = \tfrac{1}{2}$$

Adding

$$1 + 0 + \tfrac{1}{2} - \tfrac{1}{2} + \tfrac{1}{4} + 0 + \tfrac{1}{4} - \tfrac{1}{4} + \cdots \qquad = \tfrac{3}{2}$$

Thus $1+\frac{1}{2}-\frac{1}{2}+\frac{1}{4}+\frac{1}{4}-\frac{1}{4}+\frac{1}{8}+\cdots=\frac{3}{2}$. It is not difficult to check that the terms of this last series are the same $(-1)^n b_n$ but taken in a different order. There is a limit to the extent of commutativity for convergent series.

**Definition 9.5**   If $n$ is a one–one function with domain and range both $\mathbb{N}$ then the sequence $\{a_{n(k)} \mid k \in \mathbb{N}\}$ is a *rearrangement* of the sequence $\{a_n \mid n \in \mathbb{N}\}$.

Absolutely convergent series have the special property that rearranging their terms does not alter their sum.

**Theorem 9.14**   If the sequence $\{b_{n(k)}\}$ is a rearrangement of $\{b_n\}$, and $\sum b_n$ is absolutely convergent, then $\sum b_{n(k)}$ is absolutely convergent and $\sum_{n=0}^{\infty} b_n = \sum_{k=0}^{\infty} b_{n(k)}$.

*Proof*   Write $\{s_N\}$ and $\{\sigma_K\}$ for the sequences of partial sums of the series $\sum_{n=0}^{\infty} b_n$ and $\sum_{k=0}^{\infty} b_{n(k)}$ respectively. Let $\sum_{n=0}^{\infty} b_n = b$ and $\sum_{n=0}^{\infty} |b_n| = B$. Then, given $\varepsilon > 0$, we can find a positive integer $N_\varepsilon$ such that, if $N \geqslant N_\varepsilon$, both $|s_N - b| < \varepsilon/2$ and $|\sum_{n=0}^{N} |b_n| - B| < \varepsilon/2$. Now use $I$ to denote the set of integers up to $N_\varepsilon$, so that $I = \{0, 1, 2, \ldots, N_\varepsilon\}$, and $J$ to denote $n^{-1}(I)$, and consider $\sigma_K$ for $K \geqslant K_\varepsilon = \max J$. Then we have

$$\{b_n \mid n = 0, 1, \ldots, n(K_\varepsilon)\} \supset \{b_n \mid n \in I\}$$

and $\sigma_K$ is the sum of all the $b_n$ in the first of these two sets.

Thus $\sigma_K - s_{N_\varepsilon}$ is the sum of the terms $b_n$ in the first set but not in the second. Hence

$$|\sigma_K - s_{N_\varepsilon}| \leqslant \sum_{n=N_\varepsilon+1}^{\infty} |b_n| = B - \sum_{n=0}^{N_\varepsilon} |b_n| < \varepsilon/2$$

Thus $|\sigma_K - b| = |\sigma_K - s_{N_\varepsilon} + s_{N_\varepsilon} - b| \leqslant |\sigma_K - s_{N_\varepsilon}| + |s_{N_\varepsilon} - b| < \varepsilon/2 + \varepsilon/2 = \varepsilon$

provided that $K \geqslant K_\varepsilon$, that is $\lim_K \sigma_K = b$ or $\sum_{k=0}^{\infty} b_{n(k)} = b$.

Applying this result to $\sum |b_n|$ and $\sum |b_{n(k)}|$, we get $\sum_{k=0}^{\infty} |b_{n(k)}| = B$, and so the convergence is absolute.

The proof which has been given for the Leibnitz test generalizes fairly easily, giving two somewhat more powerful tests, but first it is convenient to make a definition.

**Definition 9.6**   If $\{s_N\}$ is the sequence of partial sums of the series $\sum b_n$, and $\{s_N\}$ is bounded, then the series $\sum b_n$ is said to be *bounded*.

We might have chosen to make $\{b_n\}$ bounded, but in practice it is more frequently necessary to say whether or not $\{s_N\}$ is bounded. Thus Theorem 9.5 states that if $b_n \geqslant 0$ and $\sum b_n$ is bounded then $\sum b_n$ is convergent.

**Theorem 9.15   Dirichlet's and Abel's tests**   If the terms of the sequence $\{b_n\}$ can be written in the form $B_n C_n$ where $\{B_n\}$ is a decreasing sequence of positive real numbers and either

(i)   (Dirichlet) $\lim_n B_n = 0$ and $\sum C_n$ is bounded

or

(ii)   (Abel) $\sum C_n$ is convergent

then $\sum b_n$ is convergent.

*Proof*   Let $\{s_N\}$ be the sequence of partial sums of $\sum C_n$, and take $k = 0$, as it is clear that this does not affect the argument. The following equation is known as partial summation. As $s_{n+1} - s_n = C_{n+1}$,

$$\sum_{n=0}^{N} B_n C_n = s_0 B_0 + (s_1 - s_0) B_1 + (s_2 - s_1) B_2 + \cdots + (s_N - s_{N-1}) B_N$$

$$= s_0(B_0 - B_1) + s_1(B_1 - B_2) + \cdots + s_{N-1}(B_{N-1} - B_N) + s_N B_N$$

$$= \sum_{n=0}^{N-1} s_n(B_n - B_{n+1}) + s_N B_N$$

[We show that each of these terms has a limit. Firstly, $\sum s_n(B_n - B_{n+1})$ is absolutely convergent.] As $\{B_n\}$ is real and decreasing,

$$\sum_{n=0}^{N-1} |s_n| |B_n - B_{n+1}| = \sum_{n=0}^{N-1} |s_n| (B_n - B_{n+1}) \leqslant K \sum_{n=0}^{N-1} (B_n - B_{n+1})$$

where $K$ is a bound for $\{s_N\}$ which exists either by assumption in (i) or because in (ii) $\lim_N s_N$ exists. However,

$$\sum_{n=0}^{N-1} (B_n - B_{n+1}) = B_0 - B_N \leqslant B_0$$

as $B_N \geqslant 0$. So $\sum_{n=0} |s_n| |B_n - B_{n+1}|$ is a series of positive terms and is bounded by $K B_0$. Theorem 9.5 shows that it is convergent; hence $\sum s_n(B_n - B_{n+1})$ is convergent.

Now in case (i) $\lim_N s_N B_N = 0$ by Theorem 7.5 as $|s_N B_N| = |s_N| B_N \leqslant K B_N$ and $\lim_N B_N = 0$ by assumption. In case (ii) $\lim_N s_N$ exists by the convergence of $\sum C_n$ and $\lim_N B_N$ exists as $B_N \geqslant 0$ all $N$ and $\{B_N\}$ is decreasing; thus $\lim_N s_N B_N$ exists. Hence

$$\sum_{n=0}^{\infty} B_n C_n = \sum_{n=0}^{\infty} s_n(B_n - B_{n+1}) + \lim_N s_N B_N$$

and $\sum b_n$ is convergent.

It follows immediately that for $z \neq 1$ but $|z| = 1$, as

$$\left| \sum_{n=0}^{N} z^n \right| = \left| \frac{1 - z^{N+1}}{1 - z} \right| \leqslant \frac{2}{|1 - z|}$$

the series $\sum z^n / n$ is convergent. We only have to put $B_n = 1/n$ and $C_n = z^n$.

We became interested in series by the possibility of representing some functions in the form $f(x) = \sum_{n=0}^{\infty} a_n x^n$. This is a special case of a series of functions $\sum_{n=0}^{\infty} f_n(x)$. It would be nice to be able to predict properties of the function defined by the sum from the properties of the $f_n$. Unfortunately, even if the series is convergent for each $x \in (a, b)$, it does not follow that continuity of all the $f_n$ will make the sum represent a function continuous on $(a, b)$.

To make the discussion simpler, we will look at sequences $\{g_n\}$ of functions. These then can be taken to be partial sums of a series of functions. Firstly, let us have an example of a sequence $\{g_n\}$ of functions continuous on $[0, 2]$ for which the function $g$ with $g(x) = \lim_n g_n(x)$ is defined on $[0, 2]$ but is not continuous. Define $g_n$ by

$$g_n(x) = \begin{cases} 0 & (0 \leqslant x \leqslant 1 - 1/n) \\ \frac{1}{2}n(x + 1/n - 1) & (1 - 1/n < x < 1 + 1/n) \\ 1 & (1 + 1/n \leqslant x \leqslant 2) \end{cases}$$

The graph of $g_n$ is shown in Fig. 9.2. Then we get

$$g(x) = \lim_n g_n(x) = \begin{cases} 0 & (0 \leqslant x < 1) \\ \frac{1}{2} & (x = 1) \\ 1 & (1 < x \leqslant 2) \end{cases}$$

which makes the function $g$ not continuous.

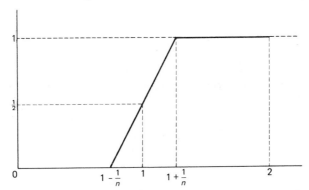

**Fig. 9.2** The function $g_n$

Indeed, for a given $x \in [0, 2]$, we can take $n$ sufficiently large so that $g_n(x)$ has the same value as the limit. However, for a given $\varepsilon > 0$, how large we need to take $n$ depends on $x$, that is $N_\varepsilon$ should be written $N_\varepsilon(x)$. Looked at another way, for any fixed $N$, by taking $x$ sufficiently close to 1 we can make $|g_N(x) - \lim_n g_n(x)|$ nearly equal to $\frac{1}{2}$ or certainly more than $\frac{1}{4}$ and less than $\frac{3}{4}$. Figure 9.3 is an attempt to show how the discontinuity of $g$ arises by putting two different functions $g_m$ and $g_n$ on the same graph.

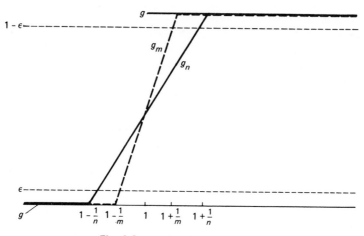

**Fig. 9.3**  The functions $g_m$ and $g_n$

**Definition 9.7**  If the sequence of functions $\{g_n\}$ all defined on a set $E$ is such that there is a function $g$ defined on $E$ with

$$\lim_n \sup_{x \in E} |g_n(x) - g(x)| = 0$$

then we say that $\lim_n g_n = g$ *uniformly on* $E$.

If the sequence of functions $\{f_n\}$ all defined on a set $E$ is such that

$$\lim_n \sum_{n=0}^{N} f_n(x) = \sum_{n=0}^{\infty} f_n(x) \text{ uniformly on } E$$

then the series of functions $\sum f_n$ is said to *converge uniformly on* $E$.

The idea of a uniform limit is illustrated in Fig. 9.4. If $\lim_n g_n = g$ uniformly on $E$ then corresponding to $\varepsilon > 0$ there is an $N_\varepsilon$ such that $\sup_{x \in E} |g_n(x) - g(x)| < \varepsilon$ for $n > N_\varepsilon$, and so for each $x \in E$, $|g_n(x) - g(x)| < \varepsilon$ if $n > N_\varepsilon$. Thus we have $\lim_n g_n(x) = g(x)$ all $x \in E$, and the same $N_\varepsilon$ can be used for all $x \in E$; that is $N_\varepsilon$ can be chosen uniformly.

The main fact we require about uniform limits is that they preserve continuity.

**Theorem 9.16**  If $\{g_n\}$ is a sequence of functions defined and continuous on a set $E$ and $\lim_n g_n = g$ uniformly on $E$ then $g$ is continuous on $E$.

*Proof*  Given $\varepsilon > 0$, there is an $N_\varepsilon$ for which

$$|g_n(x) - g(x)| \leqslant \sup_{t \in E} |g_n(t) - g(t)| < \varepsilon/3 \quad \text{for } n \geqslant N_\varepsilon, \, x \in E$$

Now for any point $\alpha \in E$ we can use the continuity of $g_{N_\varepsilon}$ to find a $\delta > 0$ such that

$$|g_{N_\varepsilon}(\alpha) - g_{N_\varepsilon}(x)| < \varepsilon/3 \quad \text{if } x \in E \cap \mathcal{N}(\alpha, \delta)$$

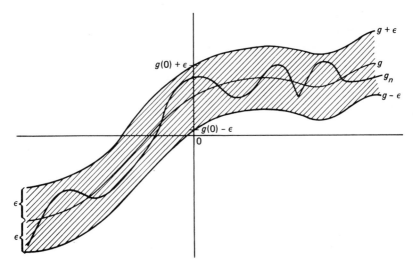

**Fig. 9.4** $\lim_n g_n = g$ uniformly

Then

$$|g(\alpha) - g(x)| \leq |g(\alpha) - g_{N_\varepsilon}(\alpha)| + |g_{N_\varepsilon}(\alpha) - g_{N_\varepsilon}(x)| + |g_{N_\varepsilon}(x) - g(x)|$$
$$< \varepsilon/3 + \varepsilon/3 + \varepsilon/3 = \varepsilon$$

provided that $x \in E \cap \mathcal{N}(\alpha, \delta)$. Thus $g$ is continuous on $\{\alpha\}$ for $\alpha \in E$, or $g$ is continuous on $E$.

**Corollary to Theorem 9.16** If $\{f_n\}$ is a sequence of functions defined and continuous on a set $E$ and $\sum f_n$ is uniformly convergent on $E$ then the function $f$, given by $f(x) = \sum_{n=k}^{\infty} f_n(x)$ for $x \in E$, is continuous on $E$.

*Proof* The partial sum $\sum_{n=k}^{N} f_n$ is continuous on $E$, being the sum of a finite number of continuous functions. Hence by the theorem $f = \sum_{n=k}^{\infty} f_n$ is continuous on $E$.

This can be applied to series of the form $\sum a_n x^n$. A series which has as its sequence of terms the sequence of functions $\{a_n x^n\}$ and $a_n \in \mathbb{R}$ all $n$ is called a *real power series*. A series with sequence of terms $\{a_n z^n\}$ where $a_n \in \mathbb{C}$ all $n$ is called a *complex power series*. A *power series* may be either a real or a complex power series.

The term 'power series' is often used for a series of numbers of the form $\sum a_n x_0^n$. For us this is the value of the power series $\sum a_n x^n$ at the point $x_0$.

If we compare the values of a power series at two points $\alpha$ and $\beta$ we get terms $a_n \alpha^n$ and $a_n \beta^n$. These differ by the factor $(\beta/\alpha)^n$, and $\{(\beta/\alpha)^n\}$ are the terms of a geometric series. This fact gives power series some very special properties.

**Theorem 9.17   Radius of convergence**   Either the series $\sum a_n z^n$ is convergent for all $z \in \mathbb{C}$ or there is some real number $r \geqslant 0$ such that $\sum a_n z^n$ is absolutely convergent for $|z| < r$ and divergent for $|z| > r$. The number $r$ is called the *radius of convergence* of the series $\sum a_n z^n$ and $\sum a_n x^n$.

[As $x^n(t) = z^n(t)$ only for real $t$, the values of $x^n$ form a subset of the values of $z^n$. Thus the values of the real power series $\sum a_n x^n$ are a subset of the values of the complex power series $\sum a_n z^n$, and in proving results in the complex case we are at the same time proving results in the real case.]

*Proof*   If there is some point $\alpha \neq 0$ such that $\sum a_n \alpha^n$ is convergent then by Theorem 9.1 $\lim_n a_n \alpha^n = 0$. However, this implies that $\{a_n \alpha^n\}$ is bounded, $|a_n \alpha^n| < K$ all $n \in \mathbb{N}'$, say. Hence for any $z \in \mathbb{C}$ with $|z| < |\alpha|$

$$|a_n z^n| = |z/\alpha|^n |a_n \alpha^n| \leqslant K|z/\alpha|^n$$

and $\{K|z/\alpha|^n\}$ is the sequence of terms of a convergent series, as $\sum |z/\alpha|^n$ is a geometric series with $|z/\alpha| < 1$. Hence applying the comparison test to $\sum |a_n z^n|$, the series $\sum a_n z^n$ is absolutely convergent.

From this it follows that if there is some point $\beta$ such that $\sum a_n \beta^n$ is divergent then $\sum a_n z^n$ diverges for all $z$ with $|\beta| < |z|$. Convergence at $z$ would imply convergence at $\beta$. Thus if there is such a $\beta$ the set $C = \{z \mid \sum a_n z^n$ is convergent$\}$ is bounded. Let $r = \sup_C |z|$. Then by the definition of $C$, $\sum a_n z^n$ diverges for any $z$ with $|z| > r$. For any $z \neq 0$ with $|z| < r$ there is a number $s = \frac{1}{2}(r + |z|) < r$ such that $\sum a_n s^n$ is convergent. As $|z| < s$, it follows that $\sum a_n z^n$ is absolutely convergent.

Thus

(i)   $\sum n! \, z^n$ converges only for $z = 0$ and so $r = 0$,

(ii)   $\sum z^n / n!$ converges for all $z$, this situation is sometimes indicated by writing $r = \infty$,

(iii)   $\sum z^n$ converges only for $|z| < 1$, so that $r = 1$,

(iv)   $\sum z^n / n$ converges only if $|z| \leqslant 1$ but $z \neq 1$, so again $r = 1$,

(v)   $\sum z^n / n^2$ converges only if $|z| \leqslant 1$ and so $r = 1$.

Notice from these last three cases that the theorem says nothing about convergence on the circle $|z| = r$, and that a wide variety of things is possible.

The circle $|z| = r$ is known as the *circle of convergence*, the disk $\{z \mid |z| < r\}$ is known as the *disk of convergence*, and for a real power series the interval $(-r, r)$ is the *interval of convergence*.

Using similar arguments to those which prove Theorem 9.17, we get the next result.

**Theorem 9.18**   The two power series $\sum a_n z^n$ and $\sum n a_n z^n$ either both converge for all $z \in \mathbb{C}$ or have the same radius of convergence. The two power series $\sum a_n x^n$ and $\sum n a_n x^n$ either both converge for all $x \in \mathbb{R}$ or have the same radius of convergence.

*Proof* If $r$ is the radius of convergence of $\sum a_n x^n$ or $\sum a_n z^n$ and $|z| < r$ then there is a real number $\zeta$ between $|z|$ and $r$ such that $\sum a_n \zeta^n$ is convergent. Now for sufficiently large $n$, $\{n(|z|/\zeta)^n\}$ is decreasing, as $(n+1)(|z|/\zeta)^{n+1} < n(|z|/\zeta)^n$ follows from $(n+1)|z|/\zeta < n$ and so from $|z|/\zeta < n/(n+1)$, and $\lim_n n/(n+1) = 1 > |z|/\zeta$. Thus $\sum |a_n| \zeta^n n(|z|/\zeta)^n = \sum n|a_n||z|^n$ is convergent by Abel's test, so that $\sum n a_n z^n$ is absolutely convergent. On the other hand, if $\sum n a_n z^n$ is convergent, since $\{1/n\}$ is decreasing with $1/n > 0$, $\sum (1/n) n a_n z^n = \sum a_n z^n$ is convergent by Abel's test.

In the case in which the series $\sum a_n z^n$ converges at all points of $\mathbb{C}$, we can simply drop the requirement on $\zeta$ that it be less than $r$.

We also have a simple way to establish uniform convergence of a power series on a suitable set.

**Theorem 9.19** If $r$ is the radius of convergence of a power series, and $0 < \rho < r$, or the power series is convergent in the whole field and $0 < \rho$, then the power series converges uniformly on $\{z \mid |z| \leq \rho\}$ if it is a complex power series, or on $[-\rho, \rho] = \{x \mid |x| \leq \rho\}$ if it is a real power series.

*Proof* The series $\sum a_n \rho^n$ is convergent. Let $E$ be either $\{z \mid |z| \leq \rho\}$ or $[-\rho, \rho]$, as appropriate. Then for $t \in E$, $|a_n t^n| \leq |a_n| \rho^n$ all $n \in \mathbb{N}$. Thus

$$\sup_{t \in E} \left| \sum_{n=0}^{\infty} a_n t^n - \sum_{n=0}^{N} a_n t^n \right| = \sup_{t \in E} \left| \sum_{N+1}^{\infty} a_n t^n \right| \leq \sum_{N+1}^{\infty} |a_n| \rho^n$$

but as $\sum |a_n| \rho^n$ is convergent, $\lim_N \sum_{N+1}^{\infty} |a_n| \rho^n = 0$. Thus

$$\lim_N \sup_{t \in E} \left| \sum_{n=0}^{\infty} a_n t^n - \sum_{n=0}^{N} a_n t^n \right| = 0$$

or $\sum a_n t^n$ is uniformly convergent on $E$.

This leads immediately to the following important result.

**Theorem 9.20** Every power series represents a continuous function within its disk or interval of convergence, or if it converges in the whole of $\mathbb{R}$ or $\mathbb{C}$ a function continuous on $\mathbb{R}$ or $\mathbb{C}$.

*Proof* The functions $x^n$ and $z^n$ are continuous on $\mathbb{R}$ and $\mathbb{C}$ respectively for all $n \in \mathbb{N}$. Thus $a_n x^n$ and $a_n z^n$ are continuous on $\mathbb{R}$ and $\mathbb{C}$ respectively. Thus by the Corollary to Theorem 9.16 and Theorem 9.19 $\sum_{n=0}^{\infty} a_n x^n$ is continuous on $[-\rho, \rho]$ and $\sum_{n=0}^{\infty} a_n z^n$ is continuous on $\{z \mid |z| \leq \rho\}$, provided that $0 \leq \rho$ and $\rho <$ the radius of convergence if there is one. Thus $\sum_{n=0}^{\infty} a_n x^n$ is continuous on $\bigcup_{0 \leq \rho < r} [-\rho, \rho]$ or, if the series is convergent everywhere, on $\bigcup_{0 \leq \rho} [-\rho, \rho]$. Also $\sum_{n=0}^{\infty} a_n z^n$ is continuous on $\bigcup_{0 \leq \rho < r} \{z \mid |z| \leq \rho\}$ or, if the series is convergent everywhere, on $\bigcup_{0 \leq \rho} \{z \mid |z| \leq \rho\}$. It remains only to observe that the unions are the sets as stated.

**Exercises**

1   By using the relation $1/n - 1/(n+1) = 1/[(n+1)n]$, or otherwise, show that $\sum 1/[n(n+1)]$ is convergent and that $\sum_{n=1}^{\infty} 1/[n(n+1)] = 1$.

2   Show that $\sum_{n=1}^{\infty} 1/[n(n+2)] = \frac{3}{4}$.

3   Let $a_n = (n+1)/(n+2)$ for $n \in \mathbb{N}$ and $b_n = (n+2)/n$ for $n \in \mathbb{N}'$. Then

$$\frac{1}{n(n+1)} = \frac{n+2}{n} \cdot \frac{1}{(n+1)(n+2)} = b_n \left( \frac{n+1}{n+2} - \frac{n}{n+1} \right) = b_n(a_n - a_{n-1})$$

and

$$\frac{2}{n(n+2)} = \frac{n+1}{n+2} \cdot \frac{2}{n(n+1)} = a_n \left( \frac{n+2}{n} - \frac{n+3}{n+1} \right) = a_n(b_n - b_{n+1})$$

Then from questions 1 and 2

$$\sum_{n=1}^{\infty} b_n(a_n - a_{n-1}) = 1 \quad \text{and} \quad \sum_{n=1}^{\infty} a_n(b_n - b_{n+1}) = \frac{3}{2}$$

However, consider the following

$$
\begin{aligned}
\sum_{n=1}^{\infty} a_n(b_n - b_{n+1}) &= a_1(b_1 - b_2) + a_2(b_2 - b_3) + a_3(b_3 - b_4) + \cdots \\
&= a_1 b_1 - a_1 b_2 + a_2 b_2 - a_2 b_3 + \cdots \\
&= b_1 a_0 + b_1(a_1 - a_0) + b_2(a_2 - a_1) + \cdots \\
&= \sum_{n=1}^{\infty} b_n(a_n - a_{n-1}) + b_1 a_0
\end{aligned}
$$

As $\frac{3}{2} \neq 1 + \frac{3}{2} = 1 + b_1 a_0$, there must be something wrong with the argument. Which step is wrong and why?

4   The sequence $\{s_N\}$ is the sequence of partial sums of the series $\sum b_n$. Show that if the series $\sum s_N$ is convergent then so is $\sum b_n$ and $\sum_{n=0}^{\infty} b_n = 0$. Prove further that if $b_n \geq 0$ for $n \in \mathbb{N}$ then $b_n = 0$ for $n \in \mathbb{N}$.

5   The function $l_2$ is monotone and $l_2(2^n) = n$, $n \in \mathbb{N}'$. Determine for what values of the real number $\alpha > 0$ (if any) the following series converge:
   (i)   $\sum 1/(n[l_2(n)]^{\alpha})$,
   (ii)  $\sum 1/(n \cdot l_2(n) \cdot [l_2 \circ l_2(n)]^{\alpha})$,
   (iii) $\sum \alpha^{-l_2(n)}$.

6   Determine whether the following series converge:
   (i)   $\sum 1/[2n^2 + (-1)^n]$,
   (ii)  $\sum 1/[2n + (-1)^n]$,
   (iii) $\sum n/(1 + \sqrt{n})$,
   (iv)  $\sum (n+1)/(n^3 + n^2 + 2)$,
   (v)   $\sum a^n/(1 + a^{2n})$ for $a \in \mathbb{R}$.

**7** Determine whether the following series converge:
  (i) $\sum n^2/n!$,
  (ii) $\sum n2^n/3^n$,
  (iii) $\sum n!/(n^2 2^n)$,
  (iv) $\sum (n!)^2/(2n)!$,
  (v) $\sum n!/(2000)^n$,
  (vi) $\sum (n!)^2/[2(n!)]$,
  (vii) $\sum [(n-2)/(2n)]^n$,
  (viii) $\sum (1+1/n)^n$.

**8** Prove that if the series $\sum a_n$ is absolutely convergent then the series $\sum a_n^2$ is convergent. Give an example of a sequence $\{b_n\}$ such that $\sum b_n$ is convergent and $\sum b_n^2$ is divergent, and an example of a sequence $\{c_n\}$ such that $\sum c_n$ is divergent and $\sum c_n^2$ is convergent.

**9** The sequence $\{b_n\}$ is defined by $p(n)=(-1)^n$ and $b_n=p(n)/n^{2+p(n)}$ for $n \in \mathbb{N}'$. Show that $b_n=(-1)^n B_n$ defines $B_n$ as a positive number for $n \in \mathbb{N}'$, where $\lim_n B_n = 0$, but that $\sum b_n$ is divergent. Which condition of the Leibnitz test is not satisfied?

**10** Find the radius of convergence, or else prove convergence for all $x$, of the following series:
  (i) $\sum (2^n/n!)x^n$,
  (ii) $\sum [1-(-1)^n]x^n$,
  (iii) $\sum (nx)^n$,
  (iv) $\sum (n^n/n!)x^n$,
  (v) $\sum [(2n+1)/(2n)]x^n$,
  (vi) $\sum (z+1)^n$.
[Note that (vi) is a power series since $(z+1)(w)=z(w+1)$, and that the limit $\lim_n [(n+1)/n]^n$ has been discussed in connection with Theorem 7.7.]

**11** The numbers $r, R>0$ are the radii of convergence of $\sum a_n x^n$ and $\sum b_n x^n$ respectively. By considering a number $z=xy$ where $0<x<r$, $0<y<R$, or otherwise, prove that the series $\sum a_n b_n x^n$ is convergent for all $x \in (-rR, rR)$.
  By taking

$$a_n = \begin{cases} 1 & (n \text{ even}) \\ 0 & (n \text{ odd}) \end{cases}$$

and making a suitable choice for $\{b_n\}$, prove that it can happen that $\sum a_n b_n x^n$ is convergent for some $x$ with $x>rR$.

# 10

# Integration

There are two ideas, which on the surface are quite different, but are both known as integration. The first is simply that of reversing the process of differentiation. The second is the definition and calculation of areas, by a type of limiting process. It is the result, at a deeper level, that these two different ideas often lead to the same answers, which makes the differential and integral calculus such a powerful tool.

## The reverse of differentiation

**Definition 10.1**   Suppose that it is possible to find a function $F$ and a set $E$ such that $F'(x) = f(x)$ if $x \in E$ and $E \subset \mathcal{D}f$. Then $F$ is a *primitive of $f$ on $E$*.

We know that if $c$ is a constant function on $E$ then $c' = 0$. Also by Theorem 8.2(i) $(F + c)'(x) = (F' + c')(x) = f(x) + 0 = f(x)$ if $x \in E$; that is if $F$ is any primitive of $f$ on $E$ then any function of the form $F + c$ is a primitive of $f$ on $E$. Now if $F$ and $G$ are both primitives of $f$ on an interval $I$ then $(F - G)' = F' - G' = f - f = 0$ on $I$, and so (because $I$ is an interval) $F - G$ is a constant function $c$, say, on $I$ by the mean value theorem, Theorem 8.4, and $G = F - c$. Thus if $F$ is a primitive of $f$ on an interval $I$ then the set of all primitives of $f$ on $I$ is $\{G \mid G(x) = F(x) + c$ if $x \in I$, for some $c \in \mathbb{R}\}$.

If $E$ is not an interval then it is not possible to find all the primitives of $f$ by adding constant functions to $F$. Thus if $E = \{x \mid 0 < |x| < 1\}$ and $f(x) = 0$ all $x \in E$ then any function $G$ given by

$$G(x) = \begin{cases} c_1 & (0 < x < 1) \\ c_2 & (-1 < x < 0) \end{cases}$$

is a primitive of $f$. So it is convenient to restrict our attention to primitives on intervals. Also it is clear that the values of a primitive outside the interval are not of interest.

**Definition 10.2**   The set of all primitives $F$ of a function $f$ on an interval $I \subset \mathcal{D}f$ such that $\mathcal{D}F = I$ is called the *indefinite integral of $f$ on $I$*, or if $I = \mathcal{D}f$, the *indefinite integral of $f$*, and is written $\int f = \int f(x)\, dx$.

Notice that if two functions $F$ and $G$ with $\mathcal{D}F = \mathcal{D}G$ are defined to be equivalent $F \sim G$, iff $F(x) = G(x) + c$ all $x \in I$ for some constant $c$, then $\sim$ is an equivalence relation, and $\int f$ an equivalence class.

Also notice that $\int f$ may be $\varnothing$. Thus if $I = [0, 1]$ and $f$ is given by

$$f(x) = \begin{cases} 0 & (0 \leqslant x < \frac{1}{2}) \\ 1 & (\frac{1}{2} \leqslant x \leqslant 1) \end{cases}$$

then if $F \in \int f$ we would have $F'(0) = 0$, $F'(1) = 1$, and so by the intermediate value theorem for derivatives, Theorem 8.7, there would have to be a point $\xi \in (0, 1)$ with $F'(\xi) = \frac{1}{3}$. However, there is no $\xi \in (0, 1)$ such that $f(\xi) = \frac{1}{3}$. In fact it is clear that if $f$ has any 'jump discontinuity' then $\int f = \varnothing$. Here $f$ has a jump discontinuity at $a$, provided that $f$ is not continuous on $\{a\}$, but for $g$ and $h$, defined by $g(x) = f(x)$ for $x \in (\cdot, a) \cap \mathcal{D}f$ and $h(x) = f(x)$ for $x \in (a, \cdot) \cap \mathcal{D}f$, both the limits $\lim_{x \to a} g(x)$ and $\lim_{x \to a} h(x)$ exist.

However, some discontinuous functions have non-empty indefinite integrals. As was indicated, the function $g$ given by

$$g(x) = \begin{cases} x + x^2 \sin(1/x) & (x \neq 0) \\ 0 & (x = 0) \end{cases}$$

has $\quad g'(x) = \begin{cases} 1 + 2x \sin(1/x) - \cos(1/x) & (x \neq 0) \\ 1 & (x = 0) \end{cases}$

Thus $g'$ is not continuous at $0$, but $g \in \int g'$.

The rules of Theorem 8.2 for calculating derivatives provide rules for calculating indefinite integrals.

**Theorem 10.1**  If at least one of $\int f$ and $\int g$ is not empty, and $\mathcal{D}f = \mathcal{D}g$ is an interval, then $\int f + \int g = \int (f + g)$.

*Proof*  If $F \in \int f$ and $G \in \int g$ then $F' = f$ and $G' = g$. By Theorem 8.2(i) $(F + G)' = F' + G' = f + g$, so that $F + G \in \int (f + g)$.

The result is still valid if one of $\int f$ or $\int g$ is empty. As then $\int f + \int g$ is empty, and $\int (f + g) = \varnothing$ because if $\int f \neq \varnothing$ then there is some $F \in \int f$ and $F' = f$, but $H' = (f + g)$ would imply $(H - F)' = H' - F' = f + g - f = g$, so that there is no $H \in \int (f + g)$.

If we wish to use this result when it is not true that $\mathcal{D}f = \mathcal{D}g$ is an interval, but there is an interval $I \subset \mathcal{D}f \cap \mathcal{D}g$, then we define $f_1$ and $g_1$ by $f_1(x) = f(x)$ iff $x \in I$ and $g_1(x) = g(x)$ iff $x \in I$.

**Theorem 10.2**  If $\alpha \neq 0$ then $\alpha \int f = \int \alpha f$.

*Proof*  If $F \in \int f$ then Theorem 8.2(ii) gives $(\alpha F)' = \alpha F' = \alpha f$. Also if $G \in \int \alpha f$ then $G' = \alpha f$, so that, since $\alpha \neq 0$, we have $(1/\alpha)G' = f$ giving $(1/\alpha)G \in \int f$ or $G \in \alpha \int f$.

If $\alpha = 0$ the result is not true, as $0 \int f = \{0\}$ unless $\int f = \varnothing$; but if $G \in \int \alpha f$ then one of $G$ or $G + 1$ is not the zero function.

With this theorem we can now evaluate indefinite integrals of the form

$$\int (a_n x^n + a_{n-1} x^{n-1} + \cdots + a_0) \ni \frac{a_n}{n+1} x^{n+1} + \frac{a_{n-1}}{n} x^n + \cdots + a_0 x$$

We also get in general $\int x^\alpha \ni x^{\alpha+1}/(\alpha+1)$, provided that $\alpha \neq -1$. Thus $\frac{1}{2} x^2 \in \int x$, $1/x \in \int -1/x^2$, $2\sqrt{x} \in \int 1/\sqrt{x}$.

**Theorem 10.3    Integration by parts**    If $\mathscr{D}f = \mathscr{D}g = I$ and $g$ is differentiable on the interval $I$ then for $F \in \int f$, $\int f \cdot g = \{F \cdot g\} - \int F \cdot g'$.

*Proof*  If $H_1 \in \int f \cdot g$ then $(F \cdot g - H_1)' = f \cdot g + F \cdot g' - f \cdot g = F \cdot g'$, and so $F \cdot g - H_1 \in \int F \cdot g'$ or

$$H_1 \in \{F \cdot g\} - \int F \cdot g'$$

If $H_2 \in \{F \cdot g\} - \int F \cdot g'$ then $H_2' = f \cdot g - F \cdot g' + F \cdot g' = f \cdot g$. Therefore

$$H_2 \in \int f \cdot g$$

Combining the results we have $\int f \cdot g = \{F \cdot g\} - \int F \cdot g'$.

The use of integration by parts is not as straightforward as the use of the product rule for derivatives, of which it is the reverse. For this reason there are many ways of using integration by parts. The following case illustrates the simplest form of use.

By differentiating $(x^2+1)^{3/2}$ we get that

$$(x^2+1)^{3/2} \in \int 3x(x^2+1)^{1/2}$$

and by differentiating $(x^2+1)^{5/2}$ we get that

$$(x^2+1)^{5/2} \in \int 5x(x^2+1)^{3/2}$$

Taking these results as known, we can integrate by parts to find

$$\int 3x^3(x^2+1)^{1/2} = \{x^2(x^2+1)^{3/2}\} - \int 2x(x^2+1)^{3/2}$$

$$\ni x^2(x^2+1)^{3/2} - \tfrac{2}{5}(x^2+1)^{5/2}$$

$$= (\tfrac{3}{5}x^2 - \tfrac{2}{5})(x^2+1)^{3/2}$$

The following theorem is a help in finding integrals like $\int 5x(x^2+1)^{5/2}$.

**Theorem 10.4    Change of variable and substitution rules**    If $\mathscr{D}f = \mathscr{R}g$ and $g$ is differentiable and if $F \in \int f$ then:
(i)    $F \circ g \in \int (f \circ g) \cdot g'$,
(ii)   if also $g$ has an inverse [since $g$ is differentiable, this is the same as $g$

being strictly monotone] and $H \in \int (f \circ g) \cdot g'$ then $F = H \circ g^{-1} + c$ for some constant function $c$.

*Note* It is an important point that $\mathcal{D}f$ and $\mathcal{D}g$ must be intervals in order that $\int f$ be defined. In use, we may have to define $f_1$ and $g_1$, as was commented after Theorem 10.1.

*Proof*
(i) If $F \in \int f$ then by Theorem 8.2(iii) $(F \circ g)' = f \circ g \cdot g'$. Hence $F \circ g \in \int f' \circ g \cdot g'$.
(ii) From (i) we have that all members of $\int f' \circ g \cdot g'$ are of the form $H + c_1$ where $c_1$ is a constant function on $\mathcal{D}g$. Suppose that $F \circ g = H + c_1$. Then $F = H \circ g^{-1} + c_1$ where we must now consider the function $c_1$ to be defined on $\mathcal{D}f$, but to take the same constant value.

The existence of functions belonging to $\int f \circ g \cdot g'$ does not ensure the existence of functions in $\int f$. Thus if $f$ is defined by

$$f(x) = \begin{cases} 1 & (x \geq 0) \\ -1 & (x < 0) \end{cases}$$

and $g$ by

$$g(x) = x^3$$

then $$\int f \circ g \cdot g' = \int f(x^3) 3x^2 \ni x^2 |x|$$

as $$\frac{d}{dx} x^2 |x| = \begin{cases} 3x^2 & (x \geq 0) \\ -3x^2 & (x < 0) \end{cases}$$

but $f$ has a jump discontinuity, and so we have $\int f = \varnothing$.

To see how Theorem 10.4 can be used, consider $\int x/(1+x^2)^{1/2}$. Put $g(x) = x^2$, so that $g'(x) = 2x$. We can write the integral as $\frac{1}{2} \int g'/(1+g)^{1/2}$. Thus if $f(t) = (1+t)^{-1/2}$, so that we can take $F(t) = 2(1+t)^{1/2}$, the integral is $\frac{1}{2} \int f \circ g \cdot g' \ni \frac{1}{2} F \circ g = (1+g)^{1/2} = (1+x^2)^{1/2}$.

As a less obvious illustration, if we take $f$ given by

$$f(x) = (x^2 - 1)^{-3/2}$$

for $x > 1$ and $g$ given by

$$g(t) = \frac{1+t^2}{1-t^2}$$

for $0 \leq t < 1$ then $g^{-1}$ is given by

$$g^{-1}(x) = \sqrt{\left(\frac{x+1}{x-1}\right)}$$

for $x > 1$. This can be found simply by solving $g(t) = x$ for $t$. Hence to find $\int (x^2 - 1)^{-3/2}$, we calculate

$$\int \left[ \left( \frac{1 + t^2}{1 - t^2} \right)^2 - 1 \right]^{-3/2} \frac{4t}{(1 - t^2)^2} \, dt$$

$$= \int \left( \frac{1 + 2t^2 + t^4}{1 - 2t^2 + t^4} - \frac{1 - 2t^2 + t^4}{1 - 2t^2 + t^4} \right)^{-3/2} \frac{4t}{(1 - t^2)^2} \, dt$$

$$= \int \left( \frac{4t^2}{(1 - t^2)^2} \right)^{-3/2} \frac{4t}{(1 - t^2)^2} \, dt = \int \frac{(1 - t^2)^3}{8t^3} \frac{4t}{(1 - t^2)^2} \, dt$$

$$= \int \frac{1 - t^2}{2t^2} \, dt = \tfrac{1}{2} \int (t^{-2} - 1) \, dt$$

$$\ni -\tfrac{1}{2}(t^{-1} + t) = H \qquad \text{(say)}$$

Then $H[g^{-1}(x)] = -\tfrac{1}{2} \left[ \sqrt{\left( \dfrac{x-1}{x+1} \right)} + \sqrt{\left( \dfrac{x+1}{x-1} \right)} \right]$

$$= -\tfrac{1}{2} \frac{(x - 1) + (x + 1)}{\sqrt{(x^2 - 1)}} = -\frac{x}{\sqrt{(x^2 - 1)}}$$

It is now easy to verify, by differentiating $H \circ g^{-1}$, that this function does in fact belong to the required indefinite integral, which is thus not empty.

As indefinite integrals often lead to functions which we have not yet discussed, the techniques of evaluation are not further developed in this chapter.

## Area—the definite integral

Every real function $f$ with an interval $[a, b]$ in its domain, which is such that $f(x) \geq 0$ for $x \in [a, b]$, defines an *ordinate set*

$$A = \{ \langle x, y \rangle \mid x \in [a, b], 0 \leq y \leq f(x) \}$$

in $\mathbb{R} \times \mathbb{R}$. It is the area of this ordinate set which corresponds to the integral. If for at least some values of $x$ in $[a, b], f(x)$ is negative then there is also a non-empty set

$$B = \{ \langle x, y \rangle \mid x \in [a, b], f(x) \leq y \leq 0 \}$$

For such a function the integral corresponds to the difference: area of $A-$ area of $B$. The sets $A$ and $B$ are illustrated in Fig. 10.1.

The case in which $f(x) \geq 0$ for all $x \in [a, b]$ corresponds most directly to the notion of area, and so it will be considered first. Although the idea that a set $A$ has an area is intuitively a very strong one, unless the graph of $f$ consists only of bits of straight lines there is no obvious simple way of calculating a number 'area of $A$'. So the problem of defining the integral as an area is the problem of assigning a number to the set $A$ in such a way that it fits with the intuitive notion of area.

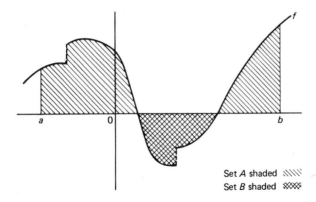

Set A shaded \\\\\
Set B shaded ✕✕✕✕

**Fig. 10.1** The sets A and B

The following properties of area will be enough to lead us to a definition. Area is given by a real valued function $\mu$ defined for some sets $A \subset \mathbb{R} \times \mathbb{R}$ and such that:

(i)  if $A$ is a set of the form $A = \{\langle x, y \rangle \mid x \in [a, b], \ m \leqslant y \leqslant M\}$—a closed bounded interval if $\mathbb{R} \times \mathbb{R}$ is thought of as $\mathbb{C}$—then $\mu(A) = (b - a) \cdot (M - m)$,

(ii)  if $A$ and $B$ are sets such that $\mu(A \cap B) = 0$ then $\mu(A \cup B) = \mu(A) + \mu(B)$,

(iii)  if $A \subset B$ then $\mu(A) \leqslant \mu(B)$.

On writing these in terms of ordinate sets, and using $\int_E f$ for $\mu(A)$ where $A = \{\langle x, y \rangle \mid x \in E, \ 0 \leqslant y \leqslant f(x)\}$, these properties give:

(1)  $\int_{[a,b]} k = k(b - a)$,

(2)  $\int_E f + \int_F f = \int_{E \cup F} f$, provided that $\int_{E \cap F} f = 0$, and as a special case, since by (1) $\int_{\{b\}} k = \int_{\{b\}} f = 0$,

$$\int_{[a,b]} f + \int_{[b,c]} f = \int_{[a,c]} f$$

(3)  $\int_E f \geqslant \int_E g$ if $f(x) \geqslant g(x)$ for all $x \in E$.

Any such function $\int$, defined for suitable pairs $\langle E, f \rangle$ and having values which are real numbers, will be called an *integral*.

We can establish some results for every such integral.

**Theorem 10.5**  If the real function $f$ is defined on $[a, b]$, and is such that $f(x) \in [0, M]$ and $F(x) = \int_{[a,x]} f$ for all $x \in [a, b]$, then the function $F$ is continuous on $[a, b]$.

*Proof*  We prove continuity at each $c \in [a, b]$.

From property (2) it follows that if $a \leqslant x_1 \leqslant x_2 \leqslant b$ then

$$F(x_2) - F(x_1) = \int_{[x_1, x_2]} f$$

Thus for any $x \in [a, b]$, $|F(x) - F(c)| = \int_I f$, where $I = [x, c] \cup [c, x]$ [that is $I$

is whichever of the intervals $[x, c]$ and $[c, x]$ is not empty]. From property (3)

$$\int_I f \le \int_I M = M|x - c|$$

by property (1). Since $\lim_{x \to c} M|x - c| = 0$, the squeeze theorem, Theorem 6.6, gives $\lim_{x \to c} \int_I f = 0$ and so $\lim_{x \to c} F(x) = F(c)$, that is $F$ is continuous on $\{c\}$.

**Theorem 10.6   Fundamental theorem of the calculus**   If the real function $f$ is defined on $[a, b]$, and is such that $f(x) \in [0, M]$ and $F(x) = \int_{[a, x]} f$ for all $x \in [a, b]$, then $F'(c) = f(c)$ at every $c \in [a, b]$ for which $f$ is continuous on $\{c\}$.

*Note*   Thus we have both the existence and the value of $F'(c)$.

*Proof*   As in Theorem 10.1, we have for $x > c$

$$F(x) - F(c) = \int_{[c, x]} f$$

and for $x < c$

$$F(x) - F(c) = -\int_{[x, c]} f$$

Thus for $x \in [a, b]$, $x \ne c$ we have

$$\frac{F(x) - F(c)}{x - c} = \frac{1}{|x - c|} \int_I f$$

where $I = [x, c] \cup [c, x]$.
   Now for $y \in I$

$$\inf\{f(t) \mid t \in I\} \le f(y) \le \sup\{f(t) \mid t \in I\}$$

where the inf and sup exist as $f([a, b]) \subset [0, M]$. Again, as in the previous theorem, we can use properties (1) and (3) to get

$$|x - c| \inf\{f(t) \mid t \in I\} \le \int_I f \le |x - c| \sup\{f(t) \mid t \in I\}$$

If $f$ is continuous on $\{c\}$ then $\lim_{x \to c} f(x) = f(c)$ and this implies that both

$$\lim_{x \to c} \inf\{f(t) \mid t \in I\} = f(c)$$

and   $$\lim_{x \to c} \sup\{f(t) \mid t \in I\} = f(c)$$

This is so because given $\varepsilon > 0$ we can choose $\delta_\varepsilon$ so that

$$|f(t) - f(c)| < \varepsilon/2 \quad \text{for } |t - c| < \delta_\varepsilon$$

and hence $\delta_\varepsilon$ is such that

$$\left|\inf\{f(t)\,|\,t\in I\}-f(c)\right|\le\varepsilon/2<\varepsilon$$

and $\quad\left|\sup\{f(t)\,|\,t\in I\}-f(c)\right|\le\varepsilon/2<\varepsilon$

for $|t-c|<\delta_\varepsilon$.

So now by the squeeze theorem, Theorem 6.6,

$$\lim_{x\to c}\frac{1}{|x-c|}\int_I f=f(c)$$

that is $F'(c)=f(c)$. [This proof is illustrated in Fig. 10.2.]

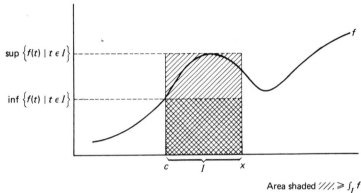

sup $\{f(t)\,|\,t\in I\}$

inf $\{f(t)\,|\,t\in I\}$

Area shaded ⁄⁄⁄ $\ge\int_I f$
Area shaded ▨▨ $\le\int_I f$

**Fig. 10.2** Fundamental theorem of the calculus

So now we have linked up the two ideas of integral. We have that if $f$ is continuous then $\int_{[a,x]}f\in\int f$.

Note that for $F'(c)$ to be defined we need $a<b$.

We seek to use property (1) to construct an integral. In order to keep the distinction between sets in $\mathbb{R}$ and $\mathbb{R}\times\mathbb{R}$ clear, we make the following definition.

**Definition 10.3** A set $E\subset\mathbb{R}\times\mathbb{R}$ of the form

$$E=\{\langle x,y\rangle\,|\,x\in[a,b],\,y\in[m,M]\}=[a,b]\times[m,M]$$

is called a *rectangle* or a *rectangular set*. The number $(b-a)(M-m)$ is written $|E|$ and called the *elementary area of E*.

One obvious way of trying to calculate $\mu(A)=\int_E f$ is to consider a collection of rectangular sets $\{E_k\}$ with $E_k=[a_k,b_k]\times[m_k,M_k]$ such that $A\subset\bigcup_k E_k$. Then we would expect that

$$\mu(A)\le\sum_k(b_k-a_k)\cdot(M_k-m_k)=\sum_k|E_k|$$

This suggests the following definition.

**Definition 10.4** For every set $A \subset \mathbb{R} \times \mathbb{R}$, a finite or countable collection of rectangular sets $\{E_k \mid k \in \mathbb{N}\}$ for which $A \subset \bigcup_{k \in \mathbb{N}} E_k$, will be called a *rectangular covering* of $A$ and denoted by $\mathbb{S}(A)$. [Notice that every collection $\{E_k \mid k \in \mathbb{N}\}$ of rectangles is a rectangular covering for some set $A$, as we may take $A = \bigcup_k E_k$.] The number $\sum_{k \in \mathbb{N}} |E_k|$ when it exists will be written $|\mathbb{S}(A)|$ and called the *sum* for $\mathbb{S}(A)$.

Suppose that we wish to restrict our considerations to only some of the possible rectangular coverings. Thus on occasion we want to use only coverings by a finite number of rectangles. Then what we have is a set $\mathbb{G}$ of rectangular coverings, that is a set $\mathbb{G}$ consisting of collections of rectangles $\{E_k \mid k \in \mathbb{N}\}$. Thus sometimes we will have $\mathbb{G}$ given by $\{E_k \mid k \in \mathbb{N}\} \in \mathbb{G}$ iff $\{E_k \mid k \in \mathbb{N}\}$ has a finite number of members $E_k$.

Only certain sets $\mathbb{G}$ are found to be really useful. Basically we need $\mathbb{G}$ to contain enough coverings.

**Definition 10.5** A set $\mathbb{G}$ consisting of finite or countable collections of rectangles will be called a *covering system* iff for every $M \geqslant 0$ and every $a, b \in \mathbb{R}$ we have $\{[a, b] \times [0, M]\} \in \mathbb{G}$. If furthermore for real numbers $\alpha$ and rectangles $E = [x_1, x_2] \times [y_1, y_2]$ we write $E^\alpha = [x_1, x_2] \times [y_1 + \alpha, y_2 + \alpha]$, and for any rectangular covering of a set $A$, $\mathbb{S}(A)$, we write $\mathbb{S}^\alpha(A) = \{E^\alpha \mid E \in \mathbb{S}(A)\}$, then a covering system $\mathbb{G}$ will be called a *regular covering system* or simply a *regular system*, provided that:
(i) whenever $A \subset [a, b] \times [0, \cdot)$, $\mathbb{S}(A) \in \mathbb{G}$ and $\alpha > 0$ then $\mathbb{S}^\alpha(A) \cup \{[a, b] \times [0, \alpha]\} \in \mathbb{G}$,
(ii) whenever $\mathbb{S}(A) \in \mathbb{G}$ and $\alpha > 0$ then $\{E^{-\alpha} \cap (\cdot, \cdot) \times [0, \cdot) \mid E \in \mathbb{S}(A)\} \in \mathbb{G}$.

The conditions (i) and (ii) are illustrated in Fig. 10.3.
Now in terms of a system of coverings $\mathbb{G}$, we can set up a definition of an integral.

**Definition 10.6** Given a covering system $\mathbb{G}$, then for any real function $f$ defined on a set $F \subset [a, b]$ with $f(F) \subset [0, M]$ and $A = \{\langle x, y\rangle \mid x \in F, y \in [0, f(x)]\}$ [so that there is at least one rectangle set $\{[a, b] \times [0, M]\}$ for which $|\mathbb{S}(A)|$ is defined] the number $\inf\{|\mathbb{S}(A)| \mid \mathbb{S}(A) \in \mathbb{G}, |\mathbb{S}(A)|$ is defined$\}$ is called the $\mathbb{G}$ *upper integral of $f$ over $F$* and written $\mathbb{G} \int_F f$.

Property (3) is immediate, as if $f(x) \geqslant g(x)$ for $x \in F$ then $B = \{\langle x, y\rangle \mid x \in F, y \in [0, g(x)]\} \subset A$, so that any $\mathbb{S}(A)$ is also an $\mathbb{S}(B)$, and the inf of all $|\mathbb{S}(B)|$ is thus at most the inf of all $|\mathbb{S}(A)|$.

The two most important upper integrals are, firstly, the one which allows the closest approximation to $A$ by putting no restriction on what rectangular coverings may be used and, secondly, the one which in many ways is the easiest to prove results about, as it only uses finite coverings. They are named

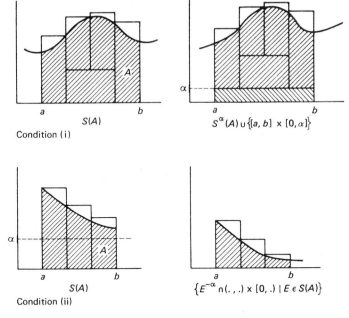

**Fig. 10.3** Conditions (i) and (ii)

after the mathematicians who were largely responsible for developing their properties and those of the associated integrals.

**Definition 10.7**
(i)  If $\mathbb{L} = \{\mathbb{S}(A) | A \subset \mathbb{R} \times \mathbb{R}, |\mathbb{S}(A)| \text{ is defined}\}$ then $\mathbb{L} \, \overline{\int}_{[a,b]} f$ is called the *Lebesgue* upper integral of $f$ over $[a, b]$.
(ii)  If $\mathbb{R}_i = \{\mathbb{S}(A) | A \subset \mathbb{R} \times \mathbb{R}, \mathbb{S}(A) \text{ has only a finite number of members}\}$ then $\mathbb{R}_i \, \overline{\int}_{[a,b]} f$ is called the *Riemann* upper integral of $f$ over $[a, b]$.

It is easy to check that both $\mathbb{L}$ and $\mathbb{R}_i$ are regular systems.

One thing to be noticed about Definition 10.4 is that $|\mathbb{S}(A)|$ need not correspond to the 'area' of $\bigcup_\mathbb{N} E_k$. The point is that the rectangular sets $E_k$ have been allowed to overlap, that is it could be that $|E_k \cap E_j| \neq 0$ for some $k \neq j$. This is illustrated in Fig. 10.4. However, this possibility is not important for the value of the upper integral, provided that $\mathbb{G}$ contains sufficient rectangular coverings of $A$ with no such overlapping.

As a first step towards seeing if we have constructed an integral, we show that the $\mathbb{G}$ upper integrals all satisfy property (1), so that we get the desired answer in the case of a constant function. However, since countable collections of rectangles are used, the proof is not quite as simple as might be expected.

**Theorem 10.7**  If the function $f$ is given by $f(x) = c \geqslant 0$ for $x \in [a, b]$ then $\mathbb{G} \, \overline{\int}_{[a,b]} f = c(b - a)$.

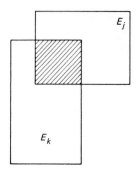

**Fig. 10.4**   $E_j$ and $E_k$ overlap

*Proof*   Let $A = [a, b] \times [0, c]$ and $\mathbb{S}(A)$ be any rectangular covering of $A$. For each rectangle $E \in \mathbb{S}(A)$ write

$$E = \{\langle x, y \rangle \mid -\alpha \leqslant x - x_0 \leqslant \alpha, -\beta \leqslant y - y_0 \leqslant \beta\}$$

so that $E$ is the rectangle centred at $\langle x_0, y_0 \rangle$ with sides $2\alpha$ and $2\beta$. Then let

$$E(\eta) = \mathcal{N}(x_0, \eta\alpha) \times \mathcal{N}(y_0, \eta\beta)$$

with $\eta > 1$ and

$$\bar{E}(\eta) = \{\langle x, y \rangle \mid -\eta\alpha \leqslant x - x_0 \leqslant \eta\alpha, -\eta\beta \leqslant y - y_0 \leqslant \eta\beta\}$$

so that $E \subset E(\eta) \subset \bar{E}(\eta)$ and $E(\eta)$ is open. Also $|E| = 4\alpha\beta$, while $|\bar{E}(\eta)| = \eta^2 4\alpha\beta$.

The collection of sets $\{E(\eta) \mid E \in \mathbb{S}(A)\}$ is an open covering of the set $A$. So by considering $\mathbb{R} \times \mathbb{R}$ to be the same as $\mathbb{C}$ the Heine–Borel theorem, Theorem 4.8, can be applied. Thus there is a finite covering $\{E_1(\eta), E_2(\eta), \ldots, E_k(\eta)\}$ of $A$.

Now if any rectangle $R = [x_1, x_2] \times [y_1, y_2]$ is written as the union of $R_1 = [x_1, x_2] \times [y_1, y_3]$ and $R_2 = [x_1, x_2] \times [y_3, y_2]$, or as the union of $R_3 = [x_1, x_3] \times [y_1, y_2]$ and $R_4 = [x_3, x_2] \times [y_1, y_2]$, then by direct calculation

$$|R| = |R_1| + |R_2| = |R_3| + |R_4|$$

So the rectangles $\bar{E}_1(\eta), \ldots, \bar{E}_k(\eta)$ can be divided up into smaller rectangles in such a way that $|A|$ is the sum of the elementary areas of just some of these smaller rectangles. [This is illustrated in Fig. 10.5.]

Thus

$$|A| \leqslant \sum_{l=1}^{k} |\bar{E}_l(\eta)| = \eta^2 \sum_{l=1}^{k} |E_l|$$

where $E_l$ is the rectangle from which $E_l(\eta)$ was constructed. Hence $|A| \leqslant \eta^2 |\mathbb{S}(A)|$, but this holds for all $\eta > 1$, and so $|A| \leqslant |\mathbb{S}(A)|$. However, we can take $\mathbb{S}(A) = \{A\}$, as we have assumed that rectangles like $A$ belong to $\mathbb{G}$.

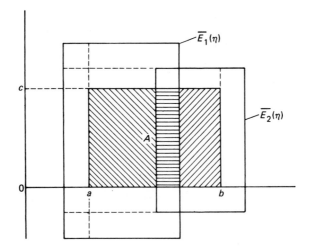

**Fig. 10.5** *A as the union of the smaller rectangles*

Thus $|\mathbb{S}(A)| = |A|$, and so

$$\mathbb{G} \int_{[a,b]}^{-} f = c(b-a)$$

**Corollary to Theorem 10.7** If $\mathbb{G}$ is a regular system then for any function $f$ with $0 \leqslant f(x) \leqslant M$ for all $x \in [a, b]$ and with $c \geqslant 0$ we have

$$\mathbb{G} \int_{[a,b]}^{-} (f+c) = \mathbb{G} \int_{[a,b]}^{-} f + c(b-a)$$

*Proof* Write

$$A = \{\langle x, y \rangle \mid y \in [0, f(x)], x \in [a, b]\}$$

and $\quad B = \{\langle x, y \rangle \mid y \in [0, f(x) + c], x \in [a, b]\}$

Then if $\mathbb{S}(A) \in \mathbb{G}$ is a covering of $A$ by rectangles there is a corresponding covering of $B$, $\mathbb{S}_1(B) = \mathbb{S}^c(A) \cup \{[a, b] \times [0, c]\} \in \mathbb{G}$, because $\mathbb{G}$ is a regular system. Now $|\mathbb{S}(A)| + c(b-a) = |\mathbb{S}_1(B)|$. Thus

$$\mathbb{G} \int_{[a,b]}^{-} (f+c) \leqslant \mathbb{G} \int_{[a,b]}^{-} f + c(b-a) \tag{1}$$

Also if $\mathbb{S}_1(B) \in \mathbb{G}$ is a covering of $B$ by rectangles then there is a corresponding covering of $A$, $\mathbb{S}(A) = \{E^{-c} \cap [(\cdot, \cdot) \times [0, \cdot)] \mid E \in \mathbb{S}_1(B)\} \in \mathbb{G}$, as $\mathbb{G}$ is a regular system. Now $\mathbb{S}^c(A)$ is a rectangular covering of $B \setminus ([a, b] \times [0, c))$, so that, using the theorem, $|\mathbb{S}_1(B)| \geq |\mathbb{S}(A)| + c(b - a)$. Thus

$$\mathbb{G} \overline{\int}_{[a,b]} (f + c) \geq \mathbb{G} \overline{\int}_{[a,b]} f + c(b - a) \tag{2}$$

By eqs (1) and (2) we must have equality.

Now we saw that if $f(x) \geq g(x)$ for all $x$ in some set $E$ then

$$A = \{\langle x, y\rangle \mid y \in [0, f(x)], x \in E\} \supset \{\langle x, y\rangle \mid y \in [0, g(x)], x \in E\} = B$$

and any rectangular covering $\mathbb{S}(A)$ is also a rectangular covering of $B$. So $\mathbb{G} \overline{\int}_E f \geq \mathbb{G} \overline{\int}_E g$ and property (3) holds. Also, provided that $\mathbb{G}$ satisfies some mild condition such as: if $\mathbb{S}(A) \in \mathbb{G}$ then $\{E \backslash ([a, b] \times \mathbb{R}) \mid E \in \mathbb{S}(A)\} \in \mathbb{G}$ for every interval $[a, b]$, it is clear that property (2) holds when the $\mathbb{G}$ upper integral is defined only over intervals.

However, we do require an integral for which property (2) holds more generally. Thus if $f$ is the characteristic function of $E \subset [a, b]$, that is

$$f(x) = \begin{cases} 1 & (x \in E) \\ 0 & (x \notin E) \end{cases}$$

and $g$ given by

$$g(x) = \begin{cases} 1 & (x \in [a, b] \backslash E) \\ 0 & (x \notin [a, b] \backslash E) \end{cases}$$

is the characteristic function of $[a, b] \backslash E$, we would expect to get $\int_E f = \int_{[a,b]} f$, $\int_{[a,b] \backslash E} 1 = \int_{[a,b] \backslash E} g = \int_{[a,b]} g$, $\int_{[a,b]} (f + g) = \int_{[a,b]} 1 = b - a$, and $\int_E 1 + \int_{[a,b] \backslash E} 1 = \int_{[a,b]} 1$.

We work towards results of this sort for fairly general sets $E$. They will be discussed in the form with all integrals taken over an interval: $\int_{[a,b]} (f + g) = \int_{[a,b]} f + \int_{[a,b]} g$. We consider not only upper integrals which supply numbers which should be greater than or equal to $\mu(A)$, but in order to approximate to $\mu(A)$ from the other side, we define also lower integrals. We make use of the fact that $f$ is assumed to be bounded on $[a, b]$, and of the Corollary to Theorem 10.7.

**Definition 10.8**  If $\mathbb{G}$ is a regular covering system then for any real function $f$ defined on an interval $[a, b]$ with $f([a, b]) \subset [0, M]$ the $\mathbb{G}$ *lower integral of $f$ over $[a, b]$* is given by

$$\mathbb{G} \underline{\int}_{[a,b]} f = M(b - a) - \mathbb{G} \overline{\int}_{[a,b]} (M - f)$$

Figure 10.6 illustrates this definition.

In the cases $\mathbb{G} = \mathbb{L}$ and $\mathbb{G} = \mathbb{R}_i$, we talk of the Lebesgue lower integral and the Riemann lower integral.

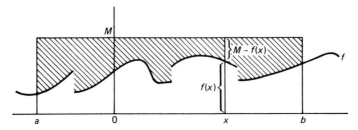

**Fig. 10.6** The lower integral

The value of the lower integral would depend on the particular choice of $M$. However, as $\mathbb{G}$ is assumed to be regular, the Corollary to Theorem 10.7 shows that we do not have dependence on $M$. It is for this reason that we here limit the definition to lower integrals over intervals.

We relate the upper and lower integrals.

**Theorem 10.8** If $\mathbb{G}$ is a regular system and $f$ is a real function defined on $[a, b]$ with $f(x) \in [0, M]$ for $x \in [a, b]$ then

$$\mathbb{G} \underline{\int}_{[a,b]} f \leqslant \mathbb{G} \overline{\int}_{[a,b]} f$$

*Proof* Let

$$A = \{\langle x, y \rangle \mid x \in [a, b], 0 \leqslant y \leqslant f(x)\}$$

and

$$B = \{\langle x, y \rangle \mid x \in [a, b], M - f(x) \leqslant y \leqslant M\}$$

Then any rectangular coverings $\mathbb{S}(A)$ of $A$ and $\mathbb{S}(B)$ of $B$ give rise to a rectangular covering $\mathbb{S}(A) \cup \mathbb{S}(B)$ of $A \cup B = [a, b] \times [0, M]$. Thus

$$|\mathbb{S}(A)| + |\mathbb{S}(B)| \geqslant |\mathbb{S}(A) \cup \mathbb{S}(B)|$$

[there might be some rectangles common to both coverings] and, as $\mathbb{S}(A) \cup \mathbb{S}(B) \in \mathbb{L}$,

$$|\mathbb{S}(A)| + |\mathbb{S}(B)| \geqslant \mathbb{L} \overline{\int}_{[a,b]} M = M(b - a)$$

by Theorem 10.7. Since this is true for all $\mathbb{S}(A) \in \mathbb{G}$ and $\mathbb{S}(B) \in \mathbb{G}$, we have

$$\mathbb{G} \overline{\int}_{[a,b]} f + \mathbb{G} \overline{\int}_{[a,b]} (M - f) \geqslant M(b - a)$$

that is

$$\mathbb{G} \overline{\int}_{[a,b]} f \geqslant \mathbb{G} \underline{\int}_{[a,b]} f$$

It would be nice if the graph of the function (the set $\{\langle x, y\rangle \mid y = f(x)\}$) had zero area, so that we always got

$$G\overline{\int}_{[a,b]} f + G\overline{\int}_{[a,b]} (M - f) = M(b - a)$$

Then we would have

$$G\overline{\int}_{[a,b]} f = G\underline{\int}_{[a,b]} f \tag{*}$$

Since the more rectangular coverings we allow ourselves to use, the smaller is the upper integral and the larger is the lower integral, we have most hope of getting eq. (*) to hold for $G = L$. Unfortunately, the problem of whether eq. (*) holds in the case $G = L$ turns out to be very deep. In order to 'construct' a function $f$ for which eq. (*) is false, it is necessary to make infinitely many arbitrary choices. Indeed, uncountably many arbitrary choices are required. So although a proof that eq. (*) holds generally is beyond us, we may expect to be able to establish it for most classes of functions we need to consider.

**Definition 10.9**   If $G$ is any regular system and $f$ is a real function defined on $[a, b]$, and if $0 \leqslant f(x) \leqslant M$ for $x \in [a, b]$ and $G\int_{[a,b]} f = G\int_{[a,b]} f$, then the common value is called the $G$ *integral of $f$ over* $[a, b]$. If $G = L$ it is the *Lebesgue integral*. If $G = R_i$ it is the *Riemann integral*.

Clearly, if $G_1 \supset G_2$ then any $G_2$ integral is also a $G_1$ integral. For we have

$$G_2\overline{\int}_{[a,b]} f \geqslant G_1\overline{\int}_{[a,b]} f \geqslant G_1\underline{\int}_{[a,b]} f \geqslant G_2\underline{\int}_{[a,b]} f$$

and so if the $G_2$ integral of $f$ over $[a, b]$ exists we must have equality throughout. In particular, every Riemann integral is also a Lebesgue integral.

The reason we consider the Riemann integral is that in many cases the nature of $R_i$ makes proofs easier. To get an idea of the way in which this works, let us consider the function $f$ given by

$$f(x) = \begin{cases} 2 & (x \text{ rational}) \\ 1 & (x \text{ irrational}) \end{cases}$$

In order to find $L\int_{[0,1]} f$, consider the rectangular covering constructed as follows. Let $\{r_k\}$ be the rational numbers in $[0, 1]$ formed into a sequence, and $\varepsilon > 0$ be any given number. Then

$$S_\varepsilon(A) = \{[0, 1] \times [0, 1]\} \cup \{[r_k - \varepsilon/2^k, r_k + \varepsilon/2^k] \times [1, 2]\}$$

is a rectangular covering of the ordinate set $A$ of $f$ over $[0, 1]$. Now $\sum_{k=1}^{\infty} 1/2^k = 1$, so that $|S_\varepsilon(A)| = 1 + 2\varepsilon$. However, $\varepsilon$ is any positive number and so we have $L\int_{[0,1]} f \leqslant 1$. Now $\{[0, 1] \times [0, 1]\}$ is a rectangular covering of the ordinate set of $2-f$, $\{\langle x, y \rangle \mid x \in [0, 1], 0 \leqslant y \leqslant 2-f(x)\}$, as

$$2-f(x) = \begin{cases} 0 & (x \text{ rational}) \\ 1 & (x \text{ irrational}) \end{cases}$$

Thus $\mathbb{L}\int_{[0,1]} f \geqslant 2-1 = 1$, so that $f$ is Lebesgue integrable over $[0, 1]$ and $\mathbb{L}\int_{[0,1]} f = 1$.

It would be nice to be able to use rectangular coverings in which all the rectangles were of the form $[x_1, x_2] \times [0, y]$. This form is suggested by the nature of an ordinate set. However, the set $F = [0, 1] \backslash \bigcup_{\mathbb{N}'} [r_k - \varepsilon/2^k, r_k + \varepsilon/2^k]$ contains no intervals other than points, and for sufficiently small $\varepsilon$ it has uncountably many points. To see this, suppose that $F$ is countable, $F = \{a_k \mid k \in \mathbb{N}'\}$, and consider the collection of intervals $\{[a_k - \varepsilon/2^k, a_k + \varepsilon/2^k]\} \cup \{[r_k - \varepsilon/2^k, r_k + \varepsilon/2^k]\}$. We have seen that for $\varepsilon$ sufficiently small the sum of the lengths of all these intervals is $<1$. But their union is supposed to contain $[0, 1]$. Hence $F$ cannot be countable, so that if we include all the rectangles $[r_k - \varepsilon/2^k, r_k + \varepsilon/2^k] \times [0, 2]$ in a rectangular covering, there have to be at least one of the rectangles $E = [r_i - \varepsilon/2^i, r_i + \varepsilon/2^i] \times [0, 2]$ and some rectangle $I$ not of this form such that $|E \cap I| \neq 0$, in the covering. So for the Lebesgue integral we have either to permit rectangles to overlap in this way, or else not to use as the standard form of rectangle $[x_1, x_2] \times [0, y]$. In contrast, the greater ease of getting results for the Riemann integral is shown by the following theorem.

**Theorem 10.9** If the real function $f$ is defined on $[a, b]$ and $f(x) \in [0, M]$ for all $x \in [a, b]$ then corresponding to any sum $|\mathbb{S}(A)|$, $\mathbb{S}(A) \in \mathbb{R}_i$ for the ordinate set $A$ of $f$ over $[a, b]$, there is a finite set of rectangles $\{[a, x_1] \times [0, y_1], [x_1, x_2] \times [0, y_2], \ldots, [x_{k-1}, b] \times [0, y_k]\} \in \mathbb{R}_i$ which is a rectangular covering of

$$A = \{\langle x, y \rangle \mid x \in [a, b], y \in [0, f(x)]\}$$

and for which, on writing $a = x_0$ and $b = x_k$,

$$\sum_{l=1}^{k} y_l(x_l - x_{l-1}) \leqslant |\mathbb{S}(A)|$$

*Proof* Suppose that $\mathbb{S}(A) = \{E_1, \ldots, E_n\}$ is a rectangular covering of $A$ and let $E_l' = E_l \cap ([a, b] \times [0, M])$. Then $\mathbb{S}_1(A) = \{E_1', \ldots, E_n'\}$ is a rectangular covering of $A$ such that $|\mathbb{S}_1(A)| \leqslant |\mathbb{S}(A)|$. Now as there is only a finite number of rectangles in $\mathbb{S}_1(A)$, we can eliminate overlapping. The process was described in the proof of Theorem 10.7. This yields an $\mathbb{S}_2(A)$ with $|\mathbb{S}_2(A)| \leqslant |\mathbb{S}_1(A)|$. Now if $[\alpha, \beta] \times [\gamma, \delta] \in \mathbb{S}_2(A)$ form $[\alpha, \beta] \times [0, \delta]$, and from each $E \in \mathbb{S}_2(A)$ for which $E \cap [\alpha, \beta] \times [0, \delta] \neq \varnothing$ form $E \backslash ([\alpha, \beta] \times [0, \delta])$ into at most two rectangles. This will require forming the union of $E \backslash ([\alpha, \beta] \times [0, \delta])$ with at most two rectangles of the form $F = [x, x] \times [\gamma_1, \delta_1]$, and $|F| = 0$. After a finite number of such steps, we get $\mathbb{S}_3(A)$ with $|\mathbb{S}_3(A)| = |\mathbb{S}_2(A)| \leqslant |\mathbb{S}(A)|$ and every rectangle of the form $[\alpha, \beta] \times [0, \delta]$, so that $\mathbb{S}_3(A)$ can be written as

$\{[a, x_1] \times [0, y_1], \ldots, [x_{k-1}, b] \times [0, y_k]\}$ and, using $a = x_0$ and $b = x_k$,

$$|S_3(A)| = \sum_{l=1}^{k} y_l(x_l - x_{l-1}) \leqslant |S(A)|$$

[Figure 10.7 illustrates two sorts of steps in the construction of $S_3(A)$.]

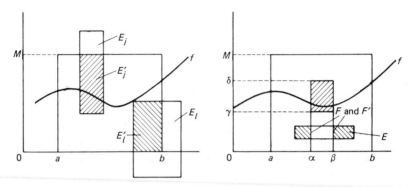

**Fig. 10.7** Construction of $S_3(A)$

**Corollary to Theorem 10.9** If the real function $f$ is defined on $[a, b]$ and $f(x) \in [0, M]$ for all $x \in [a, b]$ then corresponding to any sum $|S(B)|$, $S(B) \in \mathbb{R}_i$ for the ordinate set $B$ of $M - f$ over $[a, b]$, there is a finite set of rectangles $\{[a, x_1] \times [0, y_1], [x_1, x_2] \times [0, y_2], \ldots, [x_{k-1}, b] \times [0, y_k]\}$ for which, on writing $a = x_0$ and $b = x_k$,

$$\sum_{l=1}^{k} y_l(x_l - x_{l-1}) \geqslant M(b-a) - |S(B)|$$

*Proof* Apply the theorem to the function $M - f$. This yields the finite set of rectangles $\{[a, x_1] \times [0, M - y_1], \ldots, [x_{k-1}, b] \times [0, M - y_k]\}$, with

$$\sum_{l=1}^{k} (M - y_l)(x_l - x_{l-1}) = M(b-a) - \sum_{l=1}^{k} y_l(x_l - x_{l-1}) \leqslant |S(B)|$$

Hence

$$\sum_{l=1}^{k} y_l(x_l - x_{l-1}) \geqslant M(b-a) - |S(B)|$$

Notice that in the case of the sum approximating to the lower integral the rectangles of the type $[x_{l-1}, x_l] \times [y_l, M]$ form a covering of the set $\{\langle x, y \rangle \,|\, x \in [a, b], f(x) \leqslant y \leqslant M\}$. Hence the rectangles $[x_{l-1}, x_l] \times [0, y_l]$ are all contained in the set $A$. In particular, it follows from Theorem 10.9 and its corollary that given the set $\{a = x_0, x_1, \ldots, x_k = b\}$, and writing $m_l = \inf\{f(x) \,|\, x \in [x_{l-1}, x_l]\}$ and $M_l = \sup\{f(x) \,|\, x \in [x_{l-1}, x_l]\}$, the lower and upper Riemann integrals satisfy

$$\sum_{l=1}^{k} m_l(x_l - x_{l-1}) \leqslant \mathbb{R}_i \overline{\int}_{[a,b]} f \leqslant \mathbb{R}_i \underline{\int}_{[a,b]} f \leqslant \sum_{l=1}^{k} M_l(x_l - x_{l-1})$$

The sums at either end of this chain of inequalities are called respectively *Riemann lower* and *upper sums* for $f$ over $[a, b]$. These are illustrated in Fig. 10.8.

It is basically these inequalities which make it so much easier to prove results about the Riemann integral than about the Lebesgue integral.

The next result is valid for both Riemann and Lebesgue integrals. It is proved more generally.

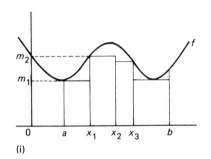 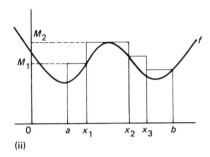

(i)                                        (ii)

**Fig. 10.8**   Riemann sums: (i) Riemann lower sum and (ii) Riemann upper sum

**Theorem 10.10**   Suppose that $\mathbb{G}$ is such that if $\{E_k\} \in \mathbb{G}$ then for every $\alpha > 0$, $\{\alpha E_k\} \in \mathbb{G}$. Then if the real function $f$ is defined on $[a, b]$ and $f(x) \in [0, M]$ for $x \in [a, b]$, for all real numbers $\alpha \geq 0$ we have

$$\mathbb{G} \overline{\int}_{[a,b]} \alpha f = \alpha \mathbb{G} \overline{\int}_{[a,b]} f$$

If furthermore $f$ has a $\mathbb{G}$ integral over $[a, b]$ then so does $\alpha f$ and

$$\mathbb{G} \int_{[a,b]} \alpha f = \alpha \mathbb{G} \int_{[a,b]} f$$

*Proof*   Let

$$A = \{\langle x, y \rangle \mid x \in [a, b], \ y \in [0, f(x)]\}$$

and    $$B = \{\langle x, y \rangle \mid x \in [a, b], \ y \in [0, \alpha f(x)]\}$$

Then if $\mathbb{S}(A)$ is a rectangular covering of $A$ and $\mathbb{S}(B) = \{E \mid E = [x_1, x_2] \times (\alpha[y_1, y_2])$ and $([x_1, x_2] \times [y_1, y_2]) \in \mathbb{S}(A)\}$, $\mathbb{S}(B)$ is a rectangular covering of $B$. Also, provided that $\alpha \neq 0$, we can start from any rectangular covering of $B$ and use the factor $1/\alpha$ in a similar construction to get a rectangular covering of $A$. Moreover $|\mathbb{S}(A)| = \alpha |\mathbb{S}(B)|$. Thus on taking the infima the required equation results. In the case $\alpha = 0$ we have $\mathbb{G} \int_{[a,b]} 0 = 0$, and so the equation also holds in this instance.

If now $\mathbb{G} \int_{[a,b]} f$ exists we have

$$\mathbb{G} \overline{\int}_{[a,b]} f = \mathbb{G} \underline{\int}_{[a,b]} f$$

but also

$$G \underline{\int}_{[a,b]} f = M \cdot (b-a) - G \overline{\int}_{[a,b]} (M-f)$$

so that

$$G \underline{\int}_{[a,b]} \alpha f = \alpha M \cdot (b-a) - \alpha G \overline{\int}_{[a,b]} (M-f) = \alpha G \underline{\int}_{[a,b]} f$$

Hence

$$G \overline{\int}_{[a,b]} \alpha f = G \underline{\int}_{[a,b]} \alpha f = \alpha G \underline{\int}_{[a,b]} f$$

The next theorem is more difficult to establish in a form suitable for the Lebesgue integral, but the result is more powerful, and is included to illustrate the greater ease with which the Lebesgue integral can be applied. The reader should not be worried if he does not fully understand the proof. A similar result with more restrictive conditions (Theorem 10.13) will be proved for the Riemann integral, and this will be adequate for the applications in this book.

A lemma is first required. It is an extension of results in Chapter 0, but could not be given there as it depends on the idea of a countable set.

**Lemma for Theorem 10.11**   If the sets $A$ and $B$, either both of real numbers or both of complex numbers, are both unions of countably many intervals then so is $A \cap B$.

*Proof*   Suppose that $\{A_n\}$ and $\{B_n\}$ are sets of intervals such that $A = \bigcup_{\mathbb{N}} A_n$ and $B = \bigcup_{\mathbb{N}} B_n$. Then $A_n \cap B_m$ is an interval for all $n, m \in \mathbb{N}$, and so, in view of Theorem 7.14, $\{A_n \cap B_m\}$ is a countable collection of intervals.

We now show that

$$\left( \bigcup_{\mathbb{N}} A_n \right) \cap \left( \bigcup_{\mathbb{N}} B_n \right) = \bigcup_{n,m \in \mathbb{N}} (A_n \cap B_m)$$

[This is a generalized distributive law.]

$$x \in \left( \bigcup_{\mathbb{N}} A_n \right) \cap \left( \bigcup_{\mathbb{N}} B_n \right)$$

$\Leftrightarrow$   $\exists n \in \mathbb{N}$ such that $x \in A_n$ and $\exists m \in \mathbb{N}$ such that $x \in B_m$

$\Leftrightarrow$   $\exists n, m \in \mathbb{N}$ such that $x \in A_n$ and $x \in B_m$

$\Leftrightarrow$   $\exists n, m \in \mathbb{N}$ such that $x \in A_n \cap B_m$

$\Leftrightarrow$   $x \in \bigcup_{n,m \in \mathbb{N}} (A_n \cap B_m)$

**Theorem 10.11**   If each of the real functions $f_n$, $n \in \mathbb{N}'$ is defined on $[a, b]$, has $f_n(x) \in [0, M]$ all $x \in [a, b]$ and has a Lebesgue integral over $[a, b]$, and if

$f_n(x) \leqslant f_{n+1}(x)$ for $n \in \mathbb{N}'$ and $x \in [a, b]$, then the function $f$ defined by $\lim_n f_n(x) = f(x)$ has a Lebesgue integral over $[a, b]$ and

$$\mathcal{L} \int_{[a,b]} f = \lim_n \mathcal{L} \int_{[a,b]} f_n$$

## Proof

(i)  We have $f_n(x) \leqslant f(x)$ all $n \in \mathbb{N}'$, $x \in [a, b]$. Hence

$$\mathcal{L} \underline{\int}_{[a,b]} f_n \leqslant \mathcal{L} \underline{\int}_{[a,b]} f$$

for $n \in \mathbb{N}'$. Now as the lower integrals of $f_n$ form an increasing sequence, the limit of these lower integrals exists and

$$\lim_n \mathcal{L} \underline{\int}_{[a,b]} f_n \leqslant \mathcal{L} \underline{\int}_{[a,b]} f \tag{1}$$

(ii)  For notational convenience write $f_0(x) = 0$ for all $x \in [a, b]$. From the rectangles used in calculating $\mathcal{L} \overline{\int}_{[a,b]} f_{n+1} - \mathcal{L} \underline{\int}_{[a,b]} f_n$, corresponding to any given $\varepsilon > 0$ we construct a rectangular covering $\mathbb{S}(D_n)$ of

$$D_n = \{\langle x, y \rangle \mid x \in [a, b], f_n(x) < y \leqslant f_{n+1}(x)\}$$

with  $\displaystyle |\mathbb{S}(D_n)| \leqslant \mathcal{L} \overline{\int}_{[a,b]} f_{n+1} - \mathcal{L} \underline{\int}_{[a,b]} f_n + \frac{\varepsilon}{2^{n+1}}$

To do this, let

$$A_n = \{\langle x, y \rangle \mid x \in [a, b], y \in [0, f_n(x)]\}$$
$$B_n = \{\langle x, y \rangle \mid x \in [a, b], y \in (f_n(x), M]\}$$

and  $C_n = \{\langle x, y \rangle \mid x \in [a, b], y \in [0, M - f_n(x)]\}$

Then for any rectangular covering $\mathbb{S}(C_n)$ of $C_n$ we can construct a rectangular covering $\mathbb{S}(B_n)$ with $|\mathbb{S}(B_n)| = |\mathbb{S}(C_n)|$ by taking each $E \in \mathbb{S}(C_n)$ and forming $E' = \{\langle x, y \rangle \mid \langle x, M - y \rangle \in E\}$. Then $\mathbb{S}(B_n) = \{E' \mid E \in \mathbb{S}(C_n)\}$.

Thus starting with rectangular coverings $\mathbb{S}(A_n)$ and $\mathbb{S}(C_n)$ such that

$$|\mathbb{S}(A_{n+1})| \leqslant \mathcal{L} \overline{\int}_{[a,b]} f_{n+1} + \frac{\varepsilon}{2^{n+2}}$$

$$|\mathbb{S}(C_n)| \leqslant \mathcal{L} \overline{\int}_{[a,b]} (M - f_n) + \frac{\varepsilon}{2^{n+2}} = M \cdot (b-a) - \mathcal{L} \underline{\int}_{[a,b]} f_n + \frac{\varepsilon}{2^{n+2}}$$

we first form the corresponding coverings $\mathbb{S}(B_n)$, and then the coverings of the rectangle $[a, b] \times [0, M]$ by countable sets of rectangles $\mathbb{S}(A_{n+1}) \cup \mathbb{S}(B_n)$.

Write $U_A$ for the union of all the rectangles in $\mathbb{S}(A_{n+1})$ and $U_B$ for the union of all the rectangles in $\mathbb{S}(B_n)$. Then by the lemma $U_A \cap U_B$ is the union of a countable number of rectangles. Since $D_n \subset U_A \cap U_B$, we can form these rectangles into a covering $\mathbb{S}(D_n)$ of $D_n$.

Now if $R_1', \ldots, R_k \in \mathbb{S}(D_n)$ then $\bigcup_{l=1}^{k} R_l \subset U_A$. So by replacing each rectangle of $\mathbb{S}(A_{n+1})$ which overlaps one of $R_1, \ldots, R_k$ by a finite number of rectangles, so that we can remove rectangles forming $\bigcup_{l=1}^{k} R_l$, we get a covering $\mathbb{S}(U_A \backslash \bigcup_{l=1}^{k} R_l)$ such that

$$\left| \mathbb{S}\left( U_A \backslash \bigcup_{l=1}^{k} R_l \right) \right| = |\mathbb{S}(A_{n+1})| - \sum_{l=1}^{k} |R_l|$$

Moreover the union of this new covering with $\mathbb{S}(B_n)$ still yields a covering of $[a, b] \times [0, M]$. Thus

$$\sum_{l=1}^{k} |R_l| \leq \mathbb{L} \int_{[a,b]}^{-} f_{n+1} - \mathbb{L} \int_{[a,b]} f_n + \frac{\varepsilon}{2^{n+1}}$$

As this holds for all $k$ and $|\mathbb{S}(D_n)| = \sum_{l=1}^{\infty} |R_l|$,

$$|\mathbb{S}(D_n)| \leq \mathbb{L} \int_{[a,b]}^{-} f_{n+1} - \mathbb{L} \int_{[a,b]} f_n + \frac{\varepsilon}{2^{n+1}}$$

So we now have a rectangular covering $\bigcup_n \mathbb{S}(D_n)$ of $\bigcup_n D_n$ which has

$$\left| \bigcup_n \mathbb{S}(D_n) \right| \leq \lim_n \mathbb{L} \int_{[a,b]}^{-} f_n + \varepsilon$$

as $\mathbb{L} \int_{[a,b]} f_n = \mathbb{L} \int_{[a,b]}^{-} f_n$ by assumption.

Unfortunately, it need not be true that $\bigcup_n D_n$ is the whole of $\{\langle x, y \rangle | x \in [a, b], 0 \leq y \leq f(x)\}$, as points where $y = f(x)$ need not be included. However, given any $\lambda > 1$, we can construct a rectangular covering $\mathbb{S}_\lambda$ from $\bigcup_n \mathbb{S}(D_n)$ by taking $[x_1, x_2] \times [\lambda y_1, \lambda y_2]$ in place of each $[x_1, x_2] \times [y_1, y_2] \in \bigcup_n \mathbb{S}(D_n)$. This covering does cover $\{\langle x, y \rangle | x \in [a, b], 0 \leq y \leq f(x)\}$.

This gives

$$|\mathbb{S}_\lambda| \leq \lim_n \mathbb{L} \int_{[a,b]}^{-} \lambda f_n + \lambda \varepsilon = \lambda \lim_n \mathbb{L} \int_{[a,b]}^{-} f_n + \lambda \varepsilon$$

for all $\varepsilon > 0$ and $\lambda > 1$. Thus

$$\mathbb{L} \int_{[a,b]} f \leq \lim_n \mathbb{L} \int_{[a,b]}^{-} f_n \tag{2}$$

As we are assuming that the integrals of the functions $f_n$ exist, eqs (1) and (2) give

$$\mathbb{L} \int_{[a,b]} f = \lim_n \mathbb{L} \int_{[a,b]} f_n$$

To establish the corresponding result for the Riemann integral, which has been mentioned, we need to use the fact that if $f$ has a Riemann integral then so does $f + c$. The relation between the two integrals is then given by the Corollary to Theorem 10.7. It is a special case of the result that the sum of two functions each of which has a Riemann integral has itself a Riemann integral.

**Theorem 10.12** If each of the functions $f$ and $g$ is defined and has a Riemann integral over $[a, b]$ then the function $f + g$ has a Riemann integral over $[a, b]$ and

$$\mathbb{R}_i \int_{[a,b]} (f + g) = \mathbb{R}_i \int_{[a,b]} f + \mathbb{R}_i \int_{[a,b]} g$$

*Proof* Suppose that the Riemann upper sums corresponding to $f$ and $g$ respectively,

$$S_f = \sum_{l=1}^{K} Y_{f,l}(X_{f,l} - X_{f,l-1})$$

and $$S_g = \sum_{l=1}^{K'} Y_{g,l}(X_{g,l} - X_{g,l-1})$$

are any such as given by Theorem 10.9 and its corollary. Suppose similarly that we have Riemann lower sums

$$s_f = \sum_{l=1}^{k} y_{f,l}(x_{f,l} - x_{f,l-1})$$

and $$s_g = \sum_{l=1}^{k'} y_{g,l}(x_{g,l} - x_{g,l-1})$$

Then let

$$\{X_l \,|\, l = 0, \ldots, K^*\} = \{X_{f,l} \,|\, l = 0, \ldots, K\} \cup \{X_{g,l} \,|\, l = 0, \ldots, K'\}$$

where $X_l \geqslant X_{l-1}$ for $l = 1, \ldots, K^*$, and

$$\{x_l \,|\, l = 0, \ldots, k^*\} = \{x_{f,l} \,|\, l = 0, \ldots, k\} \cup \{x_{g,l} \,|\, l = 0, \ldots, k'\}$$

where $x_l \geqslant x_{l-1}$ for $l = 0, \ldots, k^*$. Then $[X_{l-1}, X_l] \subset [X_{f,j-1}, X_{f,j}]$ for some $j$, so that $\sup\{f(x) \,|\, x \in [X_{l-1}, X_l]\} \leqslant Y_{f,j}$. Similarly, $\sup\{g(x) \,|\, x \in [X_{l-1}, X_l]\} \leqslant Y_{g,m}$ for some $m$. Hence $\sup\{f(x) + g(x) \,|\, x \in [X_{l-1}, X_l]\} \leqslant Y_{f,j} + Y_{g,m}$. It follows that there is a Riemann upper sum for $f + g$ over $[a, b]$ which can be written in the form

$$\sum_{l=1}^{K^*} Y_l \cdot (X_l - X_{l-1}) \leqslant \sum_{l=1}^{K} Y_{f,l} \cdot (X_{f,l} - X_{f,l-1}) + \sum_{l=1}^{K'} Y_{g,l} \cdot (X_{g,l} - X_{g,l-1})$$

$$= S_f + S_g \tag{1}$$

In an exactly similar way, we get a Riemann lower sum

$$\sum_{l=1}^{k^*} y_l \cdot (x_l - x_{l-1}) \geq \sum_{l=1}^{k} y_{f,l} \cdot (x_{f,l} - x_{f,l-1}) + \sum_{l=1}^{k'} y_{g,l} \cdot (x_{g,l} - x_{g,l-1})$$

$$= s_f + s_g \tag{2}$$

Now eqs (1) and (2) hold for all Riemann sums $S_f$, $s_f$, $S_g$ and $s_g$, and $f$ and $g$ have integrals over $[a, b]$. Thus on taking the infimum over all upper sums in eq. (1) and the supremum over all lower sums in eq. (2) it follows that

$$\mathbb{R}_i \int_{[a,b]} f + \mathbb{R}_i \int_{[a,b]} g = \mathbb{R}_i \underline{\int}_{[a,b]} (f+g) = \mathbb{R}_i \overline{\int}_{[a,b]} (f+g)$$

So the integral of $f + g$ exists and has the required value.

This result suggests a convenient way of extending the definition of integrals to functions which take both positive and negative values. In effect we just define Theorem 10.12 to hold when $f$ is such a function and $g$ is a constant function.

**Definition 10.10**   If the function $f$ is defined on $[a, b]$ and $f(x) \in [-M, M]$ for some $M$ and all $x \in [a, b]$ then the $\mathbb{G}$ integral of $f$ over $[a, b]$ is defined by

$$\mathbb{G} \int_{[a,b]} f = \mathbb{G} \int_{[a,b]} (f + M) - M(b - a)$$

If it is necessary to use lower and upper integrals, they can be defined in an analogous way.

Applying Definition 10.10, we find that Theorem 10.7 holds without restriction on $c$. Indeed properties (1) and (3) (p. 175) both hold when $f(x)$, $g(x) \in [-M, M]$ for $x \in [a, b]$. Also it is still the case that every Riemann integral is a Lebesgue integral. For positive or zero $\alpha$ and $f(x) \in [-M, M]$ Theorem 10.10 can be shown to hold using the calculation:

$$\alpha \mathbb{G} \int_{[a,b]} f = \alpha \mathbb{G} \int_{[a,b]} (f + M) - \alpha M \cdot (b - a)$$

$$= \mathbb{G} \int_{[a,b]} (\alpha f + \alpha M) - \alpha M \cdot (b - a)$$

$$= \mathbb{G} \int_{[a,b]} \alpha f$$

To show that the result also holds for negative $\alpha$, it is only necessary to prove it for $\alpha = -1$. This can be done by a straightforward combination of Definitions 10.8 and 10.10. We use the fact that the integral is equal to both the upper and the lower integrals. The calculations are as follows.

$$G\int_{[a,b]} -f = G\int_{[a,b]} \overline{(M-f)} - M\cdot(b-a)$$

$$= M\cdot(b-a) - G\int_{[a,b]} \underline{(M-M+f)} - M\cdot(b-a)$$

$$= -G\int_{[a,b]} \underline{f}$$

Also  $G\int_{[a,b]} -f = G\int_{[a,b]} \underline{(M-f)} - M\cdot(b-a)$

$$= -G\int_{[a,b]} \overline{f}$$

Whence

$$G\int_{[a,b]} \overline{f} = -G\int_{[a,b]} \underline{-f}$$

The extension of the result that property (2) holds for intervals, to functions $f$ with $f(x)\in[-M, M]$, follows by using $M(b-a) = M(b-c) + M(c-a)$. Thus $G\int$ is an integral to which Theorems 10.5 and 10.6 apply.

Finally, the extensions of Theorems 10.11 and 10.12 to $f_n(x)\in[-M, M]$ and $f(x), g(x)\in[-M, M]$ follow readily.

The result of Theorem 10.12 is now applied to prove a theorem on the integral of a limit, in a form suitable for the Riemann integral.

**Theorem 10.13**  If each of the real functions $f_n$, $n\in\mathbb{N}'$ is defined on $[a, b]$, has $f_n(x)\in[-M, M]$ all $x\in[a, b]$ and has a Riemann integral over $[a, b]$, and if $\lim_n f_n(x) = f(x)$ uniformly on $[a, b]$, then the function $f$ has a Riemann integral over $[a, b]$ and

$$\mathbb{R}_i\int_{[a,b]} f = \lim_n \mathbb{R}_i\int_{[a,b]} f_n$$

*Proof*  The limit $\lim_n f_n(x)$ is uniform over $[a, b]$, and so for any $\varepsilon > 0$ there is an $n_0\in\mathbb{N}$ such that both $f_n(x) - \varepsilon \leqslant f(x)$ and $f(x)\leqslant f_n(x) + \varepsilon$ for $n > n_0$, $n\in\mathbb{N}$ and $x\in[a, b]$. Thus

$$\mathbb{R}_i\int_{[a,b]} (f_n - \varepsilon) = \mathbb{R}_i\int_{[a,b]} f_n - \varepsilon(b-a)$$

$$\leqslant \mathbb{R}_i\int_{\underline{[a,b]}} f \leqslant \mathbb{R}_i\int_{\overline{[a,b]}} f$$

$$\leqslant \mathbb{R}_i\int_{[a,b]} (f_n + \varepsilon) = \mathbb{R}_i\int_{[a,b]} f_n + \varepsilon(b-a)$$

for $n > n_0$, $n \in \mathbb{N}$. Hence

$$-\varepsilon(b-a) \leqslant \mathbb{R}_i \int_{[a,b]} f - \mathbb{R}_i \int_{[a,b]} f_n \leqslant \mathbb{R}_i \overline{\int}_{[a,b]} f - \mathbb{R}_i \overline{\int}_{[a,b]} f_n \leqslant \varepsilon(b-a)$$

for $n > n_0$, $n \in \mathbb{N}$. Thus $f$ has a Riemann integral over $[a, b]$ and

$$\mathbb{R}_i \int_{[a,b]} f = \lim_n \mathbb{R}_i \int_{[a,b]} f_n$$

Let us now turn to the other algebraic operations between integrable functions. The next theorem shows that the product of two integrable functions is integrable. We do not get a formula for the integral of the product, and we cannot express the sums for the product in terms of the sums for the two functions.

**Theorem 10.14**  If the function $f$ defined on $[a, b]$ has a Riemann integral over that interval then the function $f^2$ has a Riemann integral over $[a, b]$.

*Proof*  Since $f$ is bounded on $[a, b]$, suppose that $|f(x)| \leqslant M$ for $x \in [a, b]$. Then $f^2(x) \leqslant M^2$ for $x \in [a, b]$.
   Now for $x, y \in [a, b]$

$$f^2(x) - f^2(y) = [f(x) - f(y)] \cdot [f(x) + f(y)]$$

Hence for any interval $I = [x_{l-1}, x_l] \subset [a, b]$

$$0 \leqslant \sup_{x \in I} f^2(x) - \inf_{x \in I} f^2(x) \leqslant \left[ \sup_{x \in I} f(x) - \inf_{x \in I} f(x) \right] 2M$$

Thus for any set $\{x_0, x_1, \ldots, x_k\}$ with $x_0 = a$, $x_k = b$ and $x_{l-1} < x_l$ for $l = 1, \ldots, k$, writing $M_l = \sup\{f(x) \,|\, x_{l-1} \leqslant x \leqslant x_l\}$, $m_l = \inf\{f(x) \,|\, x_{l-1} \leqslant x \leqslant x_l\}$, $M'_l = \sup\{f^2(x) \,|\, x_{l-1} \leqslant x \leqslant x_l\}$ and $m'_l = \inf\{f^2(x) \,|\, x_{l-1} \leqslant x \leqslant x_l\}$, we get

$$\sum_{l=1}^{k} M'_l(x_l - x_{l-1}) - \sum_{l=1}^{k} m'_l(x_l - x_{l-1})$$

$$\leqslant 2M \left( \sum_{l=1}^{k} M_l \cdot (x_l - x_{l-1}) - \sum_{l=1}^{k} m_l \cdot (x_l - x_{l-1}) \right)$$

$$0 \leqslant \mathbb{R}_i \overline{\int}_{[a,b]} f^2 - \mathbb{R}_i \underline{\int}_{[a,b]} f^2$$

$$\leqslant 2M \inf \left( \sum_{l=1}^{k} M_l \cdot (x_l - x_{l-1}) - \sum_{l=1}^{k} m_l \cdot (x_l - x_{l-1}) \right)$$

However, by Theorem 10.9 and its corollary

$$\inf\left( \sum_{l=1}^{k} M_l \cdot (x_l - x_{l-1}) - \sum_{l=1}^{k} m_l \cdot (x_l - x_{l-1}) \right)$$

taken over all such sets $\{x_0, x_1, \ldots, x_k\}$, is 0. Hence

$$\mathbb{R}_i \overline{\int}_{[a,b]} f^2 = \mathbb{R}_i \underline{\int}_{[a,b]} f^2$$

or $f^2$ is Riemann integrable over $[a, b]$.

**Corollary to Theorem 10.14**   If $f$ and $g$ have Riemann integrals over $[a, b]$ then so has $f \cdot g$.

*Proof*  By Theorem 10.12 $f + g$ is Riemann integrable over $[a, b]$, hence so is $(f + g)^2$. Then also $f^2$ and $g^2$ are Riemann integrable over $[a, b]$, and so is $f^2 + g^2$. Thus $2fg = (f + g)^2 - (f^2 + g^2)$ is Riemann integrable over $[a, b]$. Finally, $fg$ is Riemann integrable over $[a, b]$, using Theorem 10.10.

Now in order that our theory of the definite integral can be readily used, we need conditions for the existence of integrals. There are two easy and very useful conditions for the existence of the Riemann integral. The conditions, of course, also prove the existence of the Lebesgue integral. Necessary and sufficient conditions for a function to have a Riemann integral are known, and they cover a much wider class of functions. But the conditions in the following theorems have the advantage of being easy to apply.

**Theorem 10.15**   If the function $f$ is monotone on $[a, b]$ then $f$ is Riemann integrable over $[a, b]$.

*Proof*  We may assume that $f(b) \neq f(a)$, or $f$ is not a constant. We construct Riemann upper and lower sums for $f$ over $[a, b]$.
   Given $\varepsilon > 0$, let $a = x_0 < x_1 < \cdots < x_k = b$ be such that $(x_l - x_{l-1}) \leq \varepsilon [f(b) - f(a)]^{-1}$. The supremum and infimum of $f$ over $[x_{l-1}, x_l]$ are $f(x_{l-1})$ and $f(x_l)$. Which is which depends on whether $f$ is increasing or decreasing. Thus $\sum_{l=1}^{k} f(x_l) \cdot (x_l - x_{l-1})$ and $\sum_{l=1}^{k} f(x_{l-1}) \cdot (x_l - x_{l-1})$ are Riemann upper and lower sums for $f$ over $[a, b]$.
   Now because $f$ is monotone, $f(x_l) - f(x_{l-1})$ is of the same sign for all $l$. Thus

$$0 \leq \mathbb{R}_i \overline{\int}_{[a,b]} f - \mathbb{R}_i \underline{\int}_{[a,b]} f \leq \left| \sum_{l=1}^{k} [f(x_l) - f(x_{l-1})](x_l - x_{l-1}) \right|$$

$$\leq \sum_{l=1}^{k} |f(x_l) - f(x_{l-1})| \frac{\varepsilon}{|f(b) - f(a)|}$$

$$= \left| \sum_{l=1}^{k} [f(x_l) - f(x_{l-1})] \right| \frac{\varepsilon}{|f(b) - f(a)|} = \varepsilon$$

Hence

$$\mathbb{R}_i \overline{\int}_{[a,b]} f = \mathbb{R}_i \underline{\int}_{[a,b]} f$$

or $f$ has a Riemann integral over $[a, b]$.

Thus for $x \in [0, 1]$ there is only a countable set of numbers of the form $p/2^n$, where $n, p \in \mathbb{N}'$ and $p$ is odd, in $[0, x]$. If for each such $x, f(x)$ is the sum of the series formed from the corresponding numbers $1/2^{2^n}$ then $f$ is increasing. The function $f$ has a Riemann integral over $[0, 1]$ and $f$ is not continuous on each $\{p/2^n\}$.

**Theorem 10.16**    If the function $f$ is continuous on $[a, b]$ then $f$ has a Riemann integral over $[a, b]$.

*Proof*    Take a number $\varepsilon > 0$. For each $x' \in [a, b]$ there is an open interval $I_{x'}$ such that $x' \in I_{x'}$ and $|f(x) - f(x')| < \varepsilon/2$ if $x \in I_{x'}$. So $\sup\{f(x) | x \in I_{x'}\} - \inf\{f(x) | x \in I_{x'}\} \leqslant \varepsilon$. Now $[a, b] \subset \bigcup_{x' \in [a,b]} I_{x'}$. So by the Heine–Borel theorem, Theorem 4.7, there is a finite collection of intervals $\{I_{x_1'}, \ldots, I_{x_k'}\}$ with $[a, b] \subset \bigcup_{l=1}^{k} I_{x_l'}$. Clearly, we may assume that for no two such intervals do we get $I_{x_j'} \subset I_{x_i'}$, as we could certainly omit $I_{x_j'}$. Now let $x_1, \ldots, x_{n-1}$ be points in the intersections of pairs of the intervals, one in each non-empty intersection, and indexed so that $x_1 < x_2 < \cdots < x_{n-1}$. Let also $x_0 = a$ and $x_n = b$. Then each interval $[x_{l-1}, x_l]$ is contained in some $I_{x_m'}$ such that the difference between the supremum and infimum of $f(x)$ over $[x_{l-1}, x_l]$ is less than or equal to $\varepsilon$. Thus if $M_l = \sup\{f(x) | x \in [x_{l-1}, x_l]\}$ and $m_l = \inf\{f(x) | x \in [x_{l-1}, x_l]\}$ the sum $\sum_{l=1}^{n} M_l \cdot (x_l - x_{l-1})$ is a Riemann upper sum for $f$ and $\sum_{l=1}^{n} m_l \cdot (x_l - x_{l-1})$ is a Riemann lower sum for $f$. Hence

$$0 \leqslant \mathbb{R}_i \overline{\int}_{[a,b]} f - \mathbb{R}_i \underline{\int}_{[a,b]} f \leqslant \sum_{l=1}^{n} (M_l - m_l) \cdot (x_l - x_{l-1})$$

$$\leqslant \sum_{l=1}^{n} (x_l - x_{l-1})\varepsilon = (b - a)\varepsilon$$

Since given any $\eta > 0$ we may take $0 < \varepsilon \leqslant \eta/(b-a)$, we see that

$$\mathbb{R}_i \overline{\int}_{[a,b]} f = \mathbb{R}_i \underline{\int}_{[a,b]} f$$

and $f$ has a Riemann integral over $[a, b]$.

We know that continuous functions need not be monotone. Taking Theorem 10.16 together with Theorem 10.6, we see that a function $f$ continuous on an interval $[a, b]$ is a derivative. Thus for continuous functions we can use the theory of the indefinite integral to calculate the definite integral. In particular, we can use integration by parts, and substitution or change of variables. There are versions of these theorems which apply to definite integrals of functions which need not be continuous.

To facilitate the calculation of definite integrals by the use of indefinite integrals, it is customary to write the integrals without the identifying symbols, $G$, $R_i$ or $L$, the context being relied upon to show which is intended. Also we write

$$\int_{[a,b]} f = \int_a^b f = -\int_b^a f$$

Then the formula

$$F(x) - F(a) = \int_a^x f$$

holds if either $F$ is an indefinite integral of $f$ (that is $F \in \int f$) or $F(x)$ is the definite integral of $f$ over the interval with end points $a$ and $x$, with $x$ less than, equal to or greater than $a$. Thus we write, for example,

$$\tfrac{1}{2}(t^2 - 1) = \int_1^t x$$

for all $t$. Notice that all the definite integrals of $0$ are $0$, but that $1$ is an indefinite integral of $0$.

### Exercises

**1**  Find the indefinite integrals:
  (i)   $\int (1 + |x|)$,
  (ii)  $\int 2x\sqrt{(x^2 + 1)}$,
  (iii) $\int 2x^3\sqrt{(x^2 + 1)}$,
  (iv)  $\int x^3/(x^4 + 1)^2$,
  (v)   $\int x|x|$.

**2**  Starting from

$$\frac{1}{0!} \int_{x_0}^{x_0 + h} (x_0 + h - t)^0 f'(t)\, dt = f(x_0 + h) - f(x_0)$$

use integration by parts repeatedly, to arrive at a form of Taylor's theorem with an $(n-1)$th degree polynomial and remainder term

$$\frac{1}{(n-1)!} \int_{x_0}^{x_0 + h} (x_0 + h - t)^{n-1} f^{(n)}(t)\, dt$$

**3**  The function $h$ is defined by

$$h(x) = \begin{cases} 1 & (x \text{ irrational}) \\ 0 & (x \text{ rational}) \end{cases}$$

Show that $\mathrm{R_i} \, \overline{\int}_{[0,\,1]} h > \mathrm{R_i} \, \underline{\int}_{[0,1]} h$.

**4**  For some fixed number $\alpha$ the function $\theta$ is defined by $\theta(\alpha) = 1$ and $\theta(x) = 0$ if $x \neq \alpha$. Find $\mathrm{R_i} \int_0^y \theta(x) \, dx$.

**5**  Prove that if the function $f$ has a Riemann upper sum $S$ and a Riemann lower sum $s$ over $[a, b]$, and $S$ and $s$ are such that $S = s$, then $f$ is a constant on $[a, b]$.

**6**  The functions $f_n$ are defined by

$$f_n(x) = \begin{cases} 0 & (x = 0) \\ n & (0 < x < 1/n) \\ 0 & (1/n \leq x \leq 1) \end{cases}$$

Show that both $\mathrm{R_i} \int_0^1 \lim_n f_n$ and $\lim_n \mathrm{R_i} \int_0^1 f_n$ exist, but that they are not equal.

**7**  Let $B_n = \{p/2^n \mid 0 < p < 2^n, p \in \mathbb{N}', p \text{ is odd}\}$ and define $g_n$ and $g$ by

$$g_n(x) = \begin{cases} 0 & (x \notin B_n) \\ (\tfrac{1}{2})^n & (x \in B_n) \end{cases}$$

and

$$g(x) = \begin{cases} 0 & \left( x \notin \bigcup_n B_n \right) \\ (\tfrac{1}{2})^n & (x \in B_n \text{ some } n \in \mathbb{N}') \end{cases}$$

By calculating $\mathrm{R_i} \int_0^1 g_n$ and considering $\sum_{n=1}^{\infty} g_n$ show that $\mathrm{R_i} \int_0^1 g$ exists and find its value.

**8**  Let $\{r_n\}$ be the set of rational numbers in $[0, 1]$ and define $h_n$ by

$$h_n(x) = \begin{cases} r_n & (x = r_n) \\ 0 & (x \neq r_n) \end{cases}$$

Find $\mathrm{L} \int_{[0,1]} \sum_{n=0}^{\infty} h_n$.

**9**  Suppose that $f''$ is continuous on $[0, 1]$ and $f(1) = f'(1) = 0$, and that also $|f''(x)| \leq M$ for $x \in [0, 1]$. Prove by repeated integration by parts, or otherwise, that $|\int_0^1 f| \leq M/6$.

**10**  Prove that if $\mathrm{R_i} \int_{[a,b]} f$ exists and there is a $\delta > 0$ such that for all $x \in [a, b]$, $f(x) \geq \delta > 0$ then $\mathrm{R_i} \int_{[a,b]} 1/f$ exists.

**11** The functions $f^+$ and $f^-$ are defined on $\mathbb{R}$ by $f^+(x) = \max\{f(x), 0\}$ and $f^-(x) = \min\{f(x), 0\}$, so that $f = f^+ + f^-$ and $|f| = f^+ - f^-$. Prove that if $f$ is Riemann integrable over $[a, b]$ then so are $f^+$ and $f^-$ and that $\mathbb{R}_i \int_a^b f = \mathbb{R}_i \int_a^b f^+ + \mathbb{R}_i \int_a^b f^-$. Deduce that $|f|$ has a Riemann integral over $[a, b]$.

**12** Give an example of a function $f$ for which $\mathbb{R}_i \int_0^1 |f|$ exists, but for which $\mathbb{R}_i \int_0^1 f$ does not exist.

# 11

# Applications—the elementary transcendental functions

Functions are very often specified by equations involving their derivatives. Such equations are called *differential equations*. We will take a very simple one first:

$$f' = f \tag{1}$$

The equation $f' = cf$ where $c$ is some constant occurs widely and, as we shall see, once eq. (1) has been solved the solution of $f' = cf$ follows immediately. The equation $f' = cf$ describes a number of growth and decay processes. Thus with $c$ negative such an equation describes the process of radioactive decay. Since each atom is equally likely to decay at any moment, in a given interval of time there is a fixed proportion $-c$ of the atoms of a given isotope which split. Similarly, the rate of growth $f'$ of an organism can be proportional to its size $f$ at any given instant. Each part of the organism is growing at the same rate.

We have already seen at the end of Chapter 8 that if e is a function satisfying eq. (1) in some interval $[-c, c]$ then

$$e(x) = \lim_n e(0) \sum_{r=0}^{n} \frac{x^r}{r!} = e(0) \sum_{n=0}^{\infty} \frac{x^n}{n!} \tag{2}$$

But now we can see that if we use eq. (2) as a definition of e then e is in fact a solution of eq. (1). This is based on the fact that we can differentiate power series term by term.

**Theorem 11.1**   If the real function $f = \sum_{n=0}^{\infty} a_n x^n$ in $(-r, r)$ then $f$ is differentiable in $(-r, r)$ and $f' = \sum_{n=1}^{\infty} n a_n x^{n-1}$.

*Proof*   By Theorem 9.18 $\sum n a_n x^n$, and thus also $\sum n a_n x^{n-1}$, is convergent for $x \in (-r, r)$. Then by Theorems 9.19 and 9.20 $\sum n a_n x^{n-1}$ is continuous on $(-r, r)$ and, as the convergence is uniform on any $[-|\rho|, |\rho|] \subset (-r, r)$, by Theorem 10.13

$$\int_0^\rho \sum_{n=1}^{\infty} n a_n x^{n-1} \, dx = \sum_{n=1}^{\infty} a_n \rho^n = f(\rho) - f(0)$$

for $\rho \in (-r, r)$. Thus by Theorem 10.6, differentiating, we get $f' = \sum_{n=1}^{\infty} n a_n x^{n-1}$.

We use Theorem 11.1 to calculate e' from eq. (2).

$$e'(x) = e(0) \sum_{n=1}^{\infty} n \frac{x^{n-1}}{n!} = e(0) \sum_{n=1}^{\infty} \frac{x^{n-1}}{(n-1)!} = e(0) \sum_{n=0}^{\infty} \frac{x^n}{n!} = e(x)$$

Thus any function of the form $A \sum_{n=0}^{\infty} x^n/n!$ is a solution of eq. (1), and moreover we have seen that the power series is convergent for all real numbers. Indeed the corresponding complex power series converges in all of $\mathbb{C}$. The obvious choice of function to give a special name to, so that we can write the other solutions in terms of it, is the one with $A = 1$.

**Definition 11.1**   The real function e which satisfies $e' = e$ and has $e(0) = 1$ is called the *exponential function* and written exp.

Thus

$$\exp x = \sum_{n=0}^{\infty} \frac{x^n}{n!} \tag{A}$$

The solutions e to eq. (1) are $A$ exp where $A$ is any constant. Thus if $e(0) = 0$ then $e(x) = 0$ for all $x \in (\cdot, \cdot)$. However, the point 0 plays no special role in eq. (1). Consider the Taylor expansion about $x_0$. We get

$$e(x) = e(x_0) \sum_{n=0}^{N-1} \frac{(x-x_0)^n}{n!} + \frac{(x-x_0)^N}{N!} e^{(N)}(\zeta_N)$$

with $\zeta_N$ between $x_0$ and $x$. However, $e^{(N)} = e$ and, as before, e is bounded on the interval from $x$ to $x_0$, so that $\lim_N [(x-x_0)^N/N!] e(\zeta_N) = 0$, and $e(x) = e(x_0) \sum_{n=0}^{\infty} (x-x_0)^n/n!$ or $e(x) = e(x_0) \sum_{n=0}^{\infty} x^n(x-x_0)/n!$ for all $x \in (\cdot, \cdot)$. Hence if e is such that $e' = e$ and $e(x_0) = 0$ for some $x_0$ then $e(x) = 0$ for all $x$. However, $\exp 0 = 1$. Thus

$$\exp x \neq 0 \quad \text{for all } x$$

Now as exp is differentiable on $(\cdot, \cdot)$, it is continuous on $(\cdot, \cdot)$, so that $\exp a > 0$ and $\exp b < 0$ would imply the existence of $x_0$ between $a$ and $b$ such that $\exp x_0 = 0$. Hence $\exp x$ is of the same sign for all $x$ and, since $\exp 0 = 1 > 0$,

$$\exp x > 0 \quad \text{for all } x \tag{B}$$

Immediately this gives us that $\exp' x > 0$ for all $x$, and so by Theorem 8.6

$$\exp \text{ is strictly increasing on } (\cdot, \cdot) \tag{C}$$

The fact that any solution of $f' = f$ which takes the value 0 for some $x$ is the constant **0** yields an addition formula for exp.

Define a function $\varphi$ for some fixed real number $y$ by

$$\varphi(x) = \exp(x+y) - \exp x \cdot \exp y \quad \text{for } x \in (\cdot, \cdot)$$

Then

$$\varphi'(x) = \exp(x+y) - \exp x \cdot \exp y = \varphi(x)$$

which follows as the derivative with respect to $x$ of $\exp(x+y)$ works out to be $\exp(x+y)$, that is $\varphi' = \varphi$. Now

$$\varphi(0) = \exp(0+y) - \exp 0 \cdot \exp y = \exp y - \exp y = 0$$

as $\exp 0 = 1$ by the definition of exp. Thus $\varphi(x) = 0$ for all $x$, and

$$\exp(x+y) = \exp x \cdot \exp y \quad \text{for all } x, y \in (\cdot, \cdot) \tag{D}$$

Putting $y = x$ in eq. (D) gives

$$\exp 2x = (\exp x)^2$$

and then putting $y = kx$ successively for $k = 2, 3, \ldots, n-1$ we get

$$\exp nx = (\exp x)^n$$

Putting $y = -x$ in eq. (D) and using $\exp 0 = 1$ gives

$$1 = \exp x \cdot \exp -x$$

which is

$$\exp -x = (\exp x)^{-1}$$

Also $\exp x = \exp nx/n = (\exp x/n)^n$ and so

$$\exp x/n = (\exp x)^{1/n}$$

for $n \neq 0$ and $n \in \mathbb{Z}$.

We give a name to $\exp 1$.

**Definition 11.2**   The number $\exp 1 = \sum_{n=1}^{\infty} 1/n!$ is written as e.

Applying the above to e, if $r \in \mathbb{Q}$, so that $r = p/q$ for some $p, q \in \mathbb{Z}$ with $q \neq 0$, we have $e^r = (\exp 1/q)^p = \exp p/q$, so that

$$\exp r = e^r \quad \text{for } r \in \mathbb{Q} \tag{E}$$

Thus for the special case of e we have solved the problem raised near the beginning of Chapter 6 of finding a function which for rational numbers $r$ has the value $e^r$ and is continuous on $\mathbb{R}$. We can solve the problem for all $a > 0$ by noting that as exp is strictly monotone it has an inverse. However, first note that

$$\mathscr{R} \exp = (0, \cdot)$$

This follows because $e > 1$ as exp is strictly increasing, and so $\exp 1 > \exp 0$. Hence $\lim_n e^{-n} = 0$, and also, given $M$, we can find $n$ such that $e^n > M$. Thus for $x \in (0, \cdot)$ we can find $l$ and $n$ such that $e^{-l} < x < e^n$. So by the continuity of exp and the intermediate value theorem, Theorem 5.1, there is a $\xi$ such that $\exp \xi = x$.

**Definition 11.3**   The real function with domain $(0, \cdot)$ defined by $\log x = \exp^{-1} x$ is called the *logarithmic function*.

The function log is sometimes referred to as the natural or Napierian logarithm.

Since log is the inverse of exp, we have

$$a = \exp(\log a)$$

Thus for $r \in \mathbb{Q}$, $a^r = \exp(r \log a)$.

**Definition 11.4**   For any positive real number $a$ and any real $x$

$$a^x = \exp(x \log a)$$

This definition is only really needed for irrational $x$.
$a^x$ now defines a continuous and even differentiable function.
For the function log we have immediately

$$\log 1 = 0 \quad \text{and} \quad \log e = 1 \tag{F}$$

Also we have that log is strictly increasing, so that Theorem 8.2(iv) allows us to calculate that

$$\log' x = \frac{1}{x} \quad \text{or} \quad \frac{d}{dx} \log x = \frac{1}{x} \quad \text{for } x > 0 \tag{G}$$

Hence it is possible to write log as an integral. First we have

$$\log e \int \frac{1}{x}$$

but also $\log 1 = 0$, so that

$$\log x = \int_1^x \frac{1}{t}\, dt \tag{H}$$

From this we can get a simple estimate for $\log x$ in terms of $x$. Suppose that $x \geq 1$. Then $\max\{1/t \mid t \in [1, x]\} = 1$, and so

$$\log x = \int_1^x \frac{1}{t}\, dt \leq (x - 1) \cdot 1 = x - 1 \quad \text{for } x \geq 1$$

Also if $0 < x < 1$ then $\min\{1/t \mid t \in [x, 1]\} = 1$, and $(x - 1) < 0$, so that

$$\log x = \int_1^x \frac{1}{t}\, dt \leq x - 1 \quad \text{for } 0 < x < 1$$

Thus for all $x \in \mathscr{D} \log$

$$\log x \leq x - 1 \tag{I}$$

Equality occurs only for $x = 1$. This estimate is illustrated in Fig. 11.1.

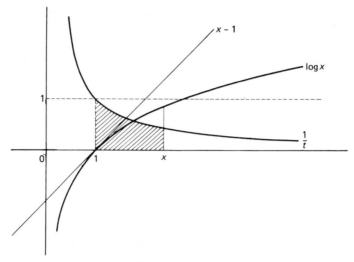

**Fig. 11.1**   $\log x \leqslant x - 1$

The problem of integrating $x^{-1}$ is not fully solved by eq. (H) itself, as it applies only for $x > 0$. But consider for $y < 0$, $y = -x$,

$$\int_{-1}^{y} \frac{1}{t}\,dt = \int_{1}^{x} \frac{1}{-s}\,(-1)\,ds \qquad \left( \text{as if } s = -t, \frac{ds}{dt} = -1 \right)$$

$$= \int_{1}^{x} \frac{1}{s}\,ds = \log x$$

that is, if $y < 0$, $\int_{-1}^{y} 1/t\,dt = \log -y$. Here $-y = |y|$, so that for the indefinite integral we can give a single formula:

$$\log \circ |x| \in \int x^{-1}$$

Translated into terms of log the addition formula for exp becomes

$$\log(a \cdot b) = \log a + \log b$$

This follows on putting $x = \log a$ and $y = \log b$, and taking the logarithms of each side. This fits in nicely with the effect of Definition 11.4 that

$$\log(a^{x}) = \log a \, \exp(x \cdot \log a) = x \log a \qquad (K)$$

The function given by $a^{x}$ is strictly increasing if $\log a > 0$ and decreasing if $\log a < 0$. If $\log a = 0$ then $a = 1$ and $1^{x} = 1$ for all $x$. Thus, unless $a = 1$, for $a > 0$ there is an inverse function to that given by $a^{x}$. This is called the *logarithm to the base a* and written $\log_{a}$. Thus $\log_{a} x = \log x / \log a$ as

$$a^{\left(\frac{\log x}{\log a}\right)} = \exp\left( \frac{\log x}{\log a} \log a \right) = \exp(\log x) = x$$

Logarithms to the base 10 have been widely used in calculations because of the convenience of

$$\log_{10}(x \cdot 10) = 1 + \log_{10} x$$

We are now in a position to complete the discussion of the derivative of $f$ given by $f(x) = x^\alpha$ for $x > 0$. Definition 11.4 gives us values for this function even when $\alpha$ is irrational. We have $f(x) = x^\alpha = \exp(\alpha \log x)$. So by Theorem 8.2(iii)

$$f'(x) = \exp(\alpha \log x)\frac{\alpha}{x} = x^\alpha \cdot \frac{\alpha}{x} = \alpha x^{\alpha-1}$$

or $\qquad \dfrac{d}{dx} x^\alpha = \alpha s^{\alpha-1} \quad$ for $x > 0$ and all real $\alpha$ $\qquad\qquad$ (L)

The inequality (I) enables us to discuss how log and exp behave for large values of the variable. We write (I) in the slightly weaker form $\log x < x$ for $x > 0$, and then replace $x$ by $x^{-a/2}$, for some $a > 0$, to get

$$-\tfrac{1}{2}a \log x < x^{-a/2}$$

Thus for $0 < x < 1$ and $a > 0$ we have

$$0 < -x^a \log x < \frac{2}{a} x^{a/2} < \frac{2}{a} x^r$$

where $r$ is any rational number $r < \tfrac{1}{2}a$. Hence by the squeeze theorem, Theorem 6.6,

$$\lim_{x\to 0} x^a \log x = 0 \quad \text{for all } a > 0 \qquad\qquad (M)$$

Again if we write inequality (I) as $y < e^y$ for all real $y$ then we get for $a > 0$ and $y > 0$

$$0 < y^a\, e^{-2ay} < y^{-a}$$

and if $x = 1/y$ then for $a > 0$ and $x > 0$

$$0 < x^{-a}\, e^{-2a/x} < x^a$$

The squeeze theorem gives

$$\lim_{x\to 0} \frac{e^{-2a/x}}{x^a} = \lim_{x\to 0} \frac{e^{-1/x}}{(x/2a)^a} = 0$$

Now $(2a)^a$ is a non-zero constant, and so

$$\lim_{x\to 0} x^{-a}\, e^{-1/x} = 0 \quad \text{for all } a > 0 \qquad\qquad (N)$$

To get results in the form of limits of sequences, put $n = 1/x$ in eqs (M) and (N). We get

$$\lim_{n} \frac{\log n}{n^a} = 0 \quad \text{and} \quad \lim_{n} n^a\, e^{-n} = 0 \quad \text{for all } a > 0 \qquad (O)$$

In the first of these, when $a = 1$, we can use the continuity of exp to get

$$\lim_n n^{1/n} = \exp 0 = 1$$

The graph of exp must be something like that shown in Fig. 11.2.

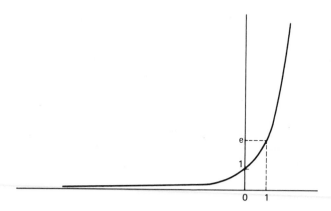

**Fig. 11.2**   exp $x$

To see how our solution to eq. (1) allows us to solve the equation $f' = cf$, first notice that if $f(x) = e^{cx}$ then $f'(x) = c\,e^{cx} = cf(x)$. Then notice that if $c \neq 0$ and $f' = cf$ we get $(d/dx)f(x/c) = (1/c)cf(x/c) = f(x/c)$. Thus the solutions of $f' = cf$ are given by $f(x) = A\,e^{cx}$. The case of $c = 0$ has already been solved by the use of the mean value theorem, Theorem 8.4. If $f' = 0$ then $f$ is constant and so can be written $f(x) = A\,e^{0x}$.

Other important functions can be found as solutions to the differential equation:

$$f'' = -f \tag{3}$$

The solutions to this equation will provide solutions to the more general equation $f'' = -cf$ where $c > 0$. The sign here is important. The solutions to $f'' = cf$ with $c > 0$ can be found in terms of $\exp(x\sqrt{c})$ and $\exp(-x\sqrt{c})$. It is easy to check that these are in fact solutions. Again in $f'' = -cf$ we have an equation which occurs widely. It describes, for example, the movement of a spring which exerts a force proportional to its displacement from rest. Here $f$ is a function of time and its values describe the position of the free end of the spring. $f''$ is then the acceleration which is proportional to the force, and so again to the value of $f$, and is directed towards eliminating the displacement.

Suppose that $g$ is a solution to eq. (3). Then we get for any even order derivative of $g$, $g^{(2n)} = (-1)^n g$. Also for any odd order derivative we get $g^{(2n+1)} = (-1)^n g'$. Now write $I$ for the closed interval with end points 0 and $x$. Then the two numbers $\sup\{|g(t)|\,|\,t \in I\}$ and $\sup\{|g'(t)|\,|\,t \in I\}$ exist as $g$ and $g'$ are continuous. Let $M$ be the larger of the two. Then the remainder term $R_n$ in the Taylor expansion of $g$ satisfies $|R_n| \leqslant |x|^n M / n!$. We have seen that

$\lim_n |x|^n/n! = 0$ for all $x$. Thus we get

$$g(x) = g(0) \sum_{n=0}^{\infty} (-1)^n \frac{x^{2n}}{(2n)!} + g'(0) \sum_{n=0}^{\infty} (-1)^n \frac{x^{2n+1}}{(2n+1)!} \quad \text{for all } x \in \mathbb{R}$$

(4)

If we calculate $g''$ from eq. (4) using Theorem 11.1, we get that $g'' = -g$, that is any function defined by eq. (4) with any pair of values for $g(0)$ and $g'(0)$ is a solution of eq. (3), and there are no others. Thus we define two new functions.

**Definition 11.5** The functions cos and sin are the solutions of the equation $f'' = -f$ for which $\cos 0 = 1$, $\cos' 0 = 0$ and $\sin 0 = 0$, $\sin' 0 = 1$. The function sin is called the *sine function*, and the function cos is called the *cosine function*.

Thus

$$\cos = \sum_{n=0}^{\infty} (-1)^n \frac{x^{2n}}{(2n)!} \quad \text{and} \quad \sin = \sum_{n=0}^{\infty} (-1)^n \frac{x^{2n+1}}{(2n+1)!}$$

(P)

These two functions may well be familiar to the reader, and may have been defined in terms of the ratios of sides of right-angled triangles. We have not yet defined 'angle' so such a definition is not available to us, although we shall see shortly that these functions are the same as the ones defined geometrically.

Equation (4) shows that any solution $g$ to $f'' = -f$ for which $g(0) = g'(0) = 0$ is the constant **0**. This again yields addition formulae. Let $\chi$ be defined for some fixed $y \in \mathbb{R}$ and all $x \in \mathbb{R}$ by

$$\chi(x) = \sin(x+y) - \sin x \cos y - \cos x \sin y$$

Notice that on applying Theorem 11.1 to the power series for cos and sin, we get

$$\sin' = \cos \quad \text{and} \quad \cos' = -\sin$$

(Q)

so that

$$\chi'(x) = \cos(x+y) - \cos x \cos y + \sin x \sin y$$

and

$$\chi''(x) = -\sin(x+y) + \sin x \cos y + \cos x \sin y = -\chi(x)$$

Thus $\chi$ is a solution of $f'' = -f$. In addition,

$$\chi(0) = \sin(0+y) - \sin 0 \cos y - \cos 0 \sin y = \sin y - \sin y = 0$$

as $\sin 0 = 0$ and $\cos 0 = 1$. Similarly,

$$\chi'(0) = \cos(0+y) - \cos 0 \cos y + \sin 0 \sin y = \cos y - \cos y = 0$$

Hence $\chi = \mathbf{0}$ or

$$\chi(x) = \sin(x+y) - \sin x \cos y - \cos x \sin y = 0$$

for all $x \in \mathbb{R}$. Now this is true for all $y \in \mathbb{R}$ and so

$$\sin(x+y) = \sin x \cos y + \cos x \sin y \quad \text{for all } x, y \in \mathbb{R}$$

(R)

Also $\chi'$ must be $\mathbf{0}$, so that

$$\cos(x+y) = \cos x \cos y - \sin x \sin y \quad \text{for all } x, y \in \mathbb{R} \tag{S}$$

In particular, putting $y = x$, $\sin 2x = 2 \sin x \cos x$ and $\cos 2x = \cos^2 x - \sin^2 x$.

By substituting $-x$ in the power series we see at once that $\cos x = \cos -x$ for all real $x$, and also that $\sin -x = -\sin x$ for all real $x$. Using this in eq. (S) we get $\cos(x - x) = \cos^2 x + \sin^2 x$ or

$$\cos^2 x + \sin^2 x = 1 \quad \text{for all } x \in \mathbb{R} \tag{T}$$

Since $\cos x$, $\sin x \in \mathbb{R}$, this implies that

$$|\cos x| \leqslant 1 \quad \text{and} \quad |\sin x| \leqslant 1 \quad \text{for all } x \in \mathbb{R} \tag{U}$$

We describe the property $\cos -x = \cos x$ by saying that cos is an *even function*, and the property $\sin -x = -\sin x$ by saying that sin is an *odd function*.

Now let us try to get a picture of sin and cos in terms of intervals over which they are increasing or decreasing. As $\sin' = \cos$ and $\cos 0 = 1$, and cos is continuous on $(\cdot, \cdot)$, there is some neighbourhood $\mathcal{N}(0, r)$ of 0 in which $\cos x > 0$ if $x \in \mathcal{N}(0, r)$. However, we have seen that as a consequence of the mean value theorem [in the form of Theorem 8.6(iii)], this means that sin is strictly increasing in $(0, r)$. In turn, this implies [Theorem 8.6(iv)] that as $\sin x > 0$ in $(0, r)$, $\cos x$ is strictly decreasing in $[0, r]$.

Suppose that sin were increasing on $[0, \cdot)$. Then write $\delta = \sin x_0$ for some convenient $x_0 > 0$. By Theorem 10.7 and property (3) (p. 175), for $x > x_0$ we can write

$$\cos x = 1 - \int_0^x \sin t \, dt = 1 - \int_0^{x_0} \sin t \, dt - \int_{x_0}^x \sin t \, dt$$

$$\leqslant 1 - \int_0^{x_0} \sin t \, dt - \delta(x - x_0) < 1 - \delta(x - x_0)$$

so that for $x > x_0 + 1/\delta$ we would have $\cos x < 0$. Now cos is continuous, $\cos 0 = 1$ and $\cos x < 0$, so that by the intermediate value theorem, Theorem 5.1, there would be at least one $\xi$, $0 < \xi < x$, such that $\cos \xi = 0$. [The occurrence of an interval in which $\cos x < 0$ shows that sin is not increasing in $[0, \cdot)$, so that this case does not occur. However, Fig. 11.3 illustrates this hypothetical situation to help to show why it is not possible.]

Suppose now that there were no $\xi > 0$ for which $\cos \xi = 0$. Then the intermediate value theorem would show that $\cos x > 0$ for all $x \geqslant 0$. This would imply, as before, that sin was increasing, but now on $[0, \cdot)$. However, this now gives a contradiction as we have seen that it implies the existence of $\xi > 0$ with $\cos \xi = 0$.

All this proves that $\exists \xi > 0$ such that $\cos \xi = 0$, so that the set $F = \{x \mid x > 0 \text{ and } \cos x = 0\}$ is not empty, and is bounded below by 0. By Theorem 2.15 there is an $x_0 = \inf F$. Now $F$ is a closed set as cos is continuous on $(0, \cdot)$ and $F = (0, \cdot)\backslash(\cos^{-1}(0, \cdot) \cup \cos^{-1}(\cdot, 0))$. Thus by Theorem 6.9 $x_0 \in F$, so that $\cos x_0 = 0$.

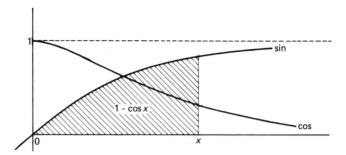

**Fig. 11.3**  sin is not increasing on $[0, \cdot)$

**Definition 11.6**  The number $\pi$ is defined by

$$\pi = 2 \inf\{x \mid x > 0 \text{ and } \cos x = 0\}$$

Thus

$$\cos(\pi/2) = 0 \tag{V}$$

In this context it would seem more natural to give a name to $x_0$, rather than $2x_0$, but the number $\pi$ occurs in such a wide variety of ways in mathematics that it does not seem worth trying also to have a special name for $\pi/2$.

Since $\pi/2$ is the smallest positive zero of cos, sin is strictly increasing on $[0, \pi/2]$. Since sin is odd, it must also be strictly increasing on $[-\pi/2, 0]$, and thus on $[-\pi/2, \pi/2]$. Now $\sin 0 = 0$, so that $\sin \pi/2 > 0$, but $\sin^2 \pi/2 = 1$, and thus $\sin \pi/2 = 1$. With this information we can discover from the addition formulae, eqs (R) and (S), exactly what happens to $\cos x$ and $\sin x$ on adding multiples of $\pi/2$ to $x$.

$$\sin\left(x + \frac{\pi}{2}\right) = \sin x \cos \frac{\pi}{2} + \cos x \sin \frac{\pi}{2} = 0 \cdot \sin x + 1 \cdot \cos x$$

and $\quad \cos\left(x + \frac{\pi}{2}\right) = \cos x \cos \frac{\pi}{2} - \sin x \sin \frac{\pi}{2} = -\sin x$

Thus for all real $x$

$$\sin\left(x + \frac{\pi}{2}\right) = \cos x \qquad \sin(x + \pi) = -\sin x \qquad \sin(x + 2\pi) = \sin x \tag{W}$$

and $\quad \cos\left(x + \frac{\pi}{2}\right) = -\sin x \qquad \cos(x + \pi) = -\cos x \qquad \cos(x + 2\pi) = \cos x$

For a function $f$ such that $f(x + \lambda) = f(x)$ for all $x \in \mathscr{D}f$ we say that $f$ is *periodic of period* $\lambda$. Thus cos and sin are periodic of period $2\pi$, so that if $I$ is any interval of length $2\pi$, and we know the values of cos or sin on $I$, then we know all the values of cos or sin. We only need to add to $x$ the appropriate number $2n\pi$, $n \in \mathbb{Z}$ to make $2n\pi + x \in I$.

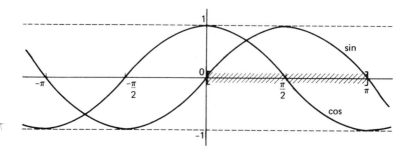

**Fig. 11.4**   The functions sin and cos

We now have a fairly full description of the behaviour of the functions cos and sin. This is shown in Fig. 11.4. Notice that the image of $[0, \pi]$ under cos is $[-1, 1] = \mathcal{R}$ cos, or the cosine takes all its values in this interval. Together with the fact that cos is an even function and sin is an odd function, this shows that for any two numbers $x$ and $y$ satisfying $x^2 + y^2 = 1$, there is some number $s$ for which cos $s = x$ and sin $s = y$.

The most basic property of angle is surely that for any two triangles with corresponding sides of equal lengths, the corresponding angles should be equal. The next most basic property is additivity. Thus in Fig. 11.5 we should have $\theta = \theta_1 + \theta_2$. It is sufficient to consider the following situation. We have the circle $\{\langle x, y \rangle \mid x^2 + y^2 = 1\}$. A triangle is defined by having as vertices the points $\langle 0, 0 \rangle$, $\langle x_1, y_1 \rangle$ and $\langle x_2, y_2 \rangle$, where these last two points are on the circle.

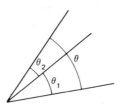

**Fig. 11.5**   $\theta = \theta_1 + \theta_2$

We require a measurement for the angle at $\langle 0, 0 \rangle$, which should depend only on the distance $d = \sqrt{[(x_1 - x_2)^2 + (y_1 - y_2)^2]}$ between $\langle x_1, y_1 \rangle$ and $\langle x_2, y_2 \rangle$. Figure 11.6 illustrates this.

Suppose that $s$ and $t$ are numbers such that cos $s = x_1$, sin $s = y_1$, cos $t = x_2$ and sin $t = y_2$. Then

$$d = \sqrt{[(\cos s - \cos t)^2 + (\sin s - \sin t)^2]}$$

$$= \sqrt{[\cos^2 s - 2 \cos s \cos t + \cos^2 t + \sin^2 s - 2 \sin s \sin t + \sin^2 t]}$$

$$= \sqrt{[2(1 - \cos s \cos t - \sin s \sin t)]} \qquad \text{[by eq. (T)]}$$

$$= \sqrt{[2(1 - \cos(t - s))]} \qquad \text{[by eq. (S)]}$$

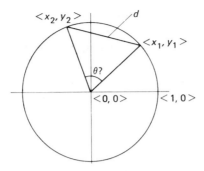

**Fig. 11.6** $\theta$ depends on $d$

This suggests that the angle should be measured by a solution $\theta$ of $\cos \theta = 1 - \frac{1}{2}d^2$. The fact that there are many such solutions is a genuine problem. We make the following definition.

**Definition 11.7** Suppose that the real numbers $x_1$, $y_1$, $x_2$ and $y_2$ are such that $x_1^2 + y_1^2 = x_2^2 + y_2^2 = 1$. The *geometric angle* between the lines from $\langle x_1, y_1 \rangle$ and $\langle x_2, y_2 \rangle$ to $\langle 0, 0 \rangle$ is the number $\theta$ in $[0, \pi]$ such that $\cos \theta = 1 - \frac{1}{2}d^2$ where $d = \sqrt{[(x_1 - x_2)^2 + (y_1 - y_2)^2]}$.

It is clear how we can use signed angles with values between $-\pi$ and $\pi$, with the sign depending on the order in which we take the two lines. If $s$ is the angle between the line from $\langle 1, 0 \rangle$ to $\langle 0, 0 \rangle$ and the line from $\langle x_1, y_1 \rangle$ to $\langle 0, 0 \rangle$, in the triangle with vertices $\langle 0, 0 \rangle$, $\langle x_1, y_1 \rangle$ and $\langle x_1, 0 \rangle$, the angle at $\langle 0, 0 \rangle$ is also $s$, the side $\langle 0, 0 \rangle$ to $\langle x_1, y_1 \rangle$ is of length 1, the side $\langle x_1, y_1 \rangle$ to $\langle x_1, 0 \rangle$ is of length $\sin s$, and the side $\langle x_1, 0 \rangle$ to $\langle 0, 0 \rangle$ is of length $\cos s$. This is in accordance with the geometric approach to the functions sin and cos. This is illustrated in Fig. 11.7.

It remains to check the scale in which our angles are measured. This is usually described in terms of lengths of the arc of the circle from the point

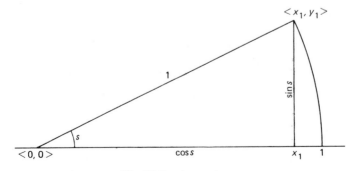

**Fig. 11.7** sin $s$ and cos $s$

$\langle x_1, y_1 \rangle$ to the point $\langle x_2, y_2 \rangle$. Since we have not defined length for general arcs, it is more convenient to use the area bounded by the two lines $\langle x_1, y_1 \rangle$ and $\langle x_2, y_2 \rangle$ to $\langle 0, 0 \rangle$ and the arc of the circle. In the geometric theory, the measure of the angle is said to be in circular measure or radians, if this area is $\frac{1}{2}$ the angle. We need only to test this for one case. So we take $s = 1$, $x_1 = \cos 1$ and $y_1 = \sin 1$, and calculate the area as shown in Fig. 11.8.

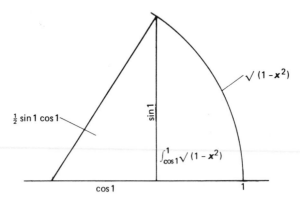

**Fig. 11.8**   Area for angle 1

The area is

$$\tfrac{1}{2}\cos 1 \sin 1 + \int_{\cos 1}^{1} \sqrt{(1-x^2)} = \tfrac{1}{2}\cos 1 \sin 1 + \int_{1}^{0} \sqrt{(1-\cos^2)}(-\sin)$$

$$= \tfrac{1}{2}\cos 1 \sin 1 + \int_{0}^{1} \sin^2 = \tfrac{1}{2}$$

as $\int_0^1 \sin^2 = [-\cos \sin]_0^1 + \int_0^1 \cos^2 = -\cos 1 \sin 1 + \int_0^1 (1-\sin^2)$, so that $2\int_0^1 \sin^2 = -\cos 1 \sin 1 + \int_0^1 1 = 1 - \cos 1 \sin 1$.

Thus our angles are measured in radians.

It should be clear that the notion of angle can be applied immediately to complex numbers.

**Definition 11.8**   If $z \in \mathbb{C}$, and $z = x + iy$ with $x, y \in \mathbb{R}$, then provided that $z \neq 0$ the angle between the lines from 1 to 0 and from $(x+iy)/|z|$ to 0 is called the *principal value of the argument* of $z$ and written arg $z$. The set $\{\theta \mid \theta = \arg z + 2n\pi, n \in \mathbb{Z}\}$ is called the *argument* of $z$ and written Arg $z$. Any member of Arg $z$ is called a value of the argument of $z$.

The argument of $z$ is sometimes called the amplitude of $z$ and written am $z$. The notations for distinguishing between the principal value of the argument and the other values, or the set of values, have no commonly accepted form.

The number 0 fails to have an argument, as the 'line from 0 to 0' does not specify a line. It will also be soon apparent that there is no way of assigning an argument to 0 which would fit with the most basic properties of the argument.

**Theorem 11.2**   If $z_1, z_2 \in \mathbb{C}$ and neither $z_1$ nor $z_2$ is 0 then:
(i)   $\text{Arg } z_1 z_2 = \text{Arg } z_1 + \text{Arg } z_2$,
(ii)   $\arg z_1 z_2 = \arg z_1 + \arg z_2 + 2n\pi$ for some $n = 0, \pm 1$.

*Proof*   Since $\arg z_1 + \arg z_2$ lies in $[-2\pi, 2\pi]$, both (i) and (ii) express the same fact.

Write $\theta_1 = \arg z_1$ and $\theta_2 = \arg z_2$. Then by the definition of arg we have

$$\frac{z_1}{|z_1|} = \cos \theta_1 + i \sin \theta_1 \quad \text{and} \quad \frac{z_2}{|z_2|} = \cos \theta_2 + i \sin \theta_2$$

Hence

$$\frac{z_1 z_2}{|z_1||z_2|} = \frac{z_1 z_2}{|z_1 z_2|} = (\cos \theta_1 + i \sin \theta_1)(\cos \theta_2 + i \sin \theta_2)$$

$$= (\cos \theta_1 \cos \theta_2 - \sin \theta_1 \sin \theta_2)$$

$$+ i(\cos \theta_1 \sin \theta_2 + \sin \theta_1 \cos \theta_2)$$

$$= \cos(\theta_1 + \theta_2) + i \sin(\theta_1 + \theta_2)$$

by eqs (R) and (S). Thus $\theta_1 + \theta_2 \in \text{Arg } z_1 z_2$.

There is an important fact which was used in this proof and deserves to be spelt out separately.

For any $z \in \mathbb{C}$, $z \neq 0$, if $\theta \in \text{Arg } z$ then

$$z = |z|(\cos \theta + i \sin \theta) \tag{X}$$

It is the fact that if $z_1 = |z_1|(\cos \theta_1 + i \sin \theta_1)$ and $z_2 = |z_2|(\cos \theta_2 + i \sin \theta_2)$ then $z_1 z_2 = |z_1||z_2|[\cos(\theta_1 + \theta_2) + i \sin(\theta_1 + \theta_2)]$ that makes the field axioms for $\mathbb{C}$ easier to check. In Chapter 1 it was remarked that Axioms (1)(ii) and (iii), (2)(ii) and (iii), and (3) remained to be checked. To draw attention to the analogy between $z = x + iy$, represented by the pair of real numbers $\langle x, y \rangle$, and $z = r(\cos \theta + i \sin \theta)$, represented by the pair of real numbers $r$ and $\theta$, for the purpose of verifying these axioms we will write $r(\cos \theta + i \sin \theta)$ as $\text{cis}(r, \theta)$. Then $\text{cis}(r_1, \theta_1) \cdot \text{cis}(r_2, \theta_2) = \text{cis}(r_1 r_2, \theta_1 + \theta_2)$.
(1)(ii)   For any two complex numbers $\langle x_1, y_1 \rangle$ and $\langle x_2, y_2 \rangle$

$$\langle x_1, y_1 \rangle + \langle x_2, y_2 \rangle$$

$$= \langle x_1 + x_2, y_1 + y_2 \rangle$$

$$= \langle x_2 + x_1, y_2 + y_1 \rangle \quad \text{(by (1)(ii) for } \mathbb{R})$$

$$= \langle x_2, y_2 \rangle + \langle x_1, y_1 \rangle$$

(2)(ii)　For any two complex numbers $\text{cis}(r_1, \theta_1)$ and $\text{cis}(r_2, \theta_2)$

$$\text{cis}(r_1, \theta_1) \cdot \text{cis}(r_2, \theta_2)$$
$$= \text{cis}(r_1 r_2, \theta_1 + \theta_2)$$
$$= \text{cis}(r_2 r_1, \theta_2 + \theta_1) \qquad \text{(by (2)(ii) and (1)(ii) for } \mathbb{R}\text{)}$$
$$= \text{cis}(r_2, \theta_2) \cdot \text{cis}(r_1, \theta_1)$$

(1)(iii)　For any three complex numbers $\langle x_j, y_j \rangle$, $j = 1, 2, 3$,

$$(\langle x_1, y_1 \rangle + \langle x_2, y_2 \rangle) + \langle x_3, y_3 \rangle$$
$$= \langle x_1 + x_2, y_1 + y_2 \rangle + \langle x_3, y_3 \rangle$$
$$= \langle (x_1 + x_2) + x_3, (y_1 + y_2) + y_3 \rangle$$
$$= \langle x_1 + (x_2 + x_3), y_1 + (y_2 + y_3) \rangle \qquad \text{(by (1)(iii) for } \mathbb{R}\text{)}$$
$$= \langle x_1, y_1 \rangle + \langle x_2 + x_3, y_2 + y_3 \rangle$$
$$= \langle x_1, y_1 \rangle + (\langle x_2, y_2 \rangle + \langle x_3, y_3 \rangle)$$

(2)(iii)　For any three complex numbers $\text{cis}(r_j, \theta_j)$, $j = 1, 2, 3$,

$$[\text{cis}(r_1, \theta_1) \cdot \text{cis}(r_2, \theta_2)] \cdot \text{cis}(r_3, \theta_3)$$
$$= \text{cis}(r_1 r_2, \theta_1 + \theta_2) \cdot \text{cis}(r_3, \theta_3)$$
$$= \text{cis}[(r_1 r_2) r_3, (\theta_1 + \theta_2) + \theta_3]$$
$$= \text{cis}[r_1 \cdot (r_2 r_3), \theta_1 + (\theta_2 + \theta_3)] \qquad \text{(by (2)(iii) and (1)(iii) for } \mathbb{R}\text{)}$$
$$= \text{cis}(r_1, \theta_1) \cdot \text{cis}(r_2 r_3, \theta_2 + \theta_3)$$
$$= \text{cis}(r_1, \theta_1) \cdot [\text{cis}(r_2, \theta_2) \cdot \text{cis}(r_3, \theta_3)]$$

For both (2)(ii) and (iii), if any one of the numbers is 0 then both sides of the equation are 0. For (3), notice that $0 = 0(\cos \theta + i \sin \theta)$ for any choice of $\theta \in \mathbb{R}$.

(3)　Suppose that the four complex numbers $r_j(\cos \theta_j + i \sin \theta_j)$ for $j = 1, 2, 3,$ 4 are such that

$$r_1(\cos \theta_1 + i \sin \theta_1) + r_2(\cos \theta_2 + i \sin \theta_2) = r_3(\cos \theta_3 + i \sin \theta_3)$$

Then　$r_1 \cos \theta_1 + r_2 \cos \theta_2 = r_3 \cos \theta_3$

and　$r_1 \sin \theta_1 + r_2 \sin \theta_2 = r_3 \sin \theta_3$

Hence multiplying the first equation by $\cos \theta_4$ and the second by $\sin \theta_4$ and subtracting

$$r_1(\cos \theta_1 \cos \theta_4 - \sin \theta_1 \sin \theta_4) + r_2(\cos \theta_2 \cos \theta_4 - \sin \theta_2 \sin \theta_4)$$
$$= r_3(\cos \theta_3 \cos \theta_4 - \sin \theta_3 \sin \theta_4)$$

that is

$$r_1 \cos(\theta_1 + \theta_4) + r_2 \cos(\theta_2 + \theta_4) = r_3 \cos(\theta_3 + \theta_4)$$

Similarly, multiplying by $\sin \theta_4$ and $\cos \theta_4$ and adding we get

$$r_1 \sin(\theta_1 + \theta_4) + r_2 \sin(\theta_2 + \theta_4) = r_3 \sin(\theta_3 + \theta_4)$$

Hence

$$r_1 r_4 [\cos(\theta_1 + \theta_4) + i \sin(\theta_1 + \theta_4)] + r_2 r_4 [\cos(\theta_2 + \theta_4) + i \sin(\theta_2 + \theta_4)]$$
$$= r_3 r_4 [\cos(\theta_3 + \theta_4) + i \sin(\theta_3 + \theta_4)]$$

or
$$r_1(\cos \theta_1 + i \sin \theta_1) \cdot r_4(\cos \theta_4 + i \sin \theta_4)$$
$$+ r_2(\cos \theta_2 + i \sin \theta_2) \cdot r_4(\cos \theta_4 + i \sin \theta_4)$$
$$= r_3(\cos \theta_3 + i \sin \theta_3) \cdot r_4(\cos \theta_4 + i \sin \theta_4)$$

This is (3). Geometrically, the result says that the triangle shown in Fig. 11.9 remains a triangle on rotating through an angle $\theta_4$ and changing scale by a factor $r_4$.

The methods that we have applied to the differential equations $f' = f$ and $f'' = -f$ can be used to develop the properties of functions satisfying many other equations.

The study in this book is all based on the assumption that $\mathbb{R}$ exists. To prove this in detail requires a lot of (often very straightforward) steps. So here only an outline of the process is suggested. First of all we must be satisfied that $\mathbb{Q}$ exists. Starting only with our basic logic and the theory of sets, we can construct $\mathbb{N}$. A set $A$ has $n$ members if there is a one–one relation with domain $A$ and range $\{0, 1, 2, \ldots, n-1\}$. If $A$ has $n$ members and $B$ has $m$ members, and $A \cap B = \varnothing$, then we can define $n + m$ as the number of members of $A \cup B$. Similarly, $n \cdot m$ is defined as the number of members of $A \times B$. We can derive the properties of $\mathbb{N}$ from the properties of sets.

We can check that the axioms for $\mathbb{Z}$ can be satisfied by adding in a number large enough to make everything positive, working in $\mathbb{N}$, and then subtracting

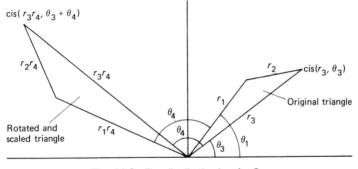

**Fig. 11.9** The distributive law for $\mathbb{C}$

the number again. A convenient way of doing this formally is to write members of $\mathbb{Z}$ as equivalence classes of pairs of members of $\mathbb{N}$. A number $l \in \mathbb{Z}$ would be written as $\langle n, m \rangle$ with $n, m \in \mathbb{N}$ where $n - m = l$. Clearly, $\langle n + k, m + k \rangle$ is equivalent to $\langle n, m \rangle$.

A similar device enables us to establish that the axioms for $\mathbb{Q}$ can be satisfied. Equivalence classes of pairs $\langle p, q \rangle$ with $p, q \in \mathbb{Z}$, $q \neq 0$ are used. A number $r = p/q \in \mathbb{Q}$ is written $\langle p, q \rangle$ where $\langle pk, qk \rangle$, $k \neq 0$ must be in the same equivalence class.

All this is lengthy but quite simple algebraic work. The key step is to see that the completeness axiom, Axiom C (p. 38), can be satisfied. To do this, we can use intervals, or perhaps equivalence classes of intervals, in $\mathbb{Q}$. For checking the properties of addition it is easier to use semi-infinite intervals; but for multiplication intervals with 0 as one end point are preferable. One technique is to construct the positive part of $\mathbb{R}$ first. Then a positive or zero real number $\alpha$ will be written as $\{r \mid r \in \mathbb{Q}, 0 \leqslant r \leqslant \alpha\}$. If $I$ and $J$ are such intervals representing real numbers $\alpha$ and $\beta$ respectively then $I + J$ represents $\alpha + \beta$. Also $I \cdot J$ represents $\alpha \cdot \beta$ and $\alpha \leqslant \beta \Leftrightarrow I \subset J$. In these terms it is not difficult to verify all the axioms, except, of course, the existence and properties of $-\alpha$. To get the whole of $\mathbb{R}$, we can now use the same device as enabled us to pass from $\mathbb{N}$ to $\mathbb{Z}$. Thus $\gamma \in \mathbb{R}$ would be treated as an equivalence class of pairs $\langle \alpha, \beta \rangle$ where $\alpha, \beta \geqslant 0$, $\alpha, \beta \in \mathbb{R}$ and $\gamma = \alpha - \beta$.

The objects which represent real numbers under this programme are extremely complicated. To take just the last stage, we get equivalence classes of pairs $\langle I, J \rangle$ of intervals in $\mathbb{Q}$. Moreover there are many possible variations in the construction, which do not affect the outcome, but change the nature of the objects representing real numbers. All we are doing is showing that there is *some* set of objects which satisfies the axioms for a complete ordered field. When we talk about $\mathbb{R}$, what we say is true of *any* such set of objects. So we do not have to be worried as to what exactly the real numbers are; all we need to know is that the axioms can be satisfied.

# Notation

# Index

Some terms which are not used in this text have been included. The exact use of such terms may differ between authors. The word indicated is equivalent to one of the usages. It should be noted that sometimes the existence of two words for the same basic idea is used to distinguish between slight variations on the same concept.